RF Coaxial Slot Radiators

Modeling, Measurements, and Applications

DISCLAIMER OF WARRANTY

For a listing of recent titles in the
Artech House Antennas and Propagation Library,
turn to the back of this book.

RF Coaxial Slot Radiators

Modeling, Measurements, and Applications

Kok Yeow You

ARTECH HOUSE
BOSTON | LONDON
artechhouse.com

Library of Congress Cataloging-in-Publication Data
A catalog record for this book is available from the U.S. Library of Congress

British Library Cataloguing in Publication Data
A catalog record for this book is available from the British Library.

ISBN-13: 978-1-60807-822-6

Cover design by John Gomes

685 Canton Street
Norwood, MA

10 9 8 7 6 5 4 3 2 1

To my dear father, mother, wife,
and sister's family.

Contents

Preface

A number of different analyses concerning coaxial radiators have been consolidated into this book. The libraries of Universiti Teknologi Malaysia (UTM) and Universiti Putra Malaysia (UPM) were valuable resources.

Chapters 1 through 5 cover the theoretical models of coaxial slot radiators. Chapters 6 and 7 present the solution of those theoretical models by numerical methods. Chapters 8 and 9 focus on the applications of the radiator, including measurements and calibration techniques for the radiators. In each chapter, the experimental results are used to confirm the theoretical computer calculations. This book is suitable for scientists and engineers in communication fields. It contains a brief theoretical background, particularly for various types of coaxial radiators. The analytical and numerical calculations of the coaxial radiator with the aid of MATLAB codes are included.

In fact, this book is different from a mathematical-based classic book. It is written using simple mathematical symbols. The mathematical symbols are as simple as possible in order to let engineers and scientists who are not from a pure mathematical background understand the content. This book contains not only mathematical formulations, but also the design guidelines, measurements, and applications for coaxial radiators. It includes both academic (theories and formulations) and industry (applications) aspects. In addition, this book contains many configurations and technical drawings to provide interpretation and explanation effectively.

The author would like to thank the International Electrotechnical Commission, *Measurement Science Review*, and *International Journal of Microwave and Optical Technology* for their permission to use copyrighted figures.

CHAPTER 1

Introduction to Coaxial Slot Radiators

This chapter focuses on wave propagation in three parts of coaxial slot radiators: (1) propagation of waves in the coaxial line, (2) wave radiation from the coaxial line, and (3) the propagation of waves on the surface of the open end of the coaxial line. The relationship between electromagnetic properties and the nature of the circuit will be described in detail. Furthermore, various types of coaxial radiators will be introduced, as well as a brief history of the development and expansion of the formulations for coaxial radiators. In this book, three kinds of coaxial radiators will be presented, namely open-ended coaxial radiator, linear bare monopole, and linear dielectric-coated monopole. The specific descriptions of the formula derivation for particular radiators will be covered in subsequent chapters: open-ended coaxial probes (Chapters 2 and 3) and monopoles (Chapters 4 and 5).

The open-ended coaxial line is a waveguide consisting of inner and outer conductors with radii a and b, respectively. In general, the medium between conductors is filled with free space or the low-loss dielectric materials. When one end of coaxial line is flatly machined as shown in Figure 1.1, it will become an open-ended coaxial radiator.

The structure of a bare monopole driven from the coaxial line is similar to an open-ended coaxial radiator, except for the protruding center of its inner conductor as shown in Figure 1.2.

A coated monopole is a bare monopole enclosed by a thin cylindrical low-loss dielectric material with an outer radius, b, as shown in Figure 1.3.

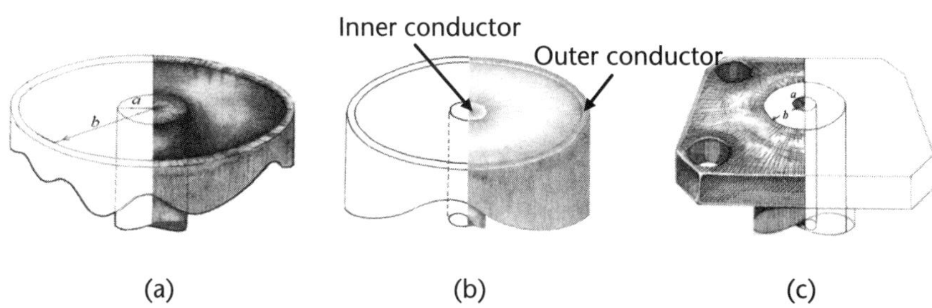

Figure 1.1 (a) Air-filled coaxial radiator. (b) Teflon-filled coaxial radiator. (c) Coaxial waveguide with open-ended flange.

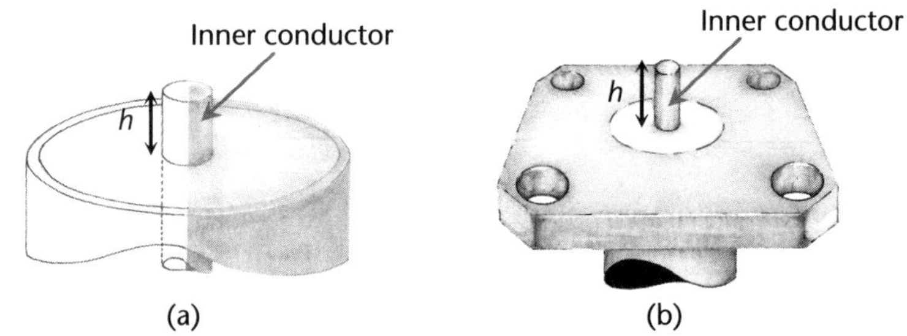

Figure 1.2 (a) Short monopole driving from coaxial line. (b) Monopole with rectangular flange.

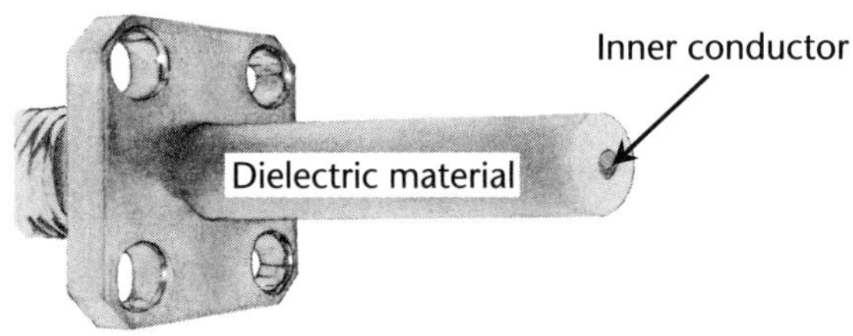

Figure 1.3 Configurations of a coated monopole.

1.1 Analytical and Numerical Analysis of Coaxial Slot Radiators

In the past 80 years, there have been many theoretical formulas that have been successfully developed and derived. Normally, the coaxial slot radiator formulations (either in differential or integration forms) have been derived from Maxwell's equations with the aid of the method of images or Gauss's theorem and Stokes' theorem. Several assumptions were made to simplify the derivation of the solution to the wave equation of the coaxial radiators. Indeed, in the past, the analytical formulas played an important role in the design of radiator due to the inadequacy of the numerical method. In other words, the insufficiency of advanced computers led to analytical formulas whereby the involved lower cost is compared to experiment design.

Since the advent of computer technology in 1960, attention to the usage of numerical methods in determining the impedance and admittance of the slot radiators has increased, including the methods of the finite-difference time-Domain (FDTD) method [1], the method of moment (MoM) [2], and the finite-element method (FEM) [3]. In order to achieve simplicity, most of the numerical methods are implemented in two dimensions. FEM and FDTD are methods used to solve the partial differential formula, such as a differential wave equation. Meanwhile, MoM is a technique used to solve the formula in the form of integration, especially for the scattering problem of unbounded wave from slot radiator.

From the author's point of view, analytical formula refers to the equation that is derived based on theoretical concepts and without any unknown variables in the equation. Although the complicated analytical formulas can be solved using numerical method, such as series expansion and trapezium's rules, it is still considered as an analytical approach. When the theoretical formula has one set of unknown variables matrix in the equation and it is solved by using modern numerical routine, such as FEM and MoM, it can be defined as a numerical approach. In fact, the modern numerical method is able to provide the results, which are closer to the practical measurement, as it will divide the problem system into smaller segments and solve them one by one.

Subsequently, in the 1970s, the first public software denominated Numerical Electromagnetics Code (NEC) emerged; it has mainly been used to solve the problem of linear wire radiators. In the 1980s, Gajda et al. [4] introduced a two-dimensional scalar FEM to solve Laplace's equation in a rotationally symmetric region of an air-filled open-ended coaxial line. However, Nevels et al. [5] had extended the MoM application to the integral admittance of open-ended coaxial aperture. In the 1990s, the vector FEM in High Frequency Structure Simulator (HFSS) was used by Blackham and Pollard [6] to improve the accuracy of the FEM in the design and analysis of Agilent probes.

Initially, the function of FEM was restricted to solve only a two-dimensional (2-D) scalar and bounded problem in order to obtain the scalar potential, V and capacitance, C. For the unbounded region under test, the volume of the finite element at the scattering region must be extended sufficiently largely and far from the source using Green's function [7]. Gajda et al. [4] reported that the unbounded region is sealed with the Neumann boundary. A hybrid method combining FEM and boundary element was also suggested as the solution to the coaxial probe problem; however, it includes particular risk of computational complexity without guaranteed accuracy improvement. Recently, interest among researchers in using FEM is gaining more support with the introduction of several commercial softwares, notably Ansoft HFSS [8] and COMSOL Multiphysics [9], whose special features include the ability to resolve the problem in the time domain.

Similar to FEM, FDTD was originally facing difficulties in solving unbounded problems; moreover, FDTD cannot be used together with the Green's function as FDTD is in the time-domain base. This was a major factor that restricted the function of the FDTD method for many years. However, this difficulty has been removed successfully with the development of perfectly matched layers for the grid, such as absorbing boundary condition (ABC) [10], as well as the Green's function, which has been formulated in the time domain recently.

Although all the numerical methods have been used to solve the problem of coaxial radiators, each method refers to the different surface problems as summarized in Figure 1.4. Figure 1.4(a) shows the MoM, which is applied to the linear monopole. The MoM has assumed that the structure of the radiator is a single linear cylindrical rod. In fact, the monopole is driving from coaxial line. Classically, the effects of fringing fields at the driving point are always corrected by the capacitance C element. The MoM in [5] only considers the wave distributed around the surface as shown in Figure 1.4(b), whereas for the extended conductor case [11], the MoM simulation surface is focused on the extended conductor and the fringing field at the driving point ($z = 0$) is normally corrected by extract capacitance terms.

The half region of the 2-D scalar or vector FEM can be used to simulate the coaxial waveguide since the distribution field for the waveguide is symmetrical as shown in Figure 1.4(c). This can sufficiently reduce the program running time and also the complication of algorithm program. The 2-D techniques are also used by FDTD as shown in Figure 1.4(d). In fact, the FDTD is easier than FEM, but it is

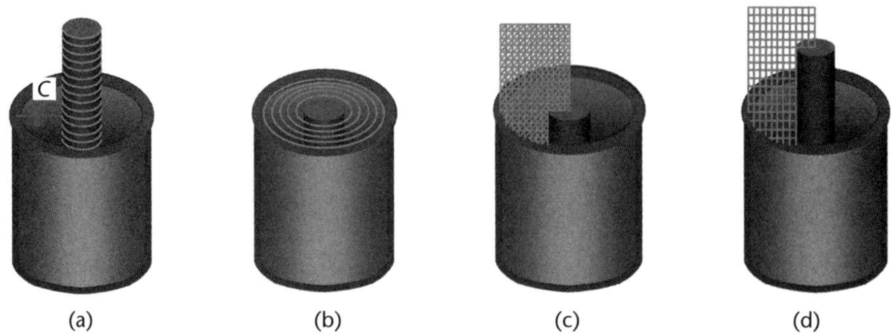

Figure 1.4 (a) The simulation surface of MoM [11]. (b) Another simulation surface of MoM [5]. (c) The symmetric 2-D FEM simulation [4, 12]. (d) The 2-D FDTD simulation [13].

only capable to solve the problem in the vector field and the mesh size normally must be consistently in a rectangular shape. The theoretical governing equations, both wave equation and scattering integral equations used in those numerical methods, will be described in Section 1.2. However, the specific descriptions of numerical calculations are covered in Chapters 6 and 7.

1.2 Wave Theory of Coaxial Radiators

1.2.1 The Propagation Wave in the Coaxial Line ($z \leq 0$ Region)

Reciprocal conversion between electrical and magnetic energy will cause waves propagating forward. The propagating wave is known as transverse electromagnetic (TEM) wave. For infinite coaxial line, the polarization of electric field, $\vec{E}$, is a function of the radius, ρ (in meters) of the cross-sectional coaxial line, whereas the polarization of magnetic field, $\vec{H}$ is only dependent on azimuth angle, ϕ (in radians). Thus, the propagation wave in the coaxial line has only field components E_ρ and H_ϕ, which are polarized in directions normal (transverse) to the direction of propagation, z (in meters), as shown in Figure 1.5.

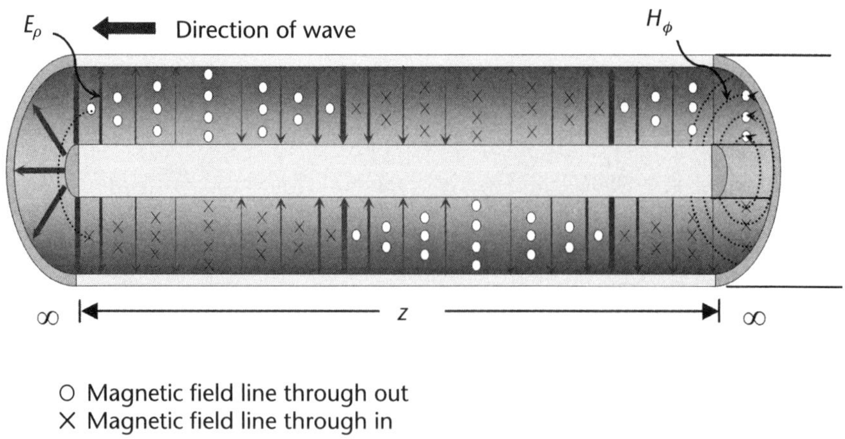

Figure 1.5 Propagation TEM wave in infinite coaxial line cross-sectional view.

In an ideal coaxial line, the magnitude of transmitted polar wave is always constant and its phase, ϕ, is shifted exponentially, $e^{-j\phi}$. The shifted phase, ϕ (in radians) can be related to the length of coaxial line, z (in meters) as:

$$\phi = \frac{2\pi f\sqrt{\varepsilon_c}}{c}z \tag{1.1}$$

where c ($\approx$ 2.99792458 ms^{-1}) and f (in hertz) are the velocity of light in free space and operation frequency, respectively. The symbol ε_c is the relative permittivity of dielectric material of the coaxial line. The incident of TEM wave field components, which are proportional to $1/\rho$, are given as:

$$H_\phi^{inc} = \frac{A_o}{\eta_c\rho}e^{-jk_cz} \tag{1.2}$$

$$E_\rho^{inc} = \frac{A_o}{\rho}e^{-jk_cz} \tag{1.3}$$

where $k_c = \omega\sqrt{\varepsilon_c}/c$ and $\eta_c = \sqrt{\mu_o/(\varepsilon_o\varepsilon_c)}$ are the propagation constant and the intrinsic wave impedance of the coaxial line, respectively. The ε_o ($\approx 8.854187817 \times 10^{-12}$ F/m) and μ_o ($\approx 4\pi \times 10^{-7}$ H/m) are the permittivity and permeability of free space. The A_o ($\approx 1/[\ln(b/a)]^{0.5}$) is the incident amplitude value of the field. In the numerical calculation, the value of A_o is assumed to be unity. The $\omega = 2\pi f$ is the angular frequency of the electromagnetic field and $j = (-1)^{1/2}$.

When one end of coaxial line is terminated into infinite half-space medium, the discontinuity at the end of the line ($z = 0$) will excite a reflected TEM wave and a series of higher-order modes, TM_{0n} in the coaxial line ($z < 0$). The reflected TEM waves are expressed as:

$$H_\phi^{ref} = \underbrace{\frac{A_o}{\eta_c\rho}\Gamma e^{jk_cz}}_{\text{Reflected TEM mode}} + \underbrace{j\omega\varepsilon_o\varepsilon_c\sum_{n=1}^{\infty}\frac{A_n}{\gamma_n}\Gamma_n e^{\gamma_n z}}_{\text{TM}_{0n}\text{ mode}} \tag{1.4}$$

$$E_\rho^{ref} = \underbrace{\frac{A_o}{\rho}\Gamma e^{jk_cz}}_{\text{Reflected TEM mode}} + \underbrace{\sum_{n=1}^{\infty}A_n\Gamma_n e^{\gamma_n z}}_{\text{TM}_{0n}\text{ mode}} \tag{1.5}$$

where Γ is the reflection coefficient for the TEM mode at $z = 0$. The reflected TEM waves are propagated in the reverse direction of the incident wave along the coaxial line. The subscript n in (1.4) and (1.5) represented the number of modes or mode orders. The symbol A_n is the radial dependence of the electric field and magnetic field of mode n, respectively, which is expressed as:

$$A_n = N_n\left[J_1(\lambda_n\rho)Y_o(\lambda_n a) - J_o(\lambda_n a)Y_1(\lambda_n\rho)\right] \tag{1.6}$$

where $J_m(x)$ is the Bessel function of the first kind and $Y_m(x)$ is the Bessel function of the second kind of order m. The factor N_n in (1.6) is defined by:

$$N_n = \frac{\pi\lambda_n}{\sqrt{2}}\left[\frac{J_o^2(\lambda_n a)}{J_o^2(\lambda_n b)} - 1\right]^{-\frac{1}{2}} \tag{1.7}$$

The Γ_n is the unknown complex reflection coefficient of mode n and $\gamma_n = \sqrt{\lambda_n^2 - k_c^2}$ is the propagation constant of mode n with $\mathrm{Re}(\gamma_n) > 0$. The λ_n is the discrete eigenvalue of the corresponding mode. The solution of Γ_n was reported in [14] and the procedure for finding the eigenvalue is available on the accompanying CD. In general, the total transverse field inside the coaxial line, ($z \le 0$), can be expressed as [5]:

$$H_\phi^- = \underbrace{\frac{A_o}{\eta_c\rho}\left(e^{-jk_cz} - \Gamma e^{jk_cz}\right)}_{\text{TEM mode}} - \underbrace{j\omega\varepsilon_o\varepsilon_c\sum_{n=1}^{\infty}\frac{A_n}{\gamma_n}\Gamma_n e^{\gamma_n z}}_{\text{TM}_{0n}\text{ mode}} \tag{1.8}$$

$$E_\rho^- = \underbrace{\frac{A_o}{\rho}\left(e^{-jk_cz} + \Gamma e^{jk_cz}\right)}_{\text{TEM mode}} + \underbrace{\sum_{n=1}^{\infty}A_n\Gamma_n e^{\gamma_n z}}_{\text{TM}_{0n}\text{ mode}} \tag{1.9}$$

1.2.2 The Scattering Wave from Coaxial Radiators ($z \ge 0$ Region)

1.2.2.1 Partial Differential Equations (Wave Equation)

All solutions to the problem of electromagnetic are always referred to as the four basic Maxwell's equations. In fact, the Maxwell's equations are mathematical expressions used to describe the law of electromagnetics. The implicit time-dependence Maxwell's equations give rise to Ampere's law, Faraday's law, and Gauss's law,

$$\vec{\nabla} \times \vec{H} = \vec{J} + j\omega\varepsilon\vec{E} \qquad \text{(Ampere's Law)} \tag{1.10}$$

$$\vec{\nabla} \times \vec{E} = -j\omega\mu\vec{H} \qquad \text{(Faraday's Law)} \tag{1.11}$$

$$\vec{\nabla} \cdot \vec{E} = \frac{Q_q}{\varepsilon} \qquad \text{(Gauss's Law)} \tag{1.12}$$

$$\vec{\nabla} \cdot \vec{H} = 0 \qquad \text{(Gauss's Law)} \tag{1.13}$$

where $\vec{E}$ is the electric field, $\vec{H}$ is the magnetic field, $\vec{J} = \sigma\vec{E}$ is the current density, Q_q is the charge density, and ω is the angular frequency. The symbols of $\varepsilon = \varepsilon_o\varepsilon_r$,

$\mu = \mu_o \mu_r$, and σ are the permittivity, permeability, and conductivity of a medium (homogeneous, isotropic, and linear), respectively.

The curl of (1.10) and the aid of identity of the vector Laplacian are given as:

$$\vec{\nabla} \times \vec{\nabla} \times \vec{H} = \vec{\nabla}\left(\vec{\nabla} \cdot \vec{H}\right) - \vec{\nabla}^2 \vec{H} = j\omega\varepsilon\left(\vec{\nabla} \times \vec{E}\right) \tag{1.14}$$

Assume that the $\vec{J} = 0$ in insulation medium or free space; replace (1.11) and (1.13) into (1.14):

$$\begin{aligned}\vec{\nabla} \times \vec{\nabla} \times \vec{H} = 0 - \vec{\nabla}^2 \vec{H} &= j\omega\varepsilon\left(-j\omega\mu\vec{H}\right) \\ &= \omega^2 \varepsilon\mu\vec{H}\end{aligned} \tag{1.15}$$

Rearranging (1.15) yields

$$\vec{\nabla}^2 \vec{H} + k^2 \vec{H} = 0 \tag{1.16}$$

For finite cylindrical symmetry cases based on the principle of TEM mode, the nonvanish components are H_ϕ, E_ρ, and E_z. Therefore, (1.16) can be simplified to a scalar equation for H_ϕ, yielded as

$$\left(\frac{\partial^2}{\partial\rho^2} + \frac{1}{\rho}\frac{\partial}{\partial\rho} - \frac{1}{\rho^2} + \frac{\partial^2}{\partial z^2} + k^2\right) H_\phi = 0 \tag{1.17}$$

The differential equation of (1.17) that governs the spatial of radiation is shown in Figure 1.6.

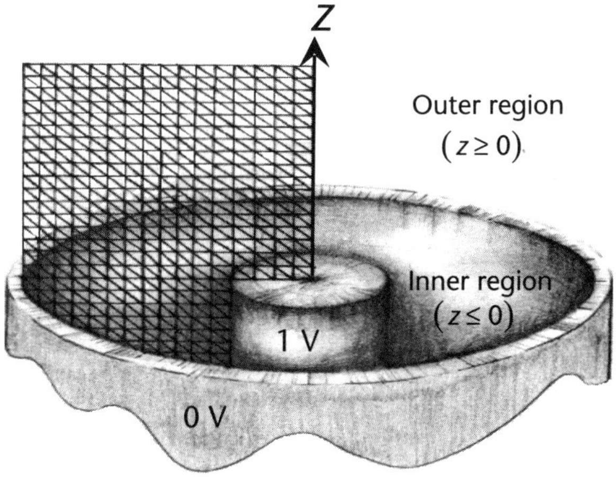

Figure 1.6 The cylindrical coordinate is independent of the angular coordinate ϕ.

1.2.2.2 Integral Equations

Double integrating both sides of (1.10) with respect to surface area, $\vec{S}$ of probe aperture is as:

$$\iint_S \vec{\nabla} \times \vec{H} \cdot d\vec{S} = j\omega\varepsilon \iint_S \vec{E} \cdot d\vec{S} \tag{1.18}$$

Using Stokes' theorem as

$$\oint \vec{H} \cdot d\vec{r} = \iint_S \vec{\nabla} \times \vec{H} \cdot d\vec{S} \tag{1.19}$$

reduces (1.18) to

$$\oint \vec{H} \cdot d\vec{r} = j\omega \iint \varepsilon \vec{E} \cdot d\vec{S} \tag{1.20}$$

The solution of scattering wave for the $\vec{E}$ field in (1.20) can be represented by

$$\begin{aligned} \vec{E} &= E_\rho \hat{\rho} + E_z \hat{z} \\ &= E_o \exp(-jkr)\hat{r} \end{aligned} \tag{1.21}$$

where k is wave propagation in the external medium at $z \geq 0$. The E_o is the magnitude of electric field. The relationship between the magnetic field and the electric field at $z \neq 0$ can be obtained from (1.20) and (1.21):

$$\int_{r_1}^{r_2} H_\phi^+ d\vec{r} = j\omega\varepsilon_o\varepsilon_r \int_a^b \int_0^\pi E_o \exp\left(-jk^+ r\right)\rho' d\phi' d\rho' \tag{1.22}$$

where the distance between two points, r, at the exterior medium is

$$r = \left|r_2 - r_1\right| = \sqrt{\rho^2 + \rho'^2 - 2\rho\rho'\cos\left(\phi' - \phi\right) + \left(z_2 - z_1\right)^2} \tag{1.23a}$$

At probe aperture, $(z_2 - z_1) = 0$, the r can be reduced as:

$$r = \sqrt{\rho^2 + \rho'^2 - 2\rho\rho'\cos\left(\phi' - \phi\right)} \tag{1.23b}$$

Upon integration of the left side of (1.22) leads to

$$H_\phi^+ r = j\omega\varepsilon_o\varepsilon_r \int_a^b \int_0^\pi E_o \exp\left(-jk^+ r\right)\rho' d\phi' d\rho' \tag{1.24}$$

The magnetic field intensity, H^+_ϕ, is the medium over the ground plane, which is a linear, isotropic, homogeneous, nonmagnetic material with relative complex permittivity; ε_r can be expressed in (1.25) as shown in Figure 1.7.

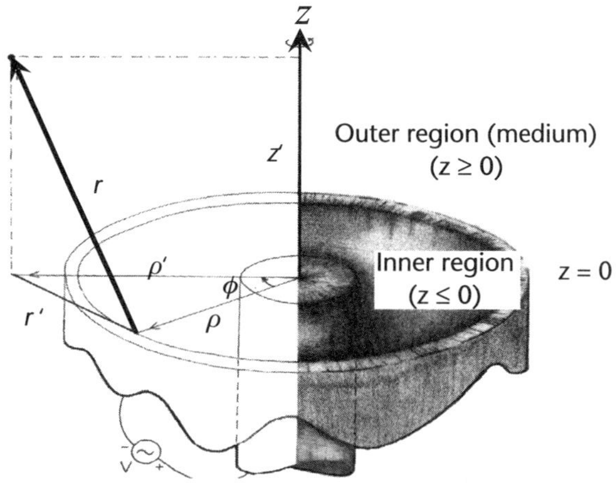

Figure 1.7 Spatial behavior of H_ϕ.

$$H_\phi^+ = \frac{j\omega\varepsilon_o\varepsilon_r}{\pi}\int_a^b\int_0^\pi E_o\frac{\exp(-jk^+r)}{r}\cos\phi'\rho' d\phi' d\rho' \qquad (1.25)$$

At discontinuity aperture, $z = 0$, the magnitude of radial electric field, E_o in (1.25) is given as:

$$E_o = \frac{V_o}{\rho'\ln(b/a)} \qquad (1.26)$$

where V_o is the voltage between the inner and outer conductors ($a < \rho < b$) at the aperture. For the V_o = 1/2 V excitation, (1.25) is rewritten as:

$$H_\phi^+ = \frac{j\omega\varepsilon_o\varepsilon_r}{2\pi\ln(b/a)}\int_a^b\int_0^\pi \frac{\exp(-jk^+r)}{r}\cos\phi' d\phi' d\rho' \qquad (1.27)$$

For the finite cylindrical symmetry cases, the divergence of (1.10) can be obtained [the derivation of (1.28) is available on the accompanying CD]

$$\frac{1}{\rho}\frac{\partial(\rho H_\phi)}{\partial\rho}\hat{z} - \frac{\partial H_\phi}{\partial z}\hat{\rho} = j\omega\varepsilon\left(E_\rho\hat{\rho} + E_z\hat{z}\right) \qquad (1.28)$$

By comparing the both sides of the unit vector for (1.28), and with the aid of (1.27), the integral formulations for E_ρ and E_z are given as:

$$E_z = \frac{1}{j\omega\varepsilon_o\varepsilon_r}\left[\frac{\partial H_\phi}{\partial\rho} + \frac{1}{\rho}H_\phi\right] \qquad (1.29a)$$

$$= \frac{1}{2\pi \ln(b/a)} \int_a^b \int_0^\pi \left\{ \begin{array}{l} \frac{\exp(-jk^+ r)}{r} \cos\phi' \\ \times \left[\frac{1}{\rho} - \left(jk^+ + \frac{1}{r} \right) \frac{(\rho - \rho' \cos\phi')}{r} \right] \end{array} \right\} d\phi' d\rho' \tag{1.29b}$$

and

$$E_\rho = \frac{-1}{j\omega\varepsilon_o\varepsilon_r} \frac{\partial H_\phi}{\partial z} \tag{1.30a}$$

$$= \frac{-1}{2\pi \ln(b/a)} \int_a^b \int_0^\pi \frac{\exp(-jk^+ r)}{r} \cos\phi' \left(jk^+ + \frac{1}{r} \right) (z_2 - z_1) d\phi' d\rho' \tag{1.30b}$$

For an open-ended coaxial radiator, the primary H_ϕ domain expressions [(1.25) or (1.27)] are often used in the derivation of the admittance aperture formula, which is further described in Chapter 2. Impedance formulations of monopole radiator that are derived from the E_z domain expression in (1.29) will be discussed in detail in Chapter 4. The MATLAB calculation program of (1.27) and (1.30b) is shown on the accompanying CD.

1.3 The Relationship Between Circuit Theory and Wave Theory

1.3.1 The Equivalent Circuit in the Coaxial Line ($z \leq 0$ Region)

The propagated electromagnetic wave along the coaxial line is caused by the mutual changing of electric energy and magnetic energy. According to circuit theory, an inductor of inductance, L_o (in H/m) is used to store or release only magnetic energy, while a capacitor of capacitance C_o (in F/m) stores or releases merely electric energy. Thus, the conversion process of electric energy and magnetic energy can be achieved by using equivalent circuit as shown in Figure 1.8. The equivalent coaxial line elements, L_o and C_o are given as:

$$L_o = \frac{\mu_o\mu_c}{2\pi} \ln(b/a) \tag{1.31}$$

$$C_o = \frac{2\pi\varepsilon_o\varepsilon_c}{\ln(b/a)} \tag{1.32}$$

whereas the electric field intensity, E_ρ, and the magnetic field intensity, H_ϕ, propagated in the coaxial line ($z < 0$, $a < \rho < b$), can be expressed in voltage, V, and current, I, terms as:

$$E_\rho = \frac{V}{\rho \ln(b/a)} \tag{1.33}$$

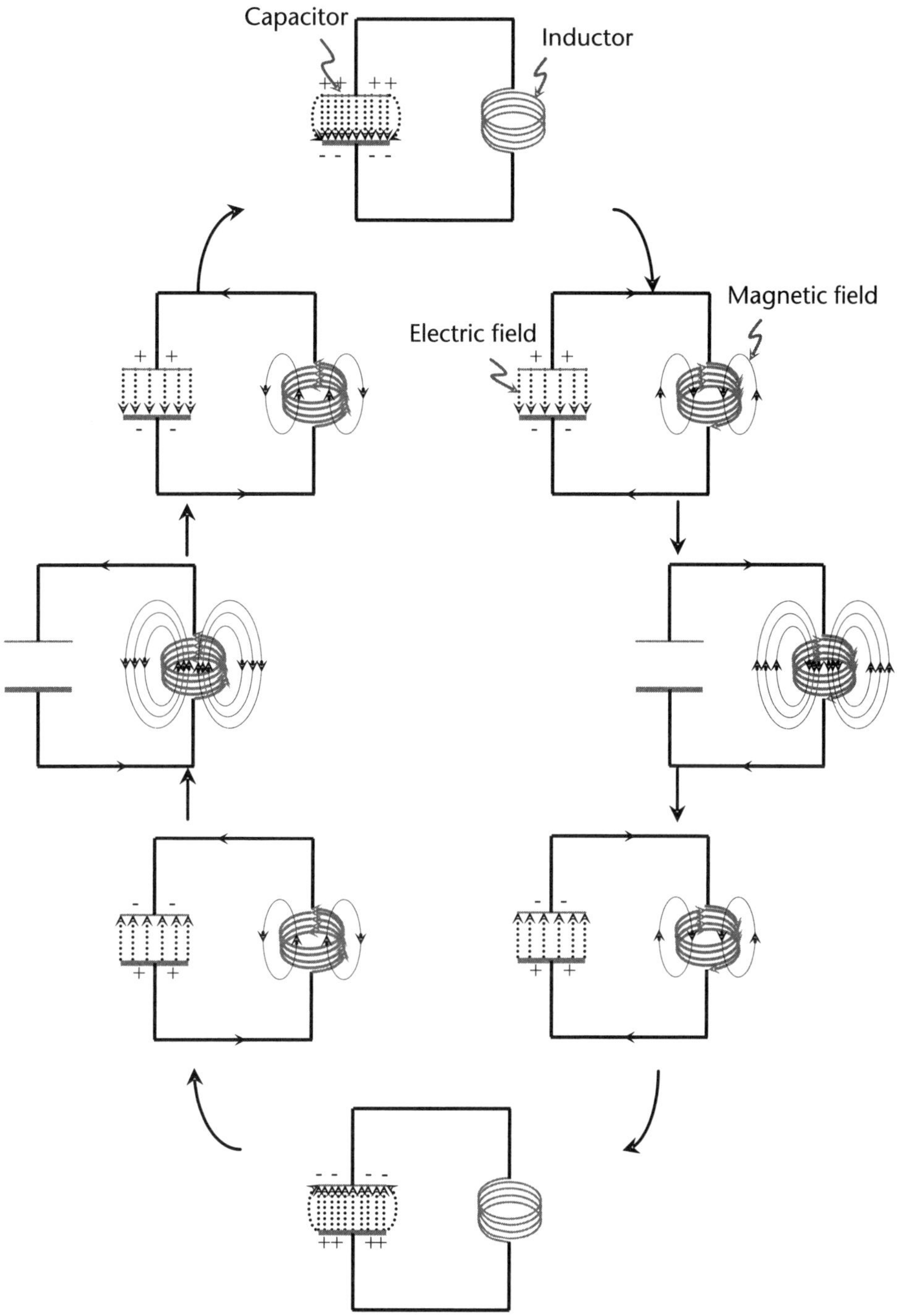

Figure 1.8 Cycle for the conversion of electric energy and magnetic energy.

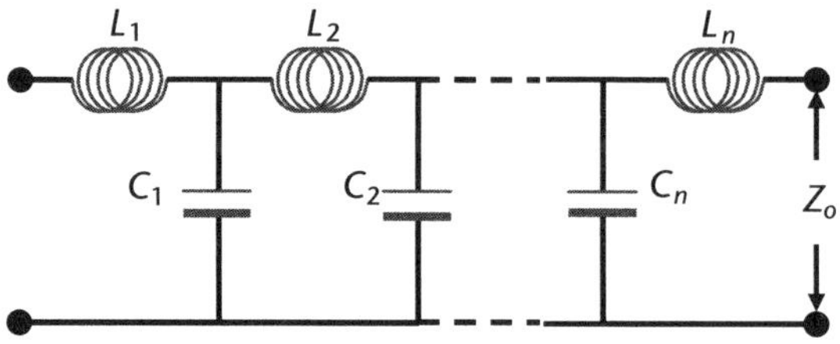

Figure 1.9 Equivalent circuit for input impedance, Z_{in}

$$H_\phi = \frac{I}{2\pi\rho} \tag{1.34}$$

The equivalent circuit in the ideal coaxial line is shown in Figure 1.9. The characteristic impedance, Z_o (in ohms), of the ideal coaxial line can be expressed as:

$$Z_o = \sqrt{\frac{L_o}{C_o}} \tag{1.35}$$

Furthermore, the propagation constant, k of the ideal coaxial line can also be expressed in L_o and C_o terms as:

$$k = j\omega\sqrt{L_o C_o} \tag{1.36}$$

In practice, the imperfect conductors and dielectric material in the coaxial line will cause attenuation in the wave during its propagation. To overcome this problem, a series resistance, R (in Ω/m), and a shunt conductance, G (in S/m) are added in the transmission equivalent circuit as shown in Figure 1.10. Based on Kirchhoff's voltage and current laws, the relations between the terminal voltage, V, and current, I, across the capacitor with capacitance, C (in F/m), and inductor with inductance, L (in H/m), respectively, in the frequency domain are given as [15, 16]:

$$\frac{\partial V}{\partial z} + (j\omega L + R)I = 0 \tag{1.37}$$

$$\frac{\partial I}{\partial z} + (j\omega C + G)V = 0 \tag{1.38}$$

The series resistance, R, of the coaxial line per unit length in (1.37) is the metal loss from the inner and outer conductors and the shunt conductance, G, of coaxial line per unit length in (1.38) represents the dielectric loss from the material filled in coaxial line. In [17], the four elements (R, L, G, and C) in (1.37) and (1.38) are expressed as:

$$R = 2\omega\delta L_o\left(1 - \frac{k^2 a^2 F_o}{2}\right)\left[\frac{1 + (b/a)}{4b\ln(b/a)}\right] \tag{1.39}$$

$$L = L_o\left\{1 + \delta\left(1 - \frac{k^2a^2F_o}{2}\right)\left[\frac{1 + (b/a)}{2b\ln(b/a)}\right]\right\} \tag{1.40}$$

$$G = \omega\delta C_o k^2a^2F_o\left[\frac{1 + (b/a)}{4b\ln(b/a)}\right] \tag{1.41}$$

$$C = C_o\left\{1 + k^2a^2F_o\left[\frac{1 + (b/a)}{4b\ln(b/a)}\right]\right\} \tag{1.42}$$

where $\delta = \sqrt{1/(\pi f\sigma\mu)}$ is the skin depth of conductor and function F_o in (1.39) through (1.42) is given as:

$$F_o = \frac{(b/a) - 1}{2\ln(b/a)} - \frac{(b/a)\ln(b/a)}{(b/a) + 1} - \frac{1}{2}\left(\frac{b}{a} - 1\right) \tag{1.43}$$

However, in [18], the characteristic impedance, Z_o is simplified by considering the different conductivity, σ between the inner and outer conductors of the coaxial line as:

$$Z_o = \frac{1}{2\pi}\sqrt{\frac{\mu_o\mu_c}{\varepsilon_o\varepsilon_c}}\ln\left(\frac{b}{a}\right) + (1 - j)\frac{1}{2\pi}\sqrt{\frac{1}{\pi\varepsilon_c\varepsilon_o f}}\left(\frac{1}{a\sqrt{\sigma_a}} + \frac{1}{b\sqrt{\sigma_b}}\right) \tag{1.44}$$

where σ_a and σ_b (in S/m) are the conductivities of the inner and outer conductors, respectively. Similar to (1.6) and (1.7) except for the higher-order modes, the voltage, V and current, I can also be expressed in terms of incident and reflected waves as:

$$V = V^+e^{-kz} + V^-e^{kz} = V^+\left(e^{-kz} + \Gamma e^{kz}\right) \tag{1.45a}$$

$$I = I^+e^{-kz} + I^-e^{kz} = I^+\left(e^{-kz} + \Gamma e^{kz}\right) \tag{1.45b}$$

where V^+ and V^- are the amplitudes of the incident and reflected voltages, respectively, and I^+ and I^- are the amplitudes of the incident and reflected currents, respectively. The propagation constant, k, for a lossy transmission line in (1.45a) is given as:

$$k = \alpha + j\beta = \sqrt{(R + j\omega L)(G + j\omega L)} \tag{1.46}$$

The attenuation constant, α, and the phase constant, β, inside the coaxial line in (1.46) are expressed as [1]:

$$\alpha = \frac{\sqrt{\pi\varepsilon_j 0\varepsilon_c f}}{2\ln(b/a)}\left(\frac{1}{a\sqrt{\sigma_a}} + \frac{1}{b\sqrt{\sigma_b}}\right) \tag{1.47a}$$

$$\beta = \frac{2\pi f}{c}\sqrt{\varepsilon_c} \tag{1.47b)]}$$

The Γ is the reflection coefficient, which is equal to $V^+/V^- = I^+/I^-$. From (1.37), (1.45b) can be rewritten in term of the voltage wave as:

$$I = \frac{V^+}{Z}e^{-kz} - \frac{V^-}{Z}e^{kz} \tag{1.48}$$

where the $Z = R + j\omega L$ is the series impedance along the transmission line as shown in Figure 1.10(a). By referring to Figure 1.10, the coaxial line terminated at surface $z = 0$ by a load with impedance Z_L. The input impedance, Z_{in} at a distance $z = -l$ from the load ($z = 0$) can be derived as:

$$Z_{\text{in}} = \frac{V(-l)}{I(-l)} = \frac{V^+(-l)\; e^{kl} + \Gamma_L e^{-kl}}{I^+(-l)\; e^{kl} - \Gamma_L e^{-kl}} = Z_o\frac{1 + \Gamma_L e^{-2kl}}{1 - \Gamma_L e^{-2kl}} \tag{1.49}$$

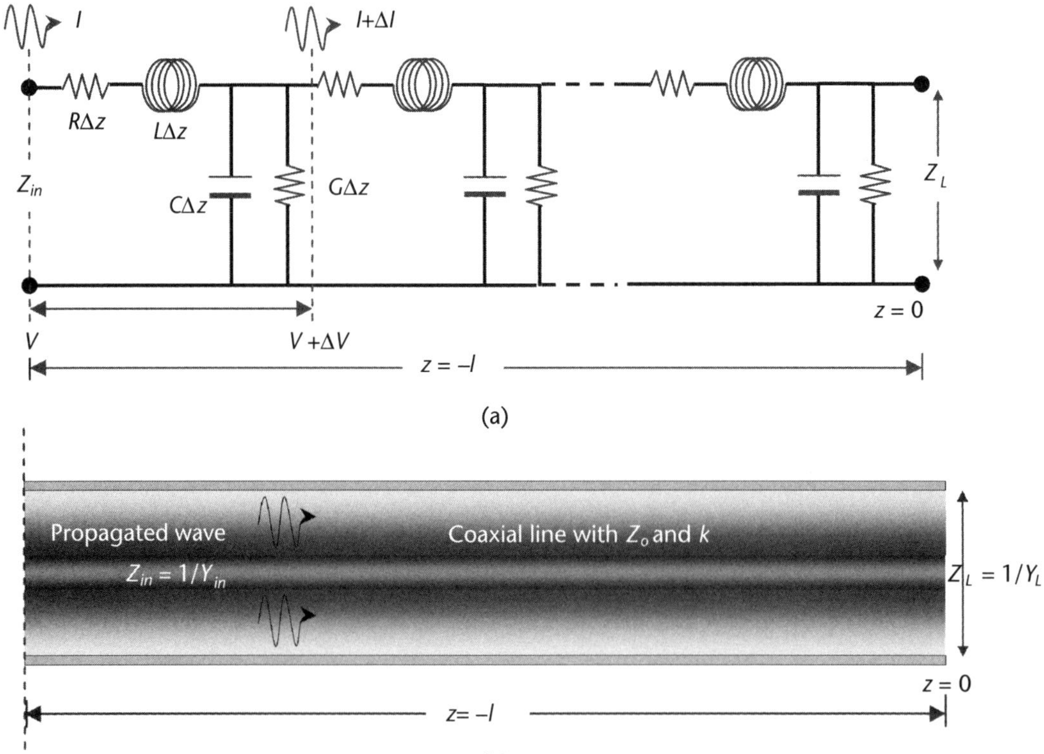

Figure 1.10 (a) Equivalent circuit of a segment of a transmission line. (b) Corresponding cross-sectional view of a physical coaxial line.

With the aid of (1.45) and (1.48), the load impedance, Z_L, at the end of the coaxial line ($z = 0$) can be expressed as

$$Z_L = \frac{V(0)}{I(0)} = Z_o \frac{V^+(0) + V^-(0)}{V^+(0) - V^-(0)} \tag{1.50}$$

From (1.50), the reflection coefficient Γ_L at $z = 0$ can be written as:

$$\Gamma_L = \frac{V^-(0)}{V^+(0)} = \frac{Z_L - Z_o}{Z_L + Z_o} \tag{1.51}$$

Then, substituting (1.51) into (1.50), it becomes:

$$\begin{aligned} Z_{\text{in}} &= Z_o \frac{(Z_L + Z_o)e^{kl} + (Z_L - Z_o)e^{-kl}}{(Z_L + Z_o)e^{kl} - (Z_L - Z_o)e^{-kl}} = Z_o \frac{Z_L \cosh(kl) + Z_o \sinh(kl)}{Z_L \sinh(kl) + Z_o \cosh(kl)} \\ &= Z_o \frac{Z_L + Z_o \tanh(kl)}{Z_o + Z_L \tanh(kl)} \end{aligned} \tag{1.52}$$

Indirectly, the admittance Y_{in} function can be written as:

$$Y_{\text{in}} = Y_o \frac{Y_L + Y_o \tanh(kl)}{Y_o + Y_L \tanh(kl)} \tag{1.53}$$

If the coaxial line in Figure 1.10(b) is lossless transmission line ($G \approx 0$ S/m and $R \approx 0$ Ω/m), the Y_{in} formula can be simplified by substituting (1.35) and (1.36) into (1.53), which yields

$$Y_{\text{in}} = \frac{Y_L + j\omega C_o l}{1 + Y_L j\omega L_o l} \tag{1.54}$$

The applications and further descriptions of (1.52), (1.53), and (1.54) are available in Chapter 8.

References

[1] Yee, K. S., "Numerical Solution of Initial Boundary Value Problems Involving Maxwell's Equations in Isotropic Media," *IEEE Trans. on Antennas Propagat.*, Vol. 14, 1966, pp. 302–307.

[2] Harrington, R. F., "Matrix Methods for Field Problems," *Proceedings of the IEEE*, Vol. 55, 1967, pp. 136–149.

[3] Silvester, P. P., "Finite Element Solution of Homogeneous Waveguide Problems," *Alta Freq.*, Vol. 38, 1969, pp. 313–317.

[4] Gajda, G. B., and S. S. Stuchly, "Numerical Analysis of Open-Ended Coaxial Lines," *IEEE Trans. on Microwave Theory Tech.*, Vol. 31, 1983, pp. 380–384.

[5] Nevels, R. D., C. M. Butler, and W. Yablon, "The Annular Slot Antenna in a Lossy Biological Medium," *IEEE Trans. on Microwave Theory Tech.*, Vol. 33, 1985, pp. 314–319.

[6] Blackham, D. V., and R. D. Pollard, "Finite Element Analysis of Open-Ended Coaxial Lines," *IEEE International Microware Symposium Digest*, Vol. 5, 1993, pp. 1247–1240.

[7] Mei, K. K., "Unimoment Method of Solving Antenna and Scattering Problems," *IEEE Trans. on Antennas Propagat.*, Vol. 22, 1974, pp. 760–766.

[8] Ansoft HFSS, *Getting Started with HFSS*, Canonsburg: ANSYS Inc., 2010.

[9] COMSOL Multiphysics, *RF Module Model Library Version 3.5a*, Sweden: COMSOL AB, TegnŽrgatan, 2008.

[10] Mur, G., "Absorbing Boundary Conditions for the Finite-Difference Approximation of the Time-Domain Electromagnetic Field Equations," *IEEE Trans. on Electromagnetic Compatibility*, Vol. 23, 1981, pp. 377–382.

[11] Misra, D. K., "A Study on Coaxial Line Excited Monopole Probes for *In-Situ* Permittivity Measurements," *IEEE Trans. on Instrum. Meas.*, Vol. 30, 1987, pp. 46–51.

[12] Sumbar, E., F. E. Vermeulen, and F. S. Chute, "Implementation of Radiation Boundary Conditions in the Finite Element Analysis of Electromagnetic Wave Propagation," *IEEE Trans. on Microwave Theory Tech.*, Vol. 39, 1991, pp. 267–273.

[13] Maloney, J. G., G. S. Smith, and W. R. Scott, "Accurate Computation of the Radiation from Simple Antennas using the Finite-Difference Time-Domain Method," *IEEE Trans. on Antennas Propagat.*, Vol. 38, 1990, pp. 1059–1068.

[14] Mosig, J. R., et al., "Reflection of an Open-Ended Coaxial Line and Application to Non-Destructive Measurement of Materials," *IEEE Trans. on Instrum. Meas.*, Vol. 30, 1981, pp. 46–51.

[15] Ishimaru, A., *Electromagnetic Wave Propagation, Radiation, and Scattering*, London: Prentice-Hall, 1991.

[16] Marcuvitz, N., *Waveguide Handbook*, Boston, MA: Boston Technical Publishers, 1964.

[17] Daywitt, W. C., "First-Order Symmetric Modes for a Slightly Lossy Coaxial Transmission Line," *IEEE Trans. on Microwave Theory Tech.*, Vol. 38, 1990, pp. 1644–1651.

[18] Agilent Technologies, Specifying Calibration Standards and Kits for Agilent Vector Network Analyzers: Application Note 1287-11, USA: Agilent Technologies Inc., 2010.

Selected Bibliography

Balanis, C. A., *Advanced Engineering Electromagnetic*, New York: John Wiley & Sons, 1989.

Barton, M. L., and Z. J. Cendes, "New Vector Finite Elements for Three-Dimensional Magnetic Field Computation," *J. Appl. Phys.*, Vol. 61, 1987, pp. 3919–3921.

Burke, G. J., E. K. Miller, and A. J. Poggio, "The Numerical Electromagnetics Code (NEC)-A Brief History," *International Symposium IEEE Antennas and Propagation Society*, Vol. 3, 2004, pp. 2871–2874.

CST Microwave Studio, *CST Microwave Studio—Workflow & Solver Overview*, Darmstadt, Germany: CST Computer Simulation Technology, 2008.

Davidson, D. B., *Computational Electromagnetic for RF and Microwave Engineering*, Cambridge, England: Cambridge University Press, 2005.

Galejs, J., *Antennas in Inhomogeneous Media*, Oxford, U.K.: Pergamon, 1969.

Garg, R., *Analytical and Computational Methods in Electromagnetic*, Norwood, MA: Artech House, 2008.

Gibson, W. C., *The Method of Moments in Electromagnetics*, London: Chapman & Hall/CRC, 2008.

Harrington, R. F., *Field Computation by Moment Methods*, New York: IEEE Press-John Wiley & Sons, 2000.

Jin, J. M., *The Finite Element Method in Electromagnetics*, New York: John Wiley & Sons, 1993.

King, R. W. P., "The Linear Antenna-Eighty Years of Progress," *Proceedings of the IEEE*, Vol. 55, 1967, pp. 2–16.

Levine, H., and C. H. Papas, "Theory of the Circular Diffraction Antenna," *J. Appl. Phys.*, Vol. 22, 1951, pp. 29–43.

Li, C. L., and K. M. Chen, "Determination of Electromagnetic Properties of Materials Using Flanged Open-Ended Coaxial Probe-Full-Wave Analysis," *IEEE Trans. on Instrum. Meas.*, Vol. 44, 1995, pp. 19–27.

Maxwell, J. C., *A Treatise on Electricity and Magnetism*, New York: Dover, 1954.

Noh, Y. C., and H. J. Eom, "Radiation from a Flanged Coaxial Line into a Dielectric Slab," *IEEE Trans. on Microwave Theory Tech.*, Vol. 47, 1999, pp. 2158–2161.

Panariello, G., L.Verolino, and G. Vitolo, "Efficient and Accurate Full-Wave Analysis of the Open-Ended Coaxial Cable," *IEEE Trans. on Microwave Theory Tech.*, Vol. 49, 2001, pp. 1304–1309.

Ridler, N. M., "Connectors, Air Lines and RF Impedance," *IEE Measurement, Sensors, Instrumentation and NDT Professional Network*, 2005.

Sadiku, M. N. O., *Numerical Techniques in Electromagnetic*, Boca Raton, FL: CRC Press, 2000.

Silvester, P. P., and R. L. Ferrari, *Finite Elements for Electrical Engineers*, England: Cambridge University Press, 1994.

Sullivan, D. M., *Electromagnetic Simulation Using the FDTD Method*, New York: IEEE Press, 2000.

Tsai, L. L., "A Numerical Solution for the Near and Far Fields of an Annular Ring of Magnetic Current," *IEEE Trans. on Antennas Propagat.*, Vol. 20, 1972, pp. 569–576.

Wadell, B. C., *Transmission Line Design Handbook*, Norwood, MA: Artech House, 1991.

Weiner, M. M., et al., *Monopole Elements on Circular Ground Plane*, Norwood, MA: Artech House, 1987.

Wexler, A., Computation of Electromagnetic Fields, *IEEE Trans. on Microwave Theory Tech.*, Vol. 17, 1969, pp. 416–439.

Yarman, B. S., *Design of Ultra Wideband Antenna Matching Networks*, New York: Springer Science, 2008.

You, K. Y., et al., "Modeling of Coaxial Slot Waveguides Using Analytical and Numerical Approaches: Revisited," *International Journal of Antennas and Propagation*, Vol. 2012, 2012, pp. 1–12.

CHAPTER 2

Analytical Modeling of Open-Ended Coaxial Line Terminated by Half-Space Infinite Media

This chapter presents the analytical modeling of open-ended coaxial probe, which is terminated by the half-space infinite lossy media. Three models are analyzed in this chapter, namely, the equivalent lumped-element circuit model, the integral admittance model, and the rational function model. The equivalent lumped-element circuit modeling is simple and easier to formulate an electromagnetic field (E_ρ and H_ϕ) at the coaxial probe aperture as compared to integral expressions. Normally, the modeling has three kinds of circuit elements, which are the resistance R, capacitance C, and inductance L. In general, the values of R, C, and L in the model are obtained either by a full-wave simulation or by measurement. This kind of semi-empirical model has a finite number of lumped elements, which is dependent on the operating frequency range with a desired accuracy. For an unbounded wave scattering case, such as radiation field at coaxial probe aperture, the analytical model in the integral form is often used. Typically, the integration of the admittance equation will contain a Green's function, G, or the kernel function for the integral formulation, which is satisfied by radiated fields at the open end of the waveguide and can be expressed as:

$$G = \frac{e^{-jk|r_2 - r_1|}}{4\pi|r_2 - r_1|}$$

Thus, the exponential properties of above expression can be found in the integral admittance equations in Section 2.3. The form of the rational function that is used to model the coaxial probe that corresponds to the lumped element equivalent circuit model will be described in detail in Section 2.4.

2.1 Theoretical Background

The aperture admittance integral equations for open-ended coaxial radiator can be derived by matching the field components (1.34) (inner region) and (1.25) (outer region), since those tangential fields are continuous across the interface aperture ($z = 0$) as shown in Figure 2.1.

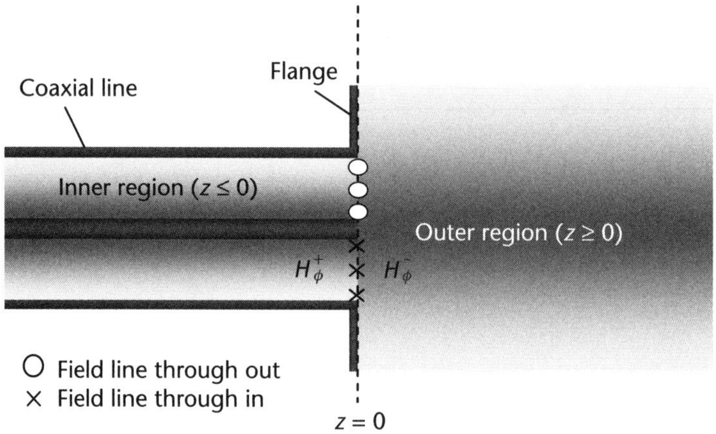

Figure 2.1 Continuity of the azimuthal magnetic field H_ϕ at the aperture surface, $z = 0$.

$$H_\phi^- = H_\phi^+ \tag{2.1a}$$

$$\frac{I}{2\pi\rho} = \frac{j\omega\varepsilon_o\varepsilon_r}{\pi}\int_a^b\int_0^\pi E_o \frac{\exp(-jk^+r)}{r}\cos\phi'\rho' d\phi' d\rho' \tag{2.1b}$$

Ignoring the effects of TM_{0n} modes in (2.1b) leads to a relatively simple mathematical solution. The current I in (2.1b) is substituted into the relationship, $I = VY$ and $V = \int_a^b E_o d\rho$, yielding

$$\frac{Y}{2\pi\rho}\int_a^b E_o d\rho = \frac{j\omega\varepsilon_o\varepsilon_r}{\pi}\int_a^b\int_0^\pi E_o \frac{\exp(-jk^+r)}{r}\cos\phi'\rho' d\phi' d\rho' \tag{2.1c}$$

where Y is the input admittance at the surface, $z = 0$. Multiplying both sides of (2.1c) by ρE_o and integrating from $\rho = a$ to b over the aperture yield

$$\frac{Y}{2\pi}\int_a^b\int_a^b E_o^2 d\rho d\rho = \frac{j\omega\varepsilon_o\varepsilon_r}{\pi}\int_a^b\int_a^b\int_0^\pi \rho E_o E_o \frac{\exp(-jk^+r)}{r}\cos\phi'\rho' d\phi' d\rho' d\rho \tag{2.2}$$

The E_o in (2.2) is assumed to be $1/\rho$. From (1.26), the double integration of E_o in (2.2) can be written as:

$$\frac{1}{|V|^2}\int_a^b\int_a^b E_o \cdot E_o d\rho d\rho' = \frac{1}{[\ln(b/a)]^2}\int_a^b\int_a^b \frac{1}{\rho\rho'} d\rho d\rho' \tag{2.3}$$

Substituting the relationship of (2.3) into (2.2) and normalizing the (2.2) by dividing by a characteristic admittance, Y_o, in (2.5). Finally, the normalized admittance, $\tilde{Y}$, is given as (2.11).

2.2 Equivalent Lumped-Element Circuits Modeling

The closed-form capacitance admittance equation (2.4) is based on a model of the discontinuity, at the termination of the coaxial line, as an equivalent lumped circuit [1–3] in Figure 2.2.

$$\tilde{Y} = j\omega\left(\varepsilon_r C_o + C_f\right)/Y_o \tag{2.4}$$

The total capacitance, C_T at the aperture probe consists of a part that is dependent on the permittivity of the test sample, C_o, and a part, C_f, which results from the fringe field inside the coaxial line filled by Teflon. The symbol Y_o is the characteristic admittance of coaxial line and is given as:

$$Y_o = \frac{2\pi}{\ln(b/a)}\sqrt{\frac{\varepsilon_o \varepsilon_c}{\mu_o}} \tag{2.5}$$

where ε_o and μ_o are the free space permittivity and permeability, respectively, ε_c is the dielectric constant of the material (Teflon) filling the coaxial line. Nevertheless, (2.4) can only be used accurately on frequency ranging from dc up to a few gigahertz. It is due to the simple capacitance circuit, which is not sufficient to model the admittance curves that exist at high frequency.

Brady et al. [4] and Stuchly et al. [5] improved (2.4) by including the conductance term, G_o, in (2.6), which is due to losses, especially for the aperture of the coaxial line, which is contacted with lossy samples.

$$\tilde{Y} = \left[j\omega\left(\varepsilon_r C_o + C_f\right) + G_o\left(\varepsilon_r\right)^{2.5}\right]/Y_o \tag{2.6}$$

From Gajda et al. [6, 7], the scalar finite element simulation results indicate that the relationship between capacitance, C, at the aperture coaxial probe, permittivity of free space, and the size of a probe is expressed as:

$$C = C_o + C_f = 2.48\varepsilon_o(b - a) \tag{2.7}$$

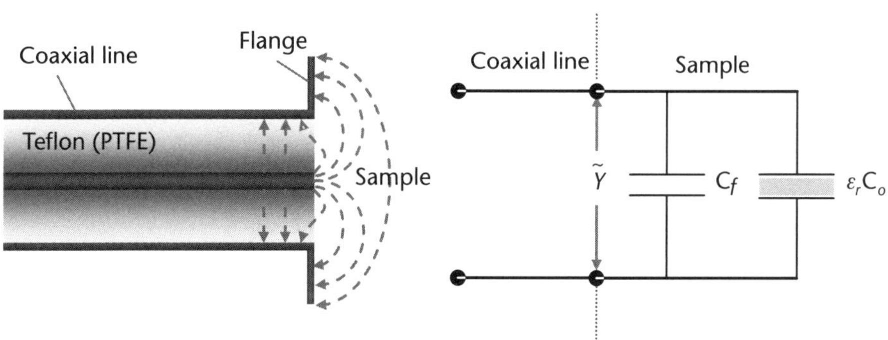

Figure 2.2 Open-ended coaxial line. (a) Cross-section geometry. (b) Equivalent circuit.

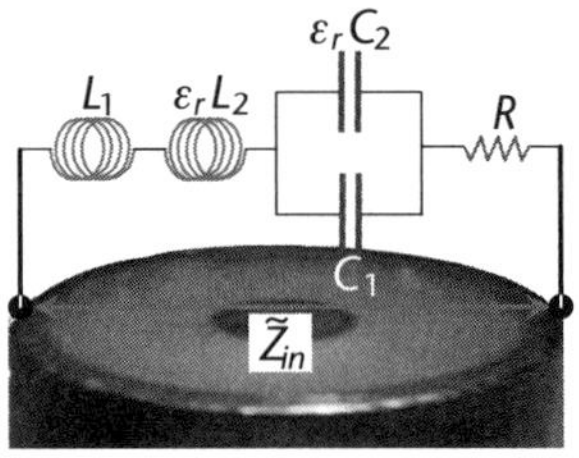

Figure 2.3 Equivalent circuit for input impedance at aperture coaxial probe.

The method of moments results obtained by [6, 7] indicated the function of capacitance C as follows:

$$C = 2.38\varepsilon_o(b - a) \tag{2.8}$$

where a is the outer radius of the inner conductor (in meters) and b is the inner radius of the outer conductor (in meters). However, the single capacitance, C, and the conductance, G_o, are not sufficient to represent the equivalent circuit for the aperture probe, which interacts with lossy materials at a high-frequency operation. You et al. [8, 9] proposed that the inductance, L, and the resistance, R, should be taken into account in the circuit model, as shown in Figure 2.3. The composition of the C, R, and L elements refer to the input impedance before it is converted to the admittance function.

The normalized input impedance, $\tilde{Z}$, and admittance, $\tilde{Y}$, are given as:

$$\tilde{Z} = Y_o\left[\frac{1}{j\omega\left(C_1 + \varepsilon_r C_2\right)} + j\omega\left(L_1 + \varepsilon_r L_2\right) + R\right] \tag{2.9a}$$

$$\tilde{Y} = \frac{1}{\tilde{Z}} \tag{2.9b}$$

The capacitances C_1 and C_2 in (2.9a) have the same meaning as in (2.4). L_1 and L_2 are the inductive properties inside the coaxial line near the aperture probe and the inductive properties in an external sample under test, respectively. R is the resistive properties at the aperture probe. The values of components C_1, L_1, L_2, and R are obtained by optimization between the normalized input impedance, $\tilde{Z}_{in}$, and the finite-element simulation results, while the C_2 values in (2.9a) are calculated from (2.7) or (2.8).

2.3 Admittance Modeling

2.3.1 Admittance Integral Modeling

The first rigorous integral of aperture admittance, $\tilde{Y}$, of the open-ended coaxial waveguide was derived using Sommerfeld's concept by Levine and Papas in 1951 [10].

$$\tilde{Y} = \frac{jk_2^2}{k_1 \ln(b/a)} \int_0^{\infty} \frac{d\zeta}{\zeta\left(\zeta^2 - k_2^2\right)^{1/2}} \left|J_o(\zeta a) - J_o(\zeta b)\right|^2 \quad (2.10)$$

where k_2 and k_1 are the propagation constant for the half-space exterior medium and coaxial line, respectively. J_o is the Bessel function of the zeroth order. The rigorous formulation (2.10) was investigated by Misra et al. [11], who successfully expressed the model as:

$$\tilde{Y} = \frac{jk_2^2}{\pi k_1 \ln(b/a)} \int_a^b \int_a^b \int_0^{\pi} \frac{\exp\left(-jk_2 r\right)}{r} \cos\phi' d\phi' d\rho' d\rho \quad (2.11)$$

In the following relationship:

$$\int_0^{\pi} \frac{\exp\left(-jk_2 r\right)}{r} \cos\phi d\phi = \pi \int_0^{\infty} \frac{J_1(\zeta\rho) J_1(\zeta\rho')\zeta}{\left(\zeta^2 - k_2^2\right)^{1/2}} d\zeta \quad (2.12)$$

where r is the distance between two points with $(z_2 - z_1) = 0$ [refer to (1.23)]. The calculated admittance of (2.11) can be solved by using the Gaussian triple integral method. However, the analytical equation is only suitable for ideal conditions in which the electric field component, E_z, has been neglected in (2.10) and (2.11) as shown in Figure 2.4(a). In practical situations, besides the components E_ρ field, the component E_z is also present on the surface $z = 0$ as shown in Figure 2.4(b).

Typically, the component E_z has been neglected to facilitate solving the problem. In order to simplify the calculation of component E_z (corresponding to the TM_{0n} modes), the component E_z can be included in the calculation of the admittance (2.10) in the form of the correction terms of C_f and G_f [9]:

$$\tilde{Y}_{\text{Corrected}} = \tilde{Y} - \frac{\omega C_f}{Y_o} \pm j\frac{\omega^{5/2}\varepsilon_r G_f}{Y_o} \quad (2.13)$$

The fringing fields C_f and G_f are obtained by fitting the normalized admittance, $\tilde{Y}$, values from (2.19) with the calculated values obtained from the finite-element method (FEM). Fitting measured values to FEM is more accurate and easier for

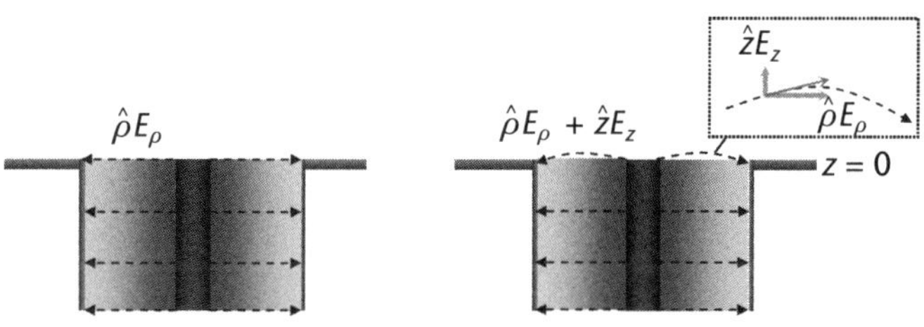

Figure 2.4 The propagated wave from the coaxial line ($z < 0$) into free space ($z > 0$) through the ground plane ($z = 0$). (a) Ideal and (b) actual.

samples with a broad range of permittivity values. For free space that is lossless ($\varepsilon_r = 1 - j0$), the capacitance used as function of the inner and outer radii a and b, respectively, the operation frequency, f, and the relative effective permittivity, ε_b, at aperture sensor are expressed as:

$$C_f = \varepsilon_o a \left[\begin{array}{l} \left(\alpha_1 \varepsilon_b^2 + \alpha_2 \varepsilon_b + \alpha_3\right)\left(\dfrac{b}{a}\right)^2 + \left(\dfrac{\alpha_4}{\varepsilon_b^2} + \dfrac{\alpha_5}{\varepsilon_b} + \alpha_6\right)\dfrac{b}{a} \\ + \left(\dfrac{\alpha_7}{\varepsilon_b^2} + \dfrac{\alpha_8}{\varepsilon_b} + \alpha_9\right) \end{array} \right] - \varepsilon_o \kappa \quad (2.14)$$

where ε_o and ε_c are the permittivities of free space and the dielectric material of the coaxial line, respectively. The effective relative permittivity ε_b at the interface ($z = 0$) between the sensor and the free space medium is given as:

$$\varepsilon_b = \left(1 + \varepsilon_c\right)/2 \quad (2.15)$$

The coefficients α_1 to α_9 and κ are listed on the accompanying CD. At $z = 0$, the propagation constant, $\gamma = \pm k_o \sqrt{\varepsilon_r - \varepsilon_c}$, where k_o is the propagation constant of free space. Obviously, γ will have different values for the two different situations $\varepsilon_r < \varepsilon_c$ and $\varepsilon_r > \varepsilon_c$. The fringing field capacitance C_f for lossy materials with permittivity ε_r that satisfy the condition ($\varepsilon_c + 1 \le \varepsilon_r \le 80$) can be expressed in the following form:

$$C_f = \varepsilon_o \left[\begin{array}{l} \left(\beta_1 b + \beta_2 a + \beta_3\right)\varepsilon_r \varepsilon_c + \left(\beta_4 b + \beta_5 a + \beta_6\right)\varepsilon_r \\ + \left(\beta_7 b + \beta_8 a + \beta_9\right)\varepsilon_c + \left(\beta_{10} b + \beta_{11} a + \beta_{12}\right) \end{array} \right] \quad (2.16)$$

The coefficients β_1 to β_{12} are given on the accompanying CD. For lossy materials, the imaginary part of relative permittivity, ε''_r, is high mainly due to high conductivity, the σ value of a sample from the relationship of $\varepsilon''_r = \sigma/(2\pi\varepsilon_o f)$. Thus, the conductive effect, G_f, becomes significant above 1.5 GHz for aperture admittance with approximately 3×10^{-31}. For the case $\varepsilon_r < \varepsilon_c$, the conductive correction term in (2.13) has a positive sign (+), and the opposite sign (–) when $\varepsilon_r > \varepsilon_c$. Both (2.14) and (2.16) are valid for coaxial sensors that satisfy the conditions: ($0.25 < a < 1.5$) mm, ($1.18 < b/a < 4$) mm, and ($1 < \varepsilon_c < 4.5$) within a range of operating frequencies, f:

$$\mathrm{DC} \le f \le \left[\frac{c}{4\pi(b - a)\sqrt{\varepsilon_r f}} \right]^2 \quad (2.17)$$

where c is the velocity of light in free space. The deviation between the measured and calculated admittances of air can be reduced to less than 1% by including the fringing field correction terms (2.14) for arbitrary ratio, b/a, and values of ε_c. Figures 2.5 and 2.6 show the relative error contour between the calculated imaginary parts of (2.13) (susceptance, $B(0)/Y_o$) by adding the fringing term of (2.16) and FEM simulation results for various sizes of sensor. The region for the relative error

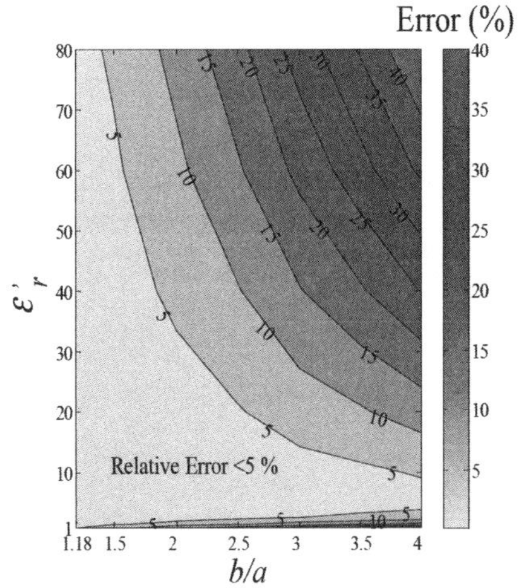

Figure 2.5 The contour for the percentage of relative error between calculated and simulated susceptances, $B(0)/Y_o$, vary with the probe ratio, b/a, and the relative permittivity, ε_r, medium (for $a = 0.65$ mm, $\varepsilon_c = 2.06$, and $f = 3$ GHz).

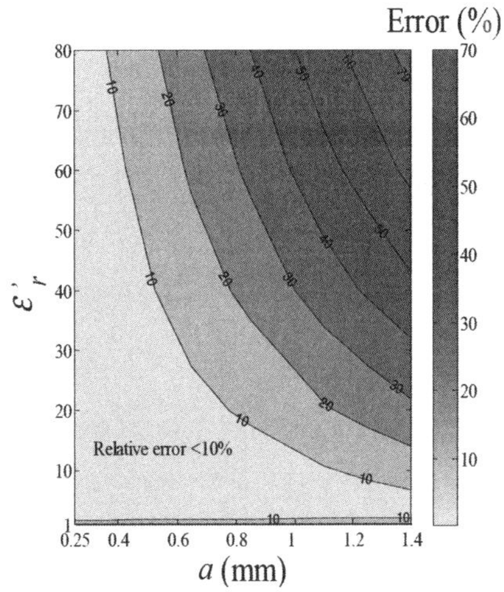

Figure 2.6 The contour for the percentage of relative error between calculated and simulated susceptances, $B(0)/Y_o$, vary with the radius of inner conductor, a, and the relative permittivity, ε_r, medium (for $b/a = 3$, $\varepsilon_c = 2.06$ and $f = 3$ GHz).

less than 10%, as shown in Figures 2.5 and 2.6, is the region where the operating frequency ranges between the calculation susceptance, $B(0)/Y_o$, without exceeding the condition of (2.17).

2.3.2 Series Admittance Solutions

The earliest derivation of the integral admittance model by Levine and Papas [10] concerned the homogeneous case, in which an air-filled coaxial line with an infinite conducting flange is radiated into the free space. The integral admittance equation was rewritten by Marcuvitz [12] for arbitrary coaxial filled material and exterior material, in real and imaginary parts of aperture admittance as:

$$\tilde{Y} = \frac{G(0)}{Y_o} + j\frac{B(0)}{Y_o} \tag{2.18}$$

The real part and imaginary part are called normalized conductance, $G(0)/Y_o$, and susceptance, $B(0)/Y_o$, respectively, which are expressed as:

$$\frac{G(0)}{Y_o} = \frac{k_2}{k_1 \ln(b/a)} \int_0^{\pi/2} \frac{d\theta}{\sin\theta} \left[J_o\left(k_2 a \sin\theta\right) - J_o\left(k_2 b \sin\theta\right) \right]^2 \tag{2.19}$$

$$\frac{B(0)}{Y_o} = \frac{k_2}{\pi k_1 \ln(b/a)} \int_0^{\pi} d\phi \begin{Bmatrix} 2Si\left(k_2 a |r|\right) - Si\left(2k_2 a |\sin(\phi/2)|\right) \\ -Si\left(2k_2 b |\sin(\phi/2)|\right) \end{Bmatrix} \tag{2.20}$$

where $r^2 = a^2 + b^2 - 2ab\cos\phi$ and Si are the sine integral or $Si(x) = \int_0^x \frac{\sin t}{t} dt$. The $G(0)/Y_o$ and $B(0)/Y_o$ can be solved analytically using an expansion method such as [13]:

$$\frac{G(0)}{Y_o} = \xi \sum_{m=0}^{N} G'_m I_m k_2^{2m+4} + \xi E'_{N+1} \left(k_2 a\right)^{2N+6} \tag{2.21}$$

$$\frac{B(0)}{Y_o} = \frac{\xi}{\pi} \sum_{m=0}^{N} B_m k_2^{2m+1} + \frac{\xi}{\pi} E''_{N+1} \left(k_2 a\right)^{2N+3} \tag{2.22}$$

where

$$I_m = \left[\frac{2^{2(m+1)} \left[(m+1)!\right]^2}{(2m+3)!} \right] \tag{2.23}$$

$$k_2 = \left(\frac{2\pi f}{c} \sqrt{\varepsilon_r} \right) \tag{2.24}$$

$$\xi = \frac{\sqrt{\varepsilon_r}}{\sqrt{\varepsilon_c} \ln(b/a)} \tag{2.25}$$

where ε_r and ε_c are the relative permittivities for the exterior sample medium and the coaxial line filled material, respectively. The conductivity in (2.25) takes into account the metal loss from the inner and outer conductors. For m = even (0, 2, 4, ...), the coefficient value in (2.21) is given as:

$$G'_0 = \left(\frac{a^2 - b^2}{2^2} \right)^2 \tag{2.26a}$$

$$G_2' = \left[\left(\frac{a^4 - b^4}{2^2 \cdot 4^2}\right)^2 + \left(\frac{a^2 - b^2}{2^2}\right)\left(\frac{a^6 - b^6}{2^2 \cdot 4^2 \cdot 6^2}\right)\right] \tag{2.26b}$$

$$G_4' = \begin{bmatrix} \left(\frac{a^6 - b^6}{2^2 \cdot 4^2 \cdot 6^2}\right) + \left(\frac{a^4 - b^4}{2^2 \cdot 4^2}\right)\left(\frac{a^8 - b^8}{2^2 \cdot 4^2 \cdot 6^2 \cdot 8^2}\right) \\ + \left(\frac{a^2 - b^2}{2^2}\right)\left(\frac{a^{10} - b^{10}}{2^2 \cdot 4^2 \cdot 6^2 \cdot 8^2 \cdot 10^2}\right) \end{bmatrix} \tag{2.26c}$$

For m = odds (1, 3, 5, ...)

$$G_1' = -\left[\left(\frac{a^2 - b^2}{2^2}\right)\left(\frac{a^4 - b^4}{2^2 \cdot 4^2}\right)\right] \tag{2.26d}$$

$$G_3' = -\left[\left(\frac{a^4 - b^4}{2^2 \cdot 4^2}\right)\left(\frac{a^6 - b^6}{2^2 \cdot 4^2 \cdot 6^2}\right) + \left(\frac{a^2 - b^2}{2^2}\right)\left(\frac{a^8 - b^8}{2^2 \cdot 4^2 \cdot 6^2 \cdot 8^2}\right)\right] \tag{2.26e}$$

$$G_5' = -\begin{bmatrix} \left(\frac{a^6 - b^6}{2^2 \cdot 4^2 \cdot 6^2}\right)\left(\frac{a^8 - b^8}{2^2 \cdot 4^2 \cdot 6^2 \cdot 8^2}\right) + \left(\frac{a^4 - b^4}{2^2 \cdot 4^2}\right)\left(\frac{a^{10} - b^{10}}{2^2 \cdot 4^2 \cdot 6^2 \cdot 8^2 \cdot 10^2}\right) \\ + \left(\frac{a^2 - b^2}{2^2}\right)\left(\frac{a^{12} - b^{12}}{2^2 \cdot 4^2 \cdot 6^2 \cdot 8^2 \cdot 10^2 \cdot 12^2}\right) \end{bmatrix} \tag{2.26f}$$

The series of (2.22) are conveniently rewritten in a single expression as:

$$B_m = \begin{Bmatrix} \left[(-1)^m \cdot \int_0^\pi \frac{2 \cdot R^{(2m+1)/2}}{(2m+1) \cdot [(2m+1)!]} d\phi\right] \\ -\left[\frac{(-1)^m \cdot (2)^{2(2m+1)}(n!)^2}{(2m+1) \cdot [(2m+1)!]^2}\right]\left(a^{2m+1} + b^{2m+1}\right) \end{Bmatrix} \tag{2.27}$$

The integration term in (2.27) can be accurately solved by numerical integral solution. Here, the integration solved by the second level of Boole's rule as:

$$\int_0^\pi R^{(2m+1)/2} d\phi = \frac{2h}{45}\begin{Bmatrix} 7R_1^{(2m+1)/2} + 32R_2^{(2m+1)/2} + 12R_3^{(2m+1)/2} \\ +32R_4^{(2m+1)/2} + 14R_5^{(2m+1)/2} + 32R_6^{(2m+1)/2} \\ +12R_7^{(2m+1)/2} + 32R_8^{(2m+1)/2} + 7R_9^{(2m+1)/2} \end{Bmatrix} \tag{2.28}$$

where $h = \pi/8$ and

$R_1 = a^2 + b^2 - 2ab\cos(0)$; $R_2 = a^2 + b^2 - 2ab\cos(\pi/8)$

$R_3 = a^2 + b^2 - 2ab\cos(\pi/4)$; $R_4 = a^2 + b^2 - 2ab\cos(3\pi/8)$

$R_5 = a^2 + b^2 - 2ab\cos(\pi/2)$; $R_6 = a^2 + b^2 - 2ab\cos(5\pi/8)$

$R_7 = a^2 + b^2 - 2ab\cos(3\pi/4)$; $R_8 = a^2 + b^2 - 2ab\cos(7\pi/8)$

$R_9 = a^2 + b^2 - 2ab\cos(\pi)$

The term of E' and E'' are the local truncation errors of the series expansion. The accuracy of the admittance calculation depends on the number of series terms, $(N + 1)$, that are plotted in Figure 2.7, for the arbitrary highest value of $\varepsilon_r = 80$ and $\varepsilon_c = 2.05$ at 20 GHz for various aspect ratios, b/a (with $a = 0.65$ mm).

Figure 2.7 shows the variation in local truncation errors, (E' and E''), of normalized conductance, $G(0)/Y_o$, and susceptance, $B(0)/Y_o$, with the number of series terms, $(N + 1)$, for the various respect ratio, b/a. The results also illustrate that the convergence rate of the series increases by increasing the number of series terms. In addition, the number of terms required for the series of normalized conductance to reach the convergence level is less than the series of normalized susceptance for all ratios, b/a. Overall, the local truncation errors, (E' and E'') of series expansions can reach convergence level using lower aspect ratio probe. The MATLAB calculation program of (2.21) and (2.22) is shown on the accompanying CD. The comparison of calculated normalized conductance, $G(0)/Y_o$ and susceptance, $B(0)/Y_o$ using (2.21) and (2.22), respectively and literature results [14, 15] are shown in Figure 2.8.

The comparison of calculated normalized conductance, $G(0)/Y_o$, and susceptance, $B(0)/Y_o$, using several models and the simulation results (using COMSOL Multiphysics simulation software), and measurements for water sample at room temperature are shown in Figure 2.9. From Figure 2.9, the closed-form equation (2.6) with $C_1 = 0.1$ pF, $C_2 = 0.0293$ pF, and $G_0 = 9 \times 10^{-29}$ is applicable up to 6 GHz, while (2.9b) shows that the calculated values are in good agreement with the corresponding simulation results up to 40 GHz. The values for the components C_1, L_1, L_2, and R in (2.6a) are given as: $C_1 = 0.1$ pF, $C_2 = 0.0293$ pF, $L_1 = 7$ pH, $L_2 = 0.9$ pH, and $R = 2.8\ \Omega$. Equations (2.11), (2.21), and (2.22) are solved analytically, but with different approaches. The calculations for admittance in (2.11)

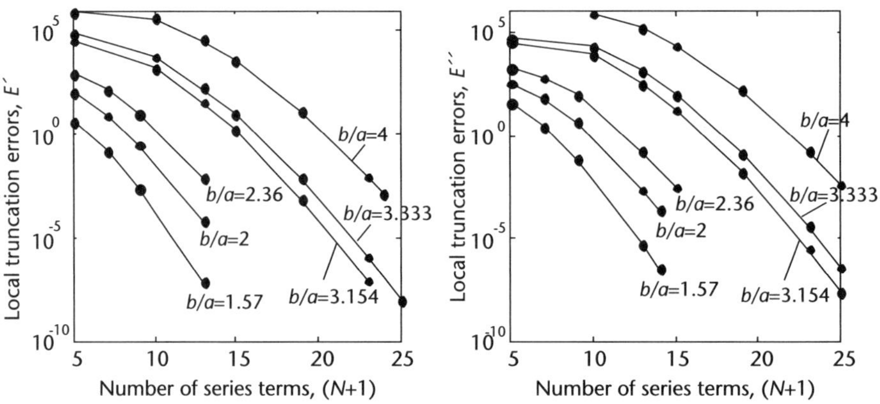

Figure 2.7 The relationship between local truncation errors of conductance, $|E'|$, and susceptance, $|E''|$, series and number of series term $(N + 1)$. (For $a = 0.65$ mm, $\varepsilon_c = 2.05$, $\varepsilon_r = 80$, and $f = 20$ GHz.)

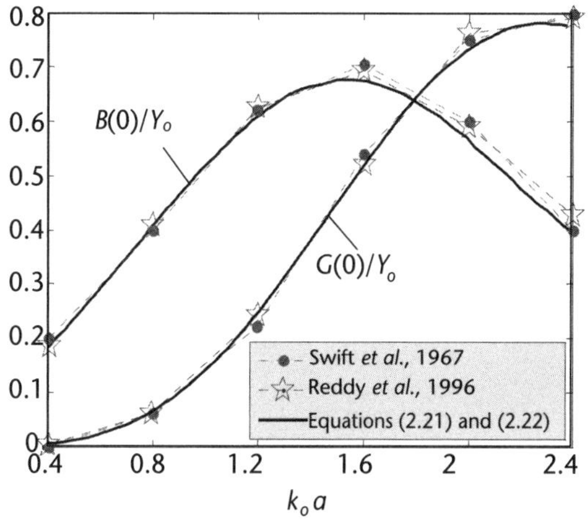

Figure 2.8 The normalized admittance as a function of k_oa (for $b/a = 1.57$, $\varepsilon_r = 1$ and $\varepsilon_c = 1$).

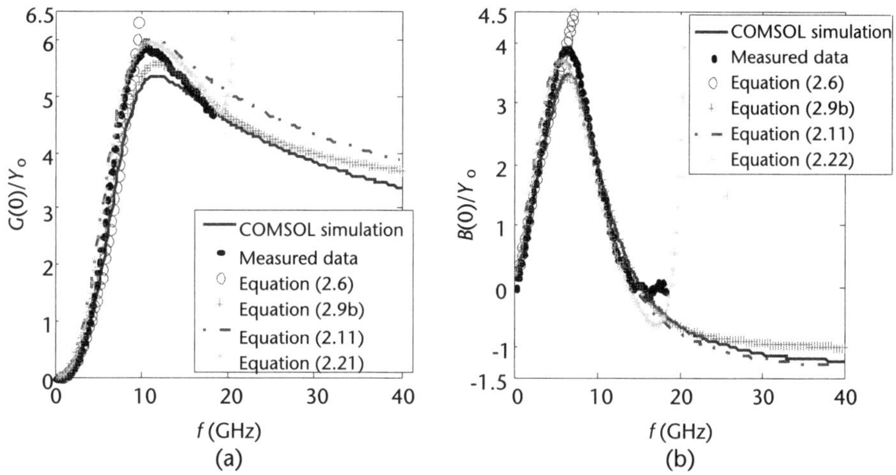

Figure 2.9 Comparison of (a) normalized conductance, $G(0)/Y_o$, and (b) normalized susceptance, $B(0)/Y_o$, for water at 25°C by considering size of probe with $a = 0.65$ mm and $b = 2.05$ mm.

are conducted by a 5 × 5 × 6-order Gaussian triple integral method, while (2.21) and (2.22) are solved by the series expansion method with 25 terms of the series. The series expansion solution of (2.21) and (2.22) agrees with those corresponding admittance results for frequencies up to 18 GHz and becomes unstable for higher frequencies. This unstable condition may be due to an inaccuracy of the decimal numeral in the terms of the series. However, the series expansion solutions can be used efficiently to minimize the running time of the program compared to the Gaussian integral method.

Similarly, the comparison of the calculated normalized conductance, $G(0)/Y_o$, and susceptance, $B(0)/Y_o$, using models (2.4), (2.11), (2.13), (2.21), and (2.22) for the simulation results and measurements for methanol at room temperature are shown in Figure 2.10. The calculations assumed that the section of coaxial line is filled with Teflon ($\varepsilon_c = 2.06$) and the value ratio of $b/a = 3.298$ with $a = 0.4545$ mm.

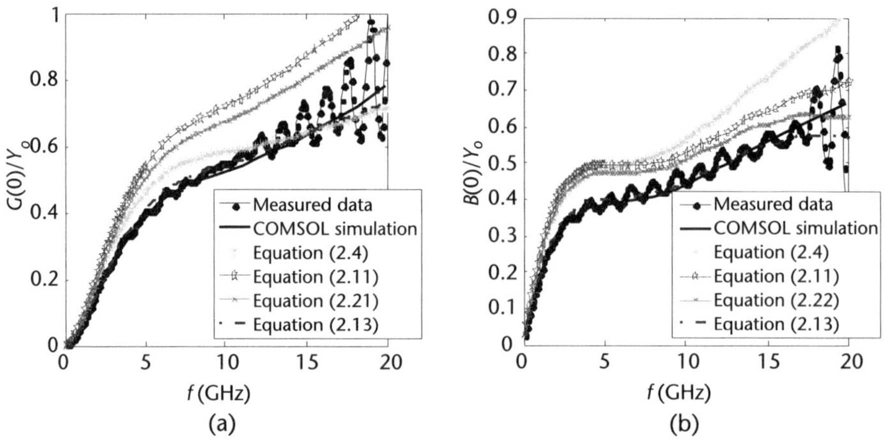

Figure 2.10 Comparison of (a) normalized conductance, $G(0)/Y_o$, and (b) normalized susceptance, $B(0)/Y_o$, for methanol at 25°C by considering size of sensor with $b/a = 3.298$ and $a = 0.4545$ mm.

Figure 2.10 is clearly shown the efficiency of (2.13) with the conductive effect, $G = 3 \times 10^{-31}$ in order to correct the fringing effect at the aperture sensor over the frequency range 0.13 GHz to 20 GHz.

2.4 Rational Expression Modeling

The universal input impedance, Z_{in}, in elements values χ_n terms when the open end of the coaxial line is terminated in unit resistance as [16, 17]:

$$Z_{in} = j\omega\chi_1 + \cfrac{1}{j\omega\chi_2 + \cfrac{1}{j\omega\chi_3 + \cfrac{1}{\ldots\ldots + \cfrac{1}{j\omega\chi_n + 1}}}}$$

$$= j\omega L_1 + \cfrac{1}{j\omega C_2 + \cfrac{1}{j\omega L_3 + \cfrac{1}{\ldots\ldots + \cfrac{1}{j\omega C_{n-1} + \cfrac{1}{j\omega L_n + 1}}}}} \quad (2.29)$$

where χ_n is a designated series inductance, L_n, for n = odd number, while the χ_n is designated a shunt capacitance, C_n, when n = even number. The equivalent circuit for input impedance, Z_{in}, is shown in Figure 1.9. However, (2.29) is only suitable for lossless circuit. When (2.29) is applied to lossy circuit, a negative impedance result is obtained. Thus, (2.29) has been reexpressed in the form of rational functions [18–20] so that the impedance results will always be positive. For instance, when $n = 1$, the Z_{in} is given as:

$$Z_{\text{in}} = j\omega\chi_1 + 1 \tag{2.30}$$

For $n = 2$,

$$Z_{\text{in}} = \frac{(j\omega)^2\chi_1\chi_2 + j\omega\chi_2 + 1}{j\omega\chi_1 + 1} \tag{2.31}$$

For $n = 5$ yields,

$$Z_{\text{in}} = \frac{a_5(j\omega)^5 + a_4(j\omega)^4 + a_3(j\omega)^3 + a_2(j\omega)^2 + a_1 j\omega + 1}{b_4(j\omega)^4 + b_3(j\omega)^3 + b_2(j\omega)^2 + b_1 j\omega + 1} = \frac{1 + \sum_{n=1}^{5} a_n(j\omega)^n}{1 + \sum_{n=1}^{4} b_n(j\omega)^n} \tag{2.32}$$

where $a_5 = \chi_1\chi_2\chi_3\chi_4\chi_5$, $a_4 = \chi_2\chi_3\chi_4\chi_5$, $a_3 = \chi_1\chi_4\chi_5 + \chi_2\chi_4\chi_5 + \chi_1\chi_2\chi_5$, $a_2 = \chi_2\chi_5 + \chi_4\chi_5$, $a_1 = \chi_1 + \chi_2 + \chi_3$, $b_4 = \chi_1\chi_2\chi_3\chi_4$, $b_3 = \chi_1\chi_4 + \chi_2\chi_4 + \chi_1\chi_2$, $b_2 = \chi_1\chi_4 + \chi_2\chi_4 + \chi_1\chi_2$, and $b_1 = \chi_2 + \chi_4$. In spite of that, the discontinuity at the open end of the coaxial line is commonly interpreted using admittance expressions $Y_{\text{in}} = 1/Z_{\text{in}}$ due to its naturally capacitive properties that dominate at the open-end surface ($z = 0$).

Stuchly et al. [19] and Anderson et al. [20] modeled the admittance aperture of open-ended coaxial radiator using an empirical rational function model as given by:

$$\tilde{Y} = \frac{\sum_{n=1}^{4}\sum_{p=1}^{8} \alpha_{np}\left(\sqrt{\varepsilon_r}\right)^p (sa)^n}{1 + \sum_{m=1}^{4}\sum_{q=0}^{8} \beta_{mq}\left(\sqrt{\varepsilon_r}\right)(sa)^m} \tag{2.33}$$

where $s = \sigma + j\omega$ and a are the complex frequency variable and the inner conductor radius of the line, respectively. α_{np} and β_{mq} in (2.33) are the unknown coefficients; both values depend on the size and type of coaxial probe, where those values are obtained by fitting to the admittance computed via the method of moments (MoM). Equation (2.33) is more accurate and easier for samples of a broad range of permittivity values.

This modeling requires that the polynomial function of the complex reflection coefficient, $\Gamma = \Gamma' + j\Gamma''$, is first obtained either by a full-wave analysis (complete EM analysis) or by measurement. Later, the polynomial function is represented in the form of a rational function (input admittance or input impedance expressions). Normally, the polynomial functions of Γ involving variable operation frequency f and complex relative permittivity ε_r as:

$$\Gamma = \sum_{m=0}^{M}\sum_{n=0}^{N} A_{mn}\varepsilon_r^m f^n \tag{2.34}$$

where A_{mn} in (2.34) are the unknown complex coefficients. In this book, the values of A_{mn} are obtained by fitting the coefficients with calculated values obtained using the FEM over a broad range of relative permittivity ε_r values. There are no typical rules to determine the order of polynomial to give a better fit with FEM simulation data. As long as the fitting line is matched with simulation data, the order of polynomial is considered as an appropriate choice.

For interpretation results, the reflection coefficient Γ is converted to normalized admittance, $\tilde{Y}$, via:

$$\tilde{Y} = \frac{1-\Gamma}{1+\Gamma} = \frac{1-\sum_{m=0}^{M}\sum_{n=0}^{N} A_{mn}\varepsilon_r^m f^n}{1+\sum_{m=0}^{M}\sum_{n=0}^{N} A_{mn}\varepsilon_r^m f^n} = \Re\text{e}\left(\tilde{Y}\right) + j\Im\text{m}\left(\tilde{Y}\right) \qquad (2.35)$$

Equation (2.35) is equivalent to (2.33). The values with 12 decimals for the complex coefficients of A_{mn} are listed on the accompanying CD.

The complex coefficient values for (2.34) or (2.35) on the accompanying CD are valid only for Teflon-filled coaxial probes with an inner conductor radius, a of 0.65 mm, and an outer conductor radius, b of 2.05 mm, satisfying the relative dielectric constant, ε_r, from 2 to 80, and the operating frequency from 0.2 to 5 GHz. The coefficient values on the accompanying CD are valid for Teflon-filled coaxial probes with a = 0.1 mm and b = 0.33 mm, satisfying the relative permittivity, ε_r from 2 to 40 and the operating frequency from 0.2 GHz to 20 GHz. Figure 2.11 shows the real part, $\Re\text{e}(\tilde{Y})$ (normalized conductance), and the imaginary part, $\Im\text{m}(\tilde{Y})$ (normalized susceptance), obtained from (2.35) using complex coefficients on the accompanying CD and FEM simulation for methanol liquid covering 0.2 to 5 GHz at room temperature. Similarly, Figure 2.12 shows the comparison between the calculation $\Re\text{e}(\tilde{Y})$ and $\Im\text{m}(\tilde{Y})$ results from (2.35), using coefficients on the accompanying CD and FEM simulation. As a result, the average percentage of relative error between the calculated and FEM-simulated values for both $\Re\text{e}(\tilde{Y})$ and $\Im\text{m}(\tilde{Y})$ were less than

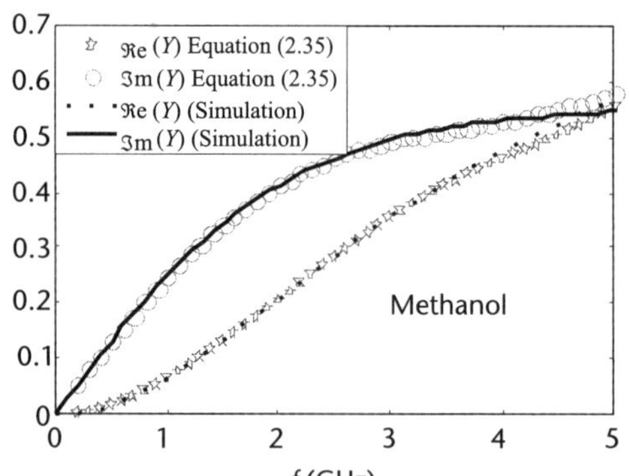

Figure 2.11 Comparison of the real part, $\Re\text{e}(\tilde{Y})$, and the imaginary part, $\Im\text{m}(\tilde{Y})$, of the normalized admittance for methanol at 25°C by considering size of sensor with b/a = 3.154 and a = 0.65 mm.

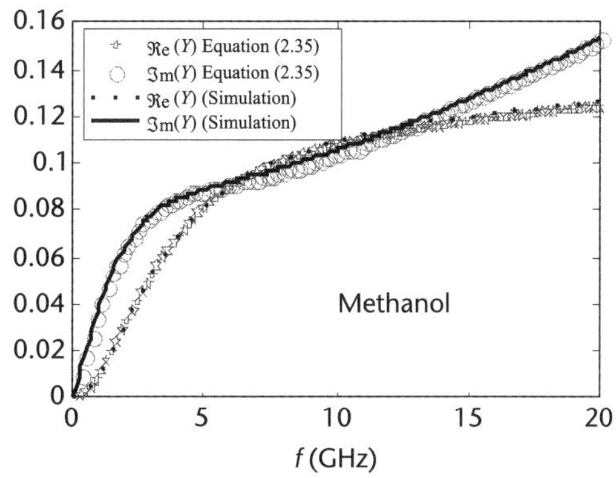

Figure 2.12 Comparison of normalized conductance, $\Re e(\tilde{Y})$, and normalized susceptance, $\Im m(\tilde{Y})$, for methanol at 25°C by considering the size of sensor with $b/a = 3.3$ and $a = 0.1$ mm.

2% in Figure 2.11 and 1% in Figure 2.12, respectively, when the FEM simulation results are used as the reference value.

References

[1] Atley, T. W., M. A. Stuchly, and S. S. Stuchly, "Measurement of Radio Frequency Permittivity of Biological Tissues with an Open-Ended Coaxial Line: Part I," *IEEE Trans. on Microwave Theory Tech.*, Vol. 30, 1982, pp. 82–86.

[2] Stuchly, M. A., and S. S. Stuchly, "Coaxial Line Reflection Methods for Measuring Dielectric Properties of Biological Substances at Radio and Microwave Frequencies—A Review," *IEEE Trans. on Instrum. Meas.*, Vol. 29, 1980, pp. 176–183.

[3] Stuchly, M. A., et al., "Equivalent Circuit of an Open-Ended Coaxial Line in a Lossy Dielectric," *IEEE Trans. on Instrum. Meas.*, Vol. 31, 1982, pp. 116–119.

[4] Brady, M. M., S. A. Symons, and S. S. Stuchly, "Dielectric Behavior of Selected Animal Tissues In Vitro at Frequencies from 2 to 4 GHz," *IEEE Trans. on Biomedical Eng.*, Vol. 28, 1981, pp. 305–307.

[5] Stuchly, M. A., et al., "Measurement of Radio Frequency Permittivity of Biological Tissues with an Open-Ended Coaxial Line: Part II—Experimental Results," *IEEE Trans. on Microwave Theory Tech.*, Vol. 30, 1982, pp. 87–91.

[6] Gajda, G. B., and S. S. Stuchly, "Numerical Analysis of Open-Ended Coaxial Lines," *IEEE Trans. Microwave Theory Tech.*, Vol. 31, 1983, pp. 380–384.

[7] Gajda, G., and S. S. Stuchly, "An Equivalent Circuit of an Open-Ended Coaxial Line," *IEEE Trans. on Instrum. Meas.*, Vol. 32, 1983, pp. 506–508.

[8] You, K. Y., and Z. Abbas, "Fringing Field Correction of Admittance Model for Open-Ended Coaxial Sensor," *IEEE Sensors Journal*, Vol. 12, 2012, pp. 1468–1469.

[9] You, K. Y., et al., "Modeling of Coaxial Slot Waveguides Using Analytical and Numerical Approaches: Revisited," *International Journal of Antennas and Propagation*, Vol. 2012, 2012, pp. 1–12.

[10] Levine, H., and C. H. Papas, "Theory of the Circular Diffraction Antenna," *J. Appl. Phys.*, Vol. 22, 1951, pp. 29–43.

[11] Misra, D., "A Quasi-Static Analysis of Open-Ended Coaxial Lines," *IEEE Trans. on Microwave Theory Tech.*, Vol. 35, 1987, pp. 925–928.

[12] Marcuvitz, N., *Waveguide Handbook*, Boston, MA: Boston Technical Publishers, 1964.

[13] You, K. Y., and Z. Abbas, *Open-Ended Coaxial Sensor Handbook: Formulations, Microwave Measurement and Applications*, Lambert Academic Publishing, 2010.

[14] Reddy, C. J., et al., "Radiation Characteristics of Cavity Backed Aperture Antennas in Finite Ground Plane Using the Hybrid FEM/MoM Technique and Geometrical Theory of Diffraction," *IEEE Trans. on Antennas Propagat.*, Vol. 44, 1996, pp. 1327–1333.

[15] Swift, C. T. "Input Admittance of a Coaxial Transmission Line Opening onto a Flat, Dielectric Covered Ground Plane," *NASA Tech. Rep.* NASA TND-4158, 1967.

[16] Cauer, W. "Die Verwirklichung von Wechselstromwiderstanden vorgeschriebener Frequenzabhangigkeit," *Arch. Electrotech*, Vol. 17, 1927, p. 355.

[17] Foster, R. M. "A Reactance Theorem," *Bell Sys. Tech. J.*, Vol. 3, 1924, pp. 259–267.

[18] Burdette, E. C., F. I. Cain, and J. Seals, "In Vivo Probe Measurement Technique at VHF Through Microwave Frequencies," *IEEE Trans. on Microwave Theory Tech.*, Vol. 28, 1980, pp. 414–427.

[19] Stuchly, S. S., C. L. Sibbald, and J. M. Anderson, "A New Aperture Admittance Model for Open-Ended Waveguide," *IEEE Trans. on Instrum. Meas.*, Vol. 42, 1994, pp. 192–198.

[20] Anderson, J. M., C. L. Sibbald, and S. S. Stuchly, "Dielectric Measurements Using a Rational Function Model," *IEEE Trans. on Microwave Theory Tech.*, Vol. 42, 1994, pp. 199–204.

Selected Bibliography

Blackham, E. V., and R. D. Pollard, "An Improved Technique for Permittivity Measurements Using a Coaxial Probe," *IEEE Trans. on Instrum. Meas.*, Vol. 46, 1997, pp. 1093–1009.

Brune, O., "Synthesis of a Finite Two-Terminal Network Whose Driving-Point Impedance Is a Prescribed Function of Frequency," *J. Math. Phys. Camb.*, Vol. 10, 1931, pp. 191–236.

Busssey, H. "Dielectric Measurement in a Shielded Open Circuit Coaxial Line," *IEEE Trans. on Instrum. Meas.*, Vol. 29, 1980, pp. 120–124.

Gabriel, C., T. Y. A. Chan, and E. H. Grant, "Admittance Models for Open Ended Coaxial Probes and Their Place in Dielectric Spectroscopy," *Phys. Med. Biol.*, Vol. 39, 1994, pp. 2183–2200.

Galejs, J., and T. W. Thompson, "Admittance of a Cavity Backed Annular Slot Antenna," *IRE Trans. on Antennas Propagat.*, Vol. 10, 1962, pp. 671–678.

Grant, J. P., et al., "A Critical Study of the Open-Ended Coaxial Line Sensor Technique for RF and Microwave Complex Permittivity Measurements," *J. Phys. E: Sci. Instrum.*, Vol. 22, 1989, pp. 757–770.

Karunanayake, P., A. P. Esselle, and S. S. Stuchly, "Capacitive Sensors for In-Vivo Measurements of the Dielectric Properties of Biological Materials," *IEEE Trans. on Instrum. Meas.*, Vol. 31, 1988, pp. 101–105.

Kraszewski, A., and S. S. Stuchly, "Capacitance of Open-Ended Dielectric-Filled Coaxial Lines-Experimental Results," *IEEE Trans. on Instrum. Meas.*, Vol. 32, 1983, pp. 517–519.

Kraszewski, A., M. A. Stuchly, and S. S. Stuchly, "ANA Calibration Method for Measurements of Dielectric Properties," *IEEE Trans. on Instrum. Meas.*, Vol. 32, 1983, pp. 385–386.

Misra, D., et al., "Noninvasive Electrical Characterization of Materials at Microwave Frequencies Using an Open-Ended Coaxial Line: Test of an Improved Calibration Technique," *IEEE Trans. on Microwave Theory Tech.*, Vol. 38, 1990, pp. 8–14.

Nyshsdham, A., C. L. Sibbald, and S. S. Stuchly, "Permittivity Measurements Using Open-Ended Sensor Reference Liquid Calibration—An Uncertainty Analysis," *IEEE Trans. on Microwave Theory Tech.*, Vol. 40, 1992, pp. 305–313.

Okoniewski, M., et al., "Further Analysis of Open-Ended Dielectric Sensors," *IEEE Trans. on Microwave Theory Tech.*, Vol. 43, 1995, pp. 1986–1989.

Pournaropoulos, C. L., and D. K. Misra, "The Co-Axial Aperture Electromagnetic Sensor and Its Application in Material Characterization," *Meas. Sci. Technol.*, Vol. 3, 1997, pp. 1191–1202.

Stuchly, S. S., A. Kraszewski, and M. A. Stuchly, "Uncertainties in Radio Frequency Dielectric Measurements of Biological Substances," *IEEE Trans. on Instrum. Meas.*, Vol. 36, 1987, pp. 67–70.

Yarman, B. S., *Design of Ultra Wideband Antenna Matching Networks*, Germany: Springer Science, 2008.

CHAPTER 3

Analytical Modeling of an Open-Ended Coaxial Line Terminated by Layered Media

This chapter presents the modeling of an open-ended coaxial probe that is terminated by layered media. For a very thin material at the coaxial probe aperture, the scattering of the wave from the probe aperture will penetrate the material and impinge on the other layer-interface media. During the past four decades, excitation of the complex waves by a probe in a layered medium was theoretically analyzed by many researchers. Various methods have been used to derive the analytical equations in order to determine the total EM fields in the layered medium. In this chapter, three types of analytical methods will be described, namely, the boundary-matching method, the images method, and the semi-empirical closed-form method. The analytical modeling mainly focuses on cases in which the aperture of the probe is placed against a two-layer or three-layer nonmagnetic medium. Furthermore, the derivation of formulas involving Bessel functions will be analyzed, since the nature of cylinder harmonics (radial harmonics) can be described using Bessel functions. In addition, many solutions of Bessel functions are aided by the Hankel transform.

3.1 Boundary-Matching Method Modeling

3.1.1 Reflection in a Stratified Medium

Let us consider the coaxial probe aperture contacts with a dielectric medium that has three layers. The first layer of the medium has a relative permittivity, ε_{r1}, and a thickness, d_1, while the second layer has a relative permittivity, ε_{r2}, and a thickness $(d_2 - d_1)$. Both layers are terminated by the third layer, which has a relative permittivity of ε_{r3} that extends to infinity as shown in Figure 3.1. An excitation wave is assumed to propagate from the coaxial line over the aperture probe and radiates into the three-layer medium.

Here, the reflection analysis is focused on the total reflection coefficient, Γ_{in}, at the boundary of the leading interface (z = 0) and on the reflection coefficient, Γ_1, at the boundary between the first and second layers of the medium ($z = d_1$), which is caused by impedance mismatched between two layers of the medium. The Hankel transform of (1.17) with respect to the radial coordinate, ρ, is given as [1]:

$$\left[\frac{d^2}{dz^2} + \left(k_i^2 - \zeta^2\right)\right]\tilde{H}_{\phi i} = 0 \tag{3.1}$$

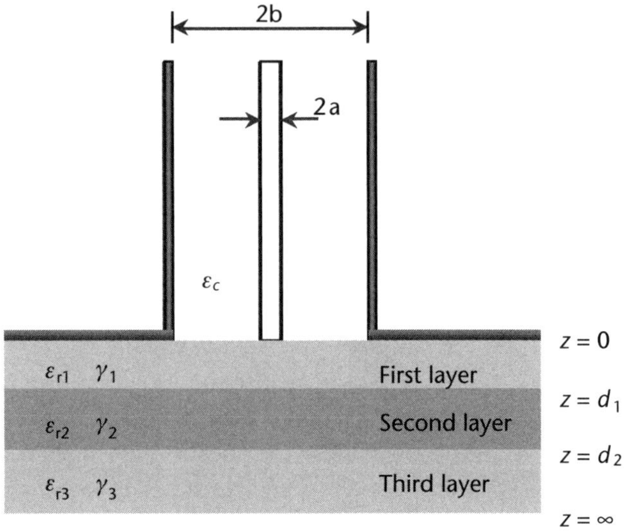

Figure 3.1 Geometry of the coaxial line and the three dielectric layers.

where $\tilde{H}_{\phi i}$ is the Hankel transform of the magnetic field in $z \geq 0$. The subscript i denotes the layer of medium (i =1, 2, 3). The k_i is the propagation constant for ith layer medium. The $\tilde{H}_{\phi i}$ in the first, second, and third layers of the medium are:

$$\tilde{H}_{\phi 1} = B_o\left(e^{-\gamma_1 z} + \Gamma_1 e^{\gamma_1 z}\right) \quad \text{for} \quad 0 \leq z \leq d_1 \tag{3.2a}$$

$$\tilde{H}_{\phi 2} = C_o\left(e^{-\gamma_2 z} + \Gamma_2 e^{\gamma_2 z}\right) \quad \text{for} \quad d_1 \leq z \leq d_2 \tag{3.2b}$$

$$\tilde{H}_{\phi 3} = D_o e^{-\gamma_3 z} \quad \text{for} \quad z \geq d_2 \tag{3.2c}$$

where the propagation constants, $\gamma_i = (\zeta^2 - k_i^2)^{1/2}$, if $\Re e(k_i) < \zeta$ and ζ is an eigenvalue that corresponds to the radial coordinate. The symbols B_o, C_o, and D_o are the unknown functions of ζ. The Hankel transform of electric field, $\tilde{E}_{\rho i}$, can be calculated from $\tilde{H}_{\phi i}$ using (1.30a) and is given as:

$$\tilde{E}_{\rho 1} = \frac{-j\gamma_1 B_o}{\omega\varepsilon_o\varepsilon_{r1}}\left(e^{-\gamma_1 z} - \Gamma_1 e^{\gamma_1 z}\right) \quad \text{for} \quad 0 \leq z \leq d_1 \tag{3.3a}$$

$$\tilde{E}_{\rho 2} = \frac{-j\gamma_2 C_o}{\omega\varepsilon_o\varepsilon_{r2}}\left(e^{-\gamma_2 z} - \Gamma_2 e^{\gamma_2 z}\right) \quad \text{for} \quad d_1 \leq z \leq d_2 \tag{3.3b}$$

$$\tilde{E}_{\rho 3} = \frac{-j\gamma_3 D_o}{\omega\varepsilon_o\varepsilon_{r3}} e^{-\gamma_3 z} \quad \text{for} \quad z \geq d_2 \tag{3.3c}$$

From Figure 3.1, the continuity transformed boundary conditions at $z = d_1$ for the tangential magnetic field, $\tilde{H}_{\phi i}$, and the tangential electric field, $\tilde{E}_{\rho i}$, are given as:

$$\begin{gathered} \tilde{H}_{\phi 1} \text{ (at } z = d_1) = \tilde{H}_{\phi 2} \text{ (at } z = d_1) \\ B_o\left(e^{-\gamma_1 d_1} + \Gamma_1 e^{\gamma_1 d_1}\right) = C_o\left(e^{-\gamma_2 d_1} + \Gamma_2 e^{\gamma_2 d_1}\right) \end{gathered} \tag{3.4}$$

and

$$\begin{gathered} \tilde{E}_{\rho 1} \text{ (at } z = d_1) = \tilde{E}_{\rho 2} \text{ (at } z = d_1) \\ \frac{j\gamma_1 B_o}{\omega\varepsilon_o\varepsilon_{r1}}\left(e^{-\gamma_1 d_1} - \Gamma_1 e^{\gamma_1 d_1}\right) = \frac{j\gamma_2 C_o}{\omega\varepsilon_o\varepsilon_{r2}}\left(e^{-\gamma_2 d_1} - \Gamma_2 e^{\gamma_2 d_1}\right) \end{gathered} \tag{3.5}$$

The unknown functions B_o and C_o in (3.5) can be eliminated by substituting (3.4) into (3.5), yielding:

$$\frac{\gamma_1}{\varepsilon_{r1}}\left(e^{-\gamma_1 d_1} - \Gamma_1 e^{\gamma_1 d_1}\right)\left(e^{-\gamma_2 d_1} + \Gamma_2 e^{\gamma_2 d_1}\right) = \frac{\gamma_2}{\varepsilon_{r2}}\left(e^{-\gamma_1 d_1} + \Gamma_1 e^{\gamma_1 d_1}\right)\left(e^{-\gamma_2 d_1} - \Gamma_2 e^{\gamma_2 d_1}\right) \tag{3.6}$$

$$\frac{\gamma_1\varepsilon_{r2}}{\gamma_2\varepsilon_{r1}} = \frac{\left(e^{-\gamma_1 d_1} + \Gamma_1 e^{\gamma_1 d_1}\right)\left(e^{-\gamma_2 d_1} - \Gamma_2 e^{\gamma_2 d_1}\right)}{\left(e^{-\gamma_1 d_1} - \Gamma_1 e^{\gamma_1 d_1}\right)\left(e^{-\gamma_2 d_1} + \Gamma_2 e^{\gamma_2 d_1}\right)} \tag{3.7}$$

Finally, the reflection coefficient, Γ_1 at $z = d_1$, is written as:

$$\Gamma_1 = \left[\frac{\varepsilon_{r2}\gamma_1\left(1 + \Gamma_2 e^{2\gamma_2 d_1}\right) - \varepsilon_{r1}\gamma_2\left(1 - \Gamma_2 e^{2\gamma_2 d_1}\right)}{\varepsilon_{r2}\gamma_1\left(1 + \Gamma_2 e^{2\gamma_2 d_1}\right) + \varepsilon_{r1}\gamma_2\left(1 - \Gamma_2 e^{2\gamma_2 d_1}\right)}\right] e^{-2\gamma_1 d_1} \tag{3.8}$$

The Γ_2 in (3.8) can be found in similar ways by equating the continuity of the tangential magnetic field, $\tilde{H}_{\phi 1}$, and tangential electric field, $\tilde{E}_{\rho i}$, at d_2 using (3.2b), (3.2c), (3.3b), and (3.3c):

$$\begin{gathered} \tilde{H}_{\phi 2} \text{ (at } z = d_2) = \tilde{H}_{\phi 3} \text{ (at } z = d_2) \\ C_o\left(e^{-\gamma_2 d_2} + \Gamma_2 e^{\gamma_2 d_2}\right) = D_o e^{-\gamma_3 d_2} \end{gathered} \tag{3.9}$$

$$\begin{gathered} \tilde{E}_{\rho 2} \text{ (at } z = d_2) = \tilde{E}_{\rho 3} \text{ (at } z = d_2) \\ \frac{j\gamma_2 C_o}{\omega\varepsilon_o\varepsilon_{r2}}\left(e^{-\gamma_2 d_2} - \Gamma_2 e^{\gamma_2 d_2}\right) = \frac{j\gamma_3 D_o}{\omega\varepsilon_o\varepsilon_{r3}} e^{-\gamma_3 d_2} \end{gathered} \tag{3.10}$$

By solving (3.9) and (3.10) simultaneously, Γ_2 can be expressed as:

$$\Gamma_2 = \frac{\gamma_2\varepsilon_{r3} - \gamma_3\varepsilon_{r2}}{\gamma_3\varepsilon_{r2} + \gamma_2\varepsilon_{r3}} e^{-2\gamma_2 d_2} \tag{3.11}$$

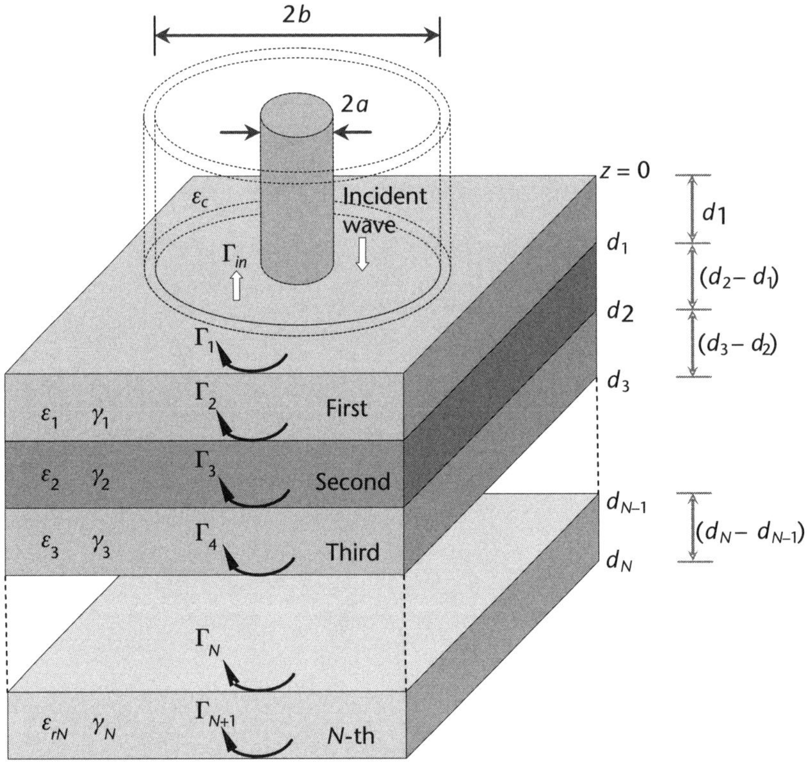

Figure 3.2 Reflection coefficients for wave propagation in stratified medium.

However, for an N-layer stratified medium as shown in Figure 3.2, a general expression of the reflection coefficient, Γ_i, in the ith layer of the medium can be written as:

$$\Gamma_i = \frac{\varepsilon_{r(i+1)}\gamma_i\left(1+\Gamma_{i+1}e^{2\gamma_{i+1}d_i}\right) - \varepsilon_{r(i+1)}\gamma_i\left(1-\Gamma_{i+1}e^{2\gamma_{i+1}d_i}\right)}{\varepsilon_{r(i+1)}\gamma_i\left(1+\Gamma_{i+1}e^{2\gamma_{i+1}d_i}\right) + \varepsilon_{r(i+1)}\gamma_i\left(1-\Gamma_{i+1}e^{2\gamma_{i+1}d_i}\right)} e^{-2\gamma_i d_i} \quad \text{for } i = 1,2,\ldots,N \tag{3.12}$$

If the Nth layer of the medium is an infinite half-space ($d_N = \infty$), then $\Gamma_{N+1} = 0$. However, $\Gamma_N = \exp(-2\gamma_N d_N)$, if the Nth layer of the medium is a finite dielectric with thickness of d_N, which is terminated by a metal plate.

3.1.2 Integral Admittance Model

From (1.8) and (1.9), the E_ρ^- and H_ϕ^- fields at the aperture probe $z = 0$ (ignoring the TM_{n0} mode) can be simplified as:

$$E_\rho^- = \frac{A_o}{\rho}\left(1+\Gamma_{\text{in}}\right) \tag{3.13}$$

$$H_\phi^- = \frac{A_o}{\eta_c\rho}\left(1-\Gamma_{\text{in}}\right) \tag{3.14}$$

The TM_{n0} mode at aperture probe is neglected in order to simplify the analytical calculations. [To be specific, the infinite higher-order series terms in (3.13) and (3.14) are neglected.] By performing the Hankel transform on (3.13) with respect to ρ, an aperture electric field can be written as:

$$\tilde{E}_\rho = \int_0^\infty E_\rho^- J_1(\zeta\rho)\rho d\rho = \int_0^\infty \frac{A_o}{\rho}\left(1+\Gamma_{\text{in}}\right) J_1(\zeta\rho)\rho d\rho \tag{3.15}$$

where $\tilde{E}_\rho$ is the Hankel transform of the aperture electric field. Since the transformed electric field can solely exist in the coaxial gap ($a < \rho < b$), (3.15) is evaluated by integrating a to b, yielding:

$$\tilde{E}_\rho = \int_a^b A_o\left(1+\Gamma_{\text{in}}\right) J_1(\zeta\rho) d\rho = A_o\left(1+\Gamma_{in}\right)\left(\frac{J_o(\zeta a) - J_o(\zeta b)}{\zeta}\right) \tag{3.16}$$

where the integration of Bessel function, $J_1(\zeta\rho)$ in (3.16) is given as [2]:

$$\int_a^b J_1(\zeta\rho) d\rho = \frac{J_o(\zeta a) - J_o(\zeta b)}{\zeta} \tag{3.17}$$

From Poynting's theorem, (3.13), and (3.14), the continuity of power P flow over the aperture surface, $d\vec{S} = \hat{z}\rho d\rho d\phi$ at $z = 0$, is expressed as:

$$\begin{aligned} P &= \frac{1}{2}\Re e\int\left(\vec{E}^* \times \vec{H}\right).\hat{z}d\vec{S} = \frac{1}{2}\int_0^{2\pi}\int_a^b E_\rho^{-*} H_\phi^- \rho d\rho d\phi \\ &= \frac{\pi\left|A_o\right|^2}{\eta_c}\left(1+\Gamma_{\text{in}}^*\right)\left(1-\Gamma_{\text{in}}\right)\ln\left(\frac{b}{a}\right) \end{aligned} \tag{3.18}$$

However, from Parseval's theorem, the complex conjugate of power P^* that is radiated by the aperture in term of the Hankel transform is expressed as [3]:

$$P^* = \pi\int_0^\infty \zeta \tilde{E}_\rho^* \tilde{H}_\phi d\zeta \tag{3.19}$$

The ratio of (3.2a) to (3.3a) at $z = 0$ is expressed as:

$$\frac{\tilde{H}_\phi}{\tilde{E}_\rho} = \frac{j\omega\varepsilon_o\varepsilon_{r1}}{\gamma_1}\frac{1+\Gamma_1}{1-\Gamma_1} \tag{3.20a}$$

Thus, $\tilde{H}_\phi$ is a function of $\tilde{E}_\rho$ and can be written as:

$$\tilde{H}_\phi = \frac{j\omega\varepsilon_o\varepsilon_{r1}}{\gamma_1}\frac{1+\Gamma_1}{1-\Gamma_1}\tilde{E}_\rho \tag{3.20b}$$

Substituting (3.16) and (3.20b) in (3.19) yields

$$P^* = \pi \int_0^\infty \zeta \left\{ A_o \left(1 + \Gamma_{\text{in}}\right) \left[\frac{J_o(\zeta a) - J_o(\zeta b)}{\zeta} \right] \right\}^2 \frac{j\omega\varepsilon_o\varepsilon_{r1}}{\left(\zeta^2 - k_1^2\right)^{1/2}} \left(\frac{1 + \Gamma_1}{1 - \Gamma_1} \right) d\zeta \tag{3.21}$$

Then equating (3.18) and (3.21) yields

$$\begin{aligned} &\frac{\pi |A_o|^2}{\eta_c} \left(1 + \Gamma_{\text{in}}^*\right)\left(1 - \Gamma_{\text{in}}\right) \ln\left(\frac{b}{a}\right) \\ &= j\omega\varepsilon_o\varepsilon_{r1}\pi \int_0^\infty A_o^2 \left(1 + \Gamma_{\text{in}}\right)^2 \frac{\left[J_o(\zeta a) - J_o(\zeta b)\right]^2}{\zeta\left(\zeta^2 - k_1^2\right)^{1/2}} \left(\frac{1 + \Gamma_1}{1 - \Gamma_1} \right) d\zeta \end{aligned} \tag{3.22}$$

Finally, the normalized admittance, $\tilde{Y}$, of an open-ended coaxial probe of layered media that is derived using Hankel transformation with respect to the radial coordinate, ρ, can be expressed in the form by [4, 5] as:

$$\tilde{Y} = \left(\frac{1 - \Gamma_{\text{in}}}{1 + \Gamma_{\text{in}}} \right) = \frac{j\omega\varepsilon_o\varepsilon_{r1}\eta_c}{\ln(b/a)} \int_0^\infty \frac{\left[J_o(\zeta a) - J_o(\zeta b)\right]^2}{\zeta\left(\zeta^2 - k_1^2\right)^{1/2}} \left(\frac{1 + \Gamma_1}{1 - \Gamma_1} \right) d\zeta \tag{3.23}$$

In addition, there is another way to derive (3.23). This can be done by expressing the fields in the first layer of the medium ($0 \geq z \geq d_1$), in the Hankel transform form with respect to ζ [4]:

$$H_{\phi 1} = \int_0^\infty B_1 \left(e^{-\gamma_1 z} + \Gamma_1 e^{\gamma_1 z} \right) J_1(\zeta\rho) \zeta d\zeta \tag{3.24}$$

$$E_{\rho 1} = \frac{1}{j\omega\varepsilon_o\varepsilon_{r1}} \int_0^\infty B_1 \gamma_1 \left(e^{-\gamma_1 z} - \Gamma_1 e^{\gamma_1 z} \right) J_1(\zeta\rho) \zeta d\zeta \tag{3.25}$$

where B_1 is an unknown function of ζ. From (2.1a), the boundary conditions at $z = 0$ are:

$$\begin{gathered} H_\phi^- \text{ (at } z = 0) = H_{\phi 1} \text{ (at } z = 0) \\ \frac{A_o}{\eta_c \rho}\left(1 - \Gamma_{\text{in}}\right) = \int_0^\infty B_1 \left(1 + \Gamma_1\right) J_1(\zeta\rho) \zeta d\zeta \end{gathered} \tag{3.26}$$

and

$$\begin{gathered} E_{\rho 1} \text{ (at } z = 0) = E_{\rho 2} \text{ (at } z = 0) \\ \frac{A_o}{\rho}\left(1 + \Gamma_{\text{in}}\right) = \frac{1}{j\omega\varepsilon_o\varepsilon_{r1}} \int_0^\infty B_1 \gamma_1 \left(1 - \Gamma_1\right) J_1(\zeta\rho) \zeta d\zeta \end{gathered} \tag{3.27}$$

The B_1 term can be solved by the inverse Hankel transform of (3.25) and given as:

$$B_1 = \frac{j\omega\varepsilon_o\varepsilon_{r1}}{\gamma_1\left(1-\Gamma_1\right)}\int_a^b E_{\rho 1}J_1(\zeta\rho)\rho d\rho \tag{3.28}$$

Substituting (3.28) into (3.26) yields:

$$\frac{A_o}{\eta_c\rho}\left(1-\Gamma_{in}\right) = j\omega\varepsilon_o\varepsilon_{r1}\int_0^\infty \frac{1}{\gamma_1}\frac{\left(1+\Gamma_1\right)}{\left(1-\Gamma_1\right)}\left[\int_a^b E_{\rho 1}J_1(\zeta\rho)\rho d\rho\right]J_1(\zeta\rho)\zeta d\zeta \tag{3.29}$$

Multiplying both sides of (3.29) with ρE_ρ and then integrating from a to b,

$$\frac{A_o}{\eta_c\rho}\left(1-\Gamma_{in}\right)\int_a^b E_\rho \rho\, d\rho = j\omega\varepsilon_o\varepsilon_{r1}\int_0^\infty \frac{1}{\gamma_1}\frac{\left(1+\Gamma_1\right)}{\left(1-\Gamma_1\right)}\left[\int_a^b E_{\rho 1}J_1(\zeta\rho)\rho d\rho\right]^2 \zeta d\zeta \tag{3.30}$$

From (3.13), the integrated E_ρ at $z = 0$ in (3.30) is given as:

$$\int_a^b E_\rho d\rho = A_o\left(1+\Gamma_{\text{in}}\right)\ln(b/a) \tag{3.31}$$

Substituting (3.31) and (3.17) into (3.30) yields:

$$\begin{aligned}&\frac{A_o}{\eta_c}\left(1-\Gamma_{\text{in}}\right)\left[A_o\left(1+\Gamma_{\text{in}}\right)\ln(b/a)\right]\\ &= j\omega\varepsilon_o\varepsilon_r\int_0^\infty \frac{1}{\gamma_1}\frac{\left(1+\Gamma_1\right)}{\left(1-\Gamma_1\right)}\left[A_o\left(1+\Gamma_{\text{in}}\right)\right]^2\left[\frac{J_o(\zeta a)-J_o(\zeta b)}{\zeta}\right]^2\zeta d\zeta\end{aligned} \tag{3.32}$$

After moving some mathematical terms between both sides of (3.32), the normalized input admittance, $\tilde{Y}$, of the probe at $z = 0$ is given as [3, 4, 6, 7]:

$$\tilde{Y} = \frac{1-\Gamma_{\text{in}}}{1+\Gamma_{\text{in}}} = \frac{j\eta_c\omega\varepsilon_o\varepsilon_{r1}}{\ln(b/a)}\int_0^\infty \frac{1}{\gamma_1}\frac{\left(1+\Gamma_1\right)}{\left(1-\Gamma_1\right)}\frac{\left[J_o(\zeta a)-J_o(\zeta b)\right]^2}{\zeta}d\zeta \tag{3.33}$$

where $\eta_c\omega\varepsilon_o\varepsilon_{r1} = (k_o\varepsilon_{r1})/(\varepsilon_c)^{0.5}$.

3.1.3 Integral Admittance Models (Practical Cases)

In practical cases, the measured thin medium typically has only a single layer, which is backed by air (second layers, $\varepsilon_{r2} = 1$) or a conducting plate (second layer, $\varepsilon_{r2} = \infty$) as shown in Figure 3.3.

For the case of a coaxial probe connected by a thin layer medium with relative permittivity of ε_{r1} and the thickness of d, which is backed by air or free space as shown in Figure 3.3(a), the Γ_1 term in (3.8) can be reduced:

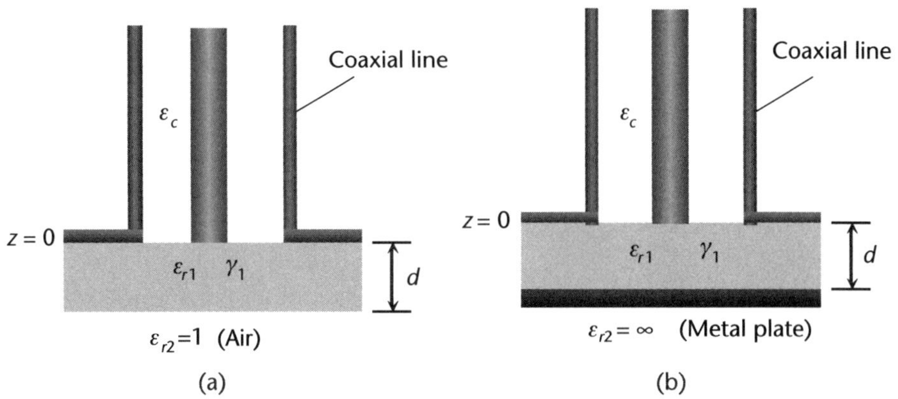

Figure 3.3 The aperture of the probe placed against the thin sample with thickness, *d*, which is backed by (a) air and (b) a metal plate.

$$\Gamma_1 = \left(\frac{\gamma_1 - \varepsilon_{r1}\gamma_2}{\gamma_1 + \varepsilon_{r1}\gamma_2}\right)e^{-2\gamma_1 d} \tag{3.34}$$

$$\Gamma_2 = 0 \tag{3.35}$$

From (3.34), the admittance model (3.23) can be decomposed into two parts. The first part is the aperture admittance for an infinite medium, while the second part of the admittance reflects the finite thickness of the medium, which is backed by free space.

$$\begin{aligned}\tilde{Y} = & \underbrace{\frac{jk_o\varepsilon_{r1}}{\sqrt{\varepsilon_c}\ln\left(\dfrac{b}{a}\right)}\int_0^\infty \frac{1}{\gamma_1}\frac{\left[J_o(\zeta a) - J_o(\zeta b)\right]^2}{\zeta}d\zeta}_{\text{Infinite thickness}} \\ & \underbrace{+\frac{jk_o\varepsilon_{r1}}{\sqrt{\varepsilon_c}\ln\left(\dfrac{b}{a}\right)}\int_0^\infty \frac{1}{\gamma_1}\frac{2(\gamma_1 - \varepsilon_{r1}\gamma_2)e^{-2\gamma_1 d}}{(\gamma_1 + \varepsilon_{r1}\gamma_2) - (\gamma_1 - \varepsilon_{r1}\gamma_2)e^{-2\gamma_1 d}}\frac{\left[J_o(\zeta a) - J_o(\zeta b)\right]^2}{\zeta}d\zeta}_{\text{Finite thickness}}\end{aligned} \tag{3.36}$$

As for the second case, the layer of air in the first case is replaced by a conducting plate, as shown in Figure 3.3(b), and its Γ_1 can be simplified as:

$$\Gamma_1 = e^{-2\gamma_1 d} \tag{3.37}$$

The skin-depth effect of the conducting plate is ignored. Thus, the normalized admittance, $\tilde{Y}$, gives [5, 7]:

$$\tilde{Y} = \frac{jk_o\varepsilon_r}{\sqrt{\varepsilon_c}\ln\left(\dfrac{b}{a}\right)}\int_0^\infty \frac{1 + e^{-2\gamma_1 d_1}}{1 - e^{-2\gamma_1 d_1}}\frac{\left[J_o(\zeta a) - J_o(\zeta b)\right]^2}{\gamma_1\zeta}d\zeta \tag{3.38a}$$

$$= \frac{jk_o\varepsilon_r}{\sqrt{\varepsilon_c}\ln\left(\frac{b}{a}\right)}\int_0^\infty \frac{\coth(\gamma_1 d)}{\gamma_1}\frac{\left[J_o(\zeta a) - J_o(\zeta b)\right]^2}{\zeta}d\zeta \tag{3.38b}$$

where $(1 + e^{-2\gamma_1 d})/(1 - e^{-2\gamma_1 d}) = \coth(\gamma_1 d)$. The solution and MATLAB calculation program of (3.38a) are shown on the accompanying CD. At the same time, the (3.38) can also be rewritten as [5]:

$$\tilde{Y} = \frac{jk_o\varepsilon_r}{\sqrt{\varepsilon_c}\ln\left(\frac{b}{a}\right)}\int_0^\infty \frac{1}{\gamma_1}\frac{\left[J_o(\zeta a) - J_o(\zeta b)\right]^2}{\zeta}d\zeta + \frac{jk_o\varepsilon_r}{\sqrt{\varepsilon_c}\ln\left(\frac{b}{a}\right)}\int_0^\infty \frac{1}{\gamma_1}\frac{2e^{-2\gamma_1 d}}{1 - e^{-2\gamma_1 d}}\frac{\left[J_o(\zeta a) - J_o(\zeta b)\right]^2}{\zeta}d\zeta \tag{3.39}$$

3.2 Image Method Modeling

The technique of images was first proposed by Lord Kelvin to determine the electric field, which is caused by a distribution of charges above a grounded boundary [8]. Anderson et al. [9] and Fan et al. [10] extended this technique to wave propagation from the coaxial line into a multilayered media problem. The analytical solution for the EM field in the layered media is assumed to be bounded by two parallel boundaries (boundary walls) at $-d$ and d on the z-axis as shown in Figure 3.4. The wall on the right side is the real configuration problem, while the wall on the left is the image wall with a mirror image along a ρ-axis plane at position $z = 0$. The boundary wall can be either the dielectric wall or conducting wall.

Let us consider that a coaxial probe aperture is terminated by a two-layered medium. The first layer and the second layer have dielectric properties of ε_{r1} and ε_{r2}, respectively. The charge, q, on the probe aperture is considered located at z' in the first layer of the medium. In fact, the distance of the boundary wall, d from the aperture probe, z', represents the actual thickness ($d' = d - z'$) of the first layer of the medium. The charge q emits an incident flux density line ψ in all radial directions from the first layer to the second layer of the medium through the boundary wall. On the wall surface, d, some fraction $K\psi$ of the flux reflects back into the first layer of the medium, while the remainder $(1 - K)\psi$ penetrates into the second layer of the medium due to the difference in dielectric characteristic between the two layers.

In order to satisfy the boundary conditions of the potential along the surface of the wall interface that is equal to zero, a virtual charge Kq is assumed to exist at $(2d - z')$ in the second layer of the medium, which will cause the reflected flux line as shown in Figure 3.4. The reflected flux lines $K\psi$ are further reflected partially into the first layer and penetrate into the second layer of medium and subsequently

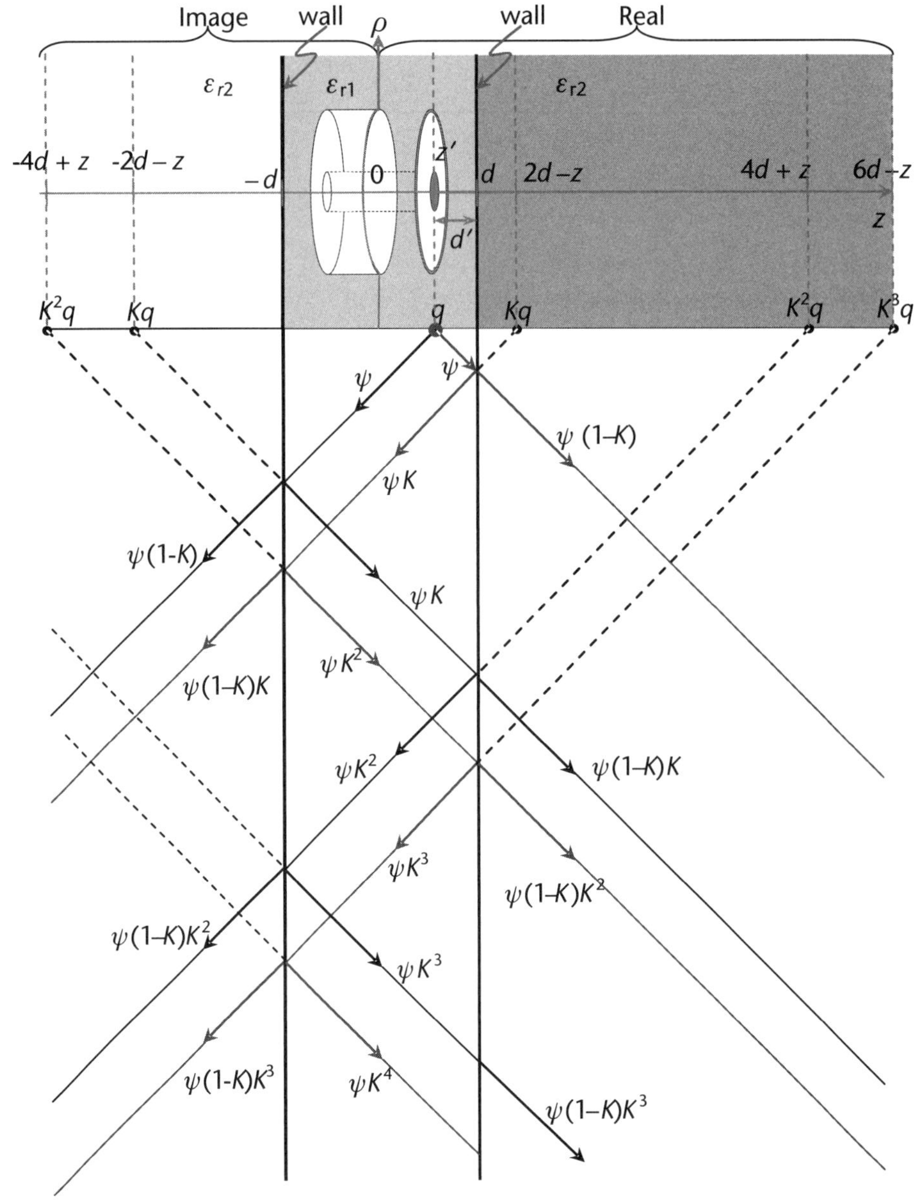

Figure 3.4 Configuration of the image model and distribution of flux lines in the finite-layer medium.

give rise to $K^2\psi$ and $\psi(1 - K)K$, respectively, at the left interface wall, $-d$. To offset the reflected flux lines, $K^2\psi$, another virtual charge, K^2q, must exist on the left at $(-4d + z')$ and so on ad infinitum between the two walls. In fact, the charge distribution around the surface of the inner conductor at the probe aperture is in the form of a ring with a radius of ρ'. Thus, the potential at point (ρ, z), due to a symmetry ring charge that is located at $(\rho', z'$ within the first layer of the medium (medium between two bounded walls), can be expressed as [9]:

$$V = \frac{1}{4\pi\varepsilon_o\varepsilon_{r1}} \int_0^{2\pi} \left\{ \begin{array}{l} \dfrac{q}{\sqrt{r^2 + (z - z')^2}} \\ + \displaystyle\sum_{n=1}^{\infty} \dfrac{qK^n}{\sqrt{r^2 + \left[z - 2nd + (-1)^{n+1} z'\right]^2}} \\ + \displaystyle\sum_{n=1}^{\infty} \dfrac{qK^n}{\sqrt{r^2 + \left[z + 2nd + (-1)^{n+1} z'\right]^2}} \end{array} \right\} d\phi \tag{3.40}$$

where r is the distance between two points [refer to (1.23b)]. The expression of fraction K in (3.40) can be found by electrostatic continuity rules of material interfaces or Snell's law. Figure 3.5 shows the normal and tangential components of the flux density, ψ, at the boundary wall, d. As is well known, only the tangential component of electric field, $E_t = (\psi \cos \theta)/\varepsilon_r$ (in V/m), is continuous at the boundary of the interface between the two layers. The angles θ of incidence and reflection are assumed to be equal. Thus, the continuity of the total tangential electric field E_t at point P on the interface of the wall can be written as [11]:

$$\begin{aligned} E_t(1 - K) &= E_t + E_t K \\ \frac{\psi\cos\theta}{\varepsilon_{r2}}(1 - K) &= \frac{\psi\cos\theta}{\varepsilon_{r1}} + \frac{\psi K\cos\theta}{\varepsilon_{r1}} \\ \varepsilon_{r1}(1 - K) &= \varepsilon_{r2}(1 + K) \end{aligned} \tag{3.41}$$

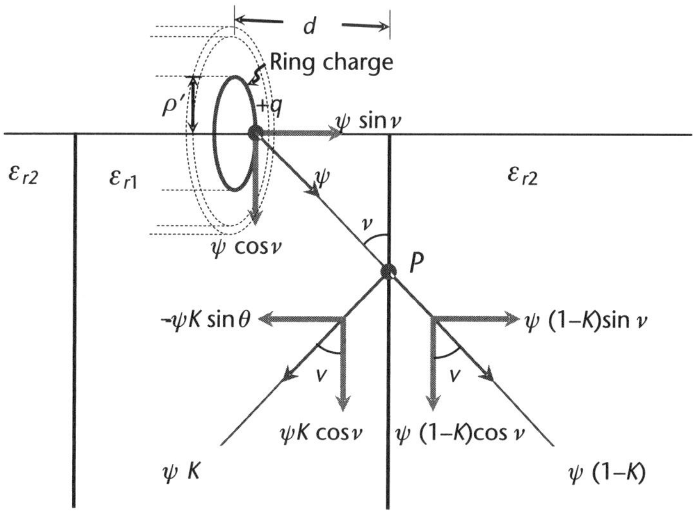

Figure 3.5 Part of the enlarged Figure 3.4: normal component and tangential component of the flux lines associated with a ring charge.

From (3.41), K, the image coefficient, can be expressed as:

$$K = \frac{\varepsilon_{r1} - \varepsilon_{r2}}{\varepsilon_{r1} + \varepsilon_{r2}} \tag{3.42}$$

If the bounded wall is perfectly electrically conducting (PEC) metal, then the value of ε_{r2} is infinite and results in the coefficient $K = -1$. If the second layer of medium is the air or free space, the image coefficient is $K = (\varepsilon_{r1} - 1)/(\varepsilon_{r1} + 1)$. For excitation of the uniform ring of charge, q is perfectly located at the mirror image plane ($z = z' = 0$), and the thickness of the first layer of the medium is equal to the distance to the wall ($d' = d$) from the mirror plane. In addition, the flux lines, ψ in the real and image regions, are symmetrical as shown in Figure 3.6. Thus, total voltages of (3.40) at probe aperture, $z = 0$, can be simplified as:

$$V = \frac{1}{4\pi\varepsilon_o\varepsilon_{r1}} \int_0^{2\pi} \left\{ \frac{q}{r} + \sum_{n=1}^{\infty} \frac{2qK^n}{\sqrt{r^2 + (2nd)^2}} \right\} d\phi \tag{3.43}$$

Fan et al. [10] and Misra et al. [12] determined the field in the dielectric layer using image analysis and the aperture admittance was written as:

$$\begin{aligned} Y = {} & \frac{j2\omega\varepsilon_o\varepsilon_{r1}}{[\ln(b/a)]^2} \int_a^b \int_a^b \int_0^{\pi} \frac{\exp\left(-jk_1\sqrt{r^2 + z^2}\right)}{\sqrt{r^2 + z^2}} \cos\phi' d\phi' d\rho' d\rho \\ & + \frac{j2\omega\varepsilon_o\varepsilon_{r1}}{\left[\ln(b/a)\right]^2} \sum_{n=1}^{\infty} \left(\frac{\varepsilon_{r2} - \varepsilon_{r1}}{\varepsilon_{r2} + \varepsilon_{r1}} \right)^n \int_a^b \int_a^b \int_0^{\pi} \left\{ \begin{aligned} & \frac{\exp\left(-jk_1\sqrt{r^2 + (2nd - z)^2}\right)}{\sqrt{r^2 + (2nd - z)^2}} \\ & + \frac{\exp\left(-jk_1\sqrt{r^2 + (2nd + z)^2}\right)}{\sqrt{r^2 + (2nd + z)^2}} \end{aligned} \right\} \cos\phi' d\phi' d\rho' d\rho \end{aligned} \tag{3.44}$$

For the sample under test that is backed by a conducting medium ($\varepsilon_{r2} \to \infty$), (3.44) at $z = 0$ can be reduced as:

$$\begin{aligned} Y = {} & \frac{j2\omega\varepsilon_o\varepsilon_{r1}}{[\ln(b/a)]^2} \int_a^b \int_a^b \int_0^{2\pi} \frac{\exp(-jk_1 r)}{r} \cos\phi' d\phi' d\rho' d\rho \\ & + \frac{j4\omega\varepsilon_o\varepsilon_{r1}}{[\ln(b/a)]^2} \sum_{n=1}^{\infty} \int_a^b \int_a^b \int_0^{2\pi} \frac{\exp\left(-jk_1\sqrt{r^2 + 4n^2d^2}\right)}{\sqrt{r^2 + 4n^2d^2}} \cos\phi' d\phi' d\rho' d\rho \end{aligned} \tag{3.45}$$

Equation (3.45) can be normalized by dividing the characteristic admittance, Y_o [refer to (2.5)], and is given as:

$$\begin{aligned} \tilde{Y} = {} & \frac{jk_o\varepsilon_{r1}}{\sqrt{\varepsilon_c}\ln(b/a)} \int_a^b \int_a^b \int_0^{2\pi} \frac{\exp(-jk_1 r)}{r} \cos\phi' d\phi' d\rho' d\rho \\ & + \frac{2jk_o\varepsilon_{r1}}{\sqrt{\varepsilon_c}\ln(b/a)} \sum_{n=1}^{\infty} \int_a^b \int_a^b \int_0^{2\pi} \frac{\exp\left(-jk_1\sqrt{r^2 + 4n^2d^2}\right)}{\sqrt{r^2 + 4n^2d^2}} \cos\phi' d\phi' d\rho' d\rho \end{aligned} \tag{3.46}$$

where k_o is the propagation constant of free space. The ε_c is the relative permittivity for the medium filled in the coaxial line of the probe. The multi-integral of first term in (3.46) can be reduced to a single Bessel function integral:

$$\int_a^b \int_a^b \int_0^{2\pi} \frac{\exp(-jk_1 r)}{r} \cos\phi' d\phi' d\rho' d\rho = \pi \int_0^\infty \frac{1}{\gamma_1} \frac{\left[J_o(\zeta a) - J_o(\zeta b)\right]^2}{\zeta} d\zeta \quad (3.47)$$

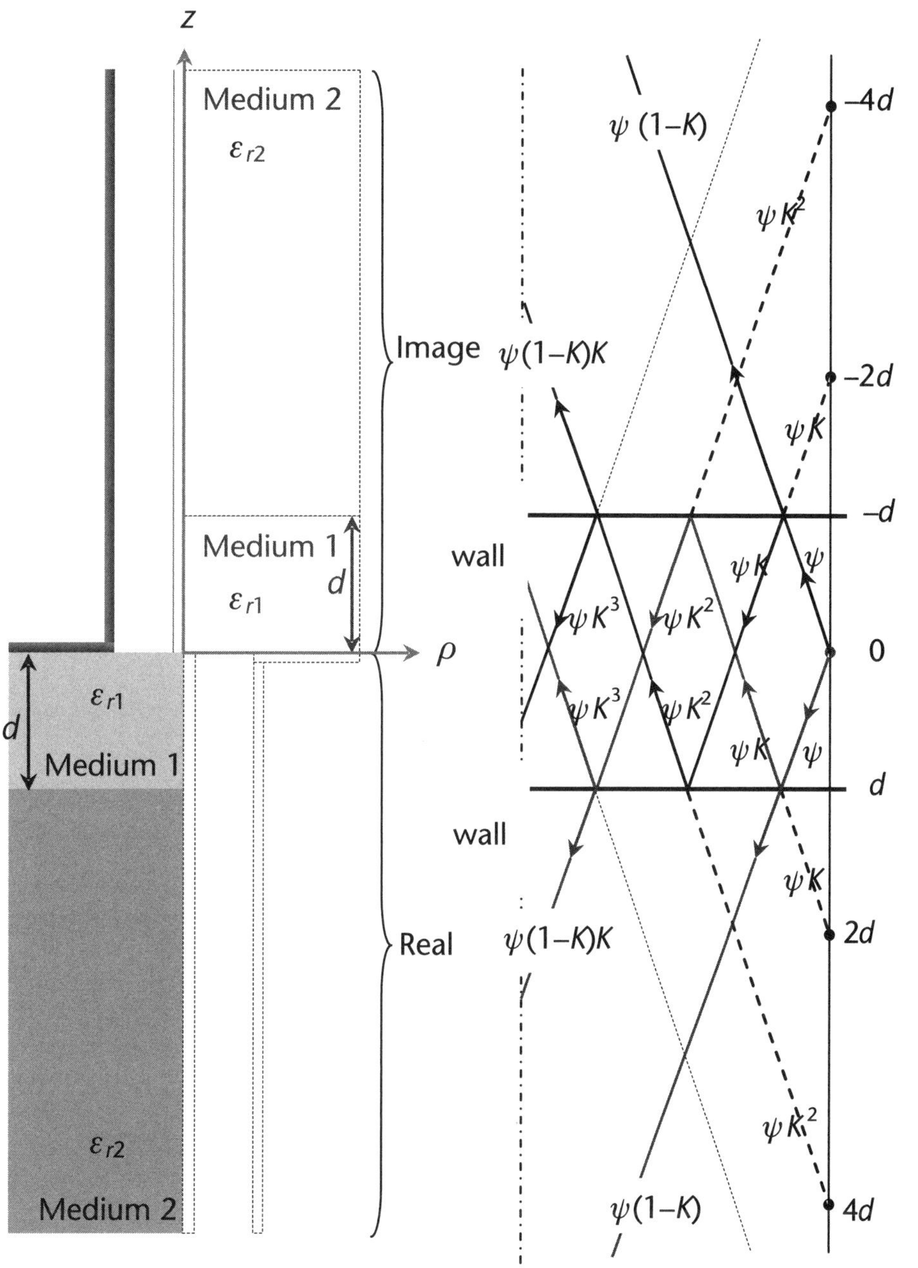

Figure 3.6 Configurations of image model and the flux lines in finite layer of the medium.

Thus, (3.46) can also be rewritten as:

$$\tilde{Y} = \frac{jk_o\varepsilon_{r1}}{\sqrt{\varepsilon_c}\ln(b/a)}\int_0^{\infty}\frac{1}{\gamma_1}\frac{\left[J_o(\zeta a)-J_o(\zeta b)\right]^2}{\zeta}d\zeta$$
$$+\frac{2jk_o\varepsilon_{r1}}{\sqrt{\varepsilon_c}\ln(b/a)}\sum_{n=1}^{\infty}\int_a^b\int_a^b\int_0^{2\pi}\frac{\exp\left(-jk_1\sqrt{r^2+4n^2d^2}\right)}{\sqrt{r^2+4n^2d^2}}\cos\phi' d\phi' d\rho' d\rho \quad (3.48)$$

The MATLAB calculation program of (3.48) is shown on the accompanying CD. Similar to (3.36) and (3.39), the first term in (3.48) is exactly equal to the aperture admittance for an infinite medium, while the second term corresponds to the thickness of the finite medium. Subsequently, the expression (3.45) can be specialized for the static capacitance C_o over the aperture probe as [10, 13, 14]:

$$C_o = C_1 + C_2 \quad (3.49)$$

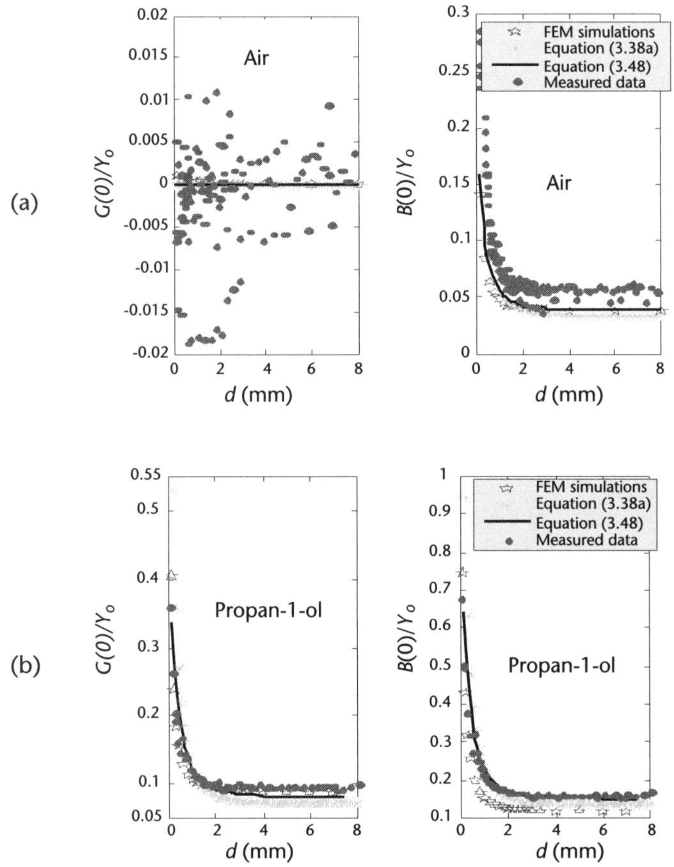

Figure 3.7 Variation in normalized admittance $\tilde{Y} = G(0)/Y_o + j\,B(0)/Y_o$ with the thickness layer, at room temperature for 4 GHz, of (a) air, (b) propan-1-ol, (c) ethanol, and (d) methanol.

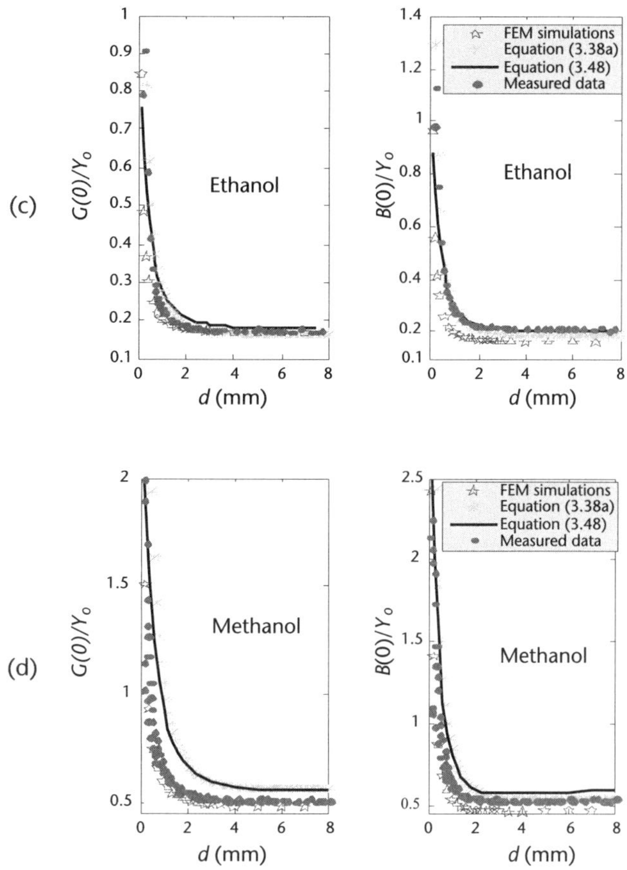

Figure 3.7 *(Continued)*

where

$$C_1 = \frac{2\varepsilon_o \varepsilon_{r1}}{\left(\ln(b/a)\right)^2} \int_a^b \int_a^b \int_0^{2\pi} \frac{\cos\phi'}{r} \, d\phi' \, d\rho' \, d\rho \tag{3.50a}$$

$$C_2 = \frac{4\varepsilon_o \varepsilon_{r1}}{\left[\ln(b/a)\right]^2} \sum_{n=1}^{\infty} \left(\frac{\varepsilon_{r2} - \varepsilon_{r1}}{\varepsilon_{r2} + \varepsilon_{r1}} \right)^n \int_a^b \int_a^b \int_0^{2\pi} \frac{\cos\phi'}{\sqrt{r^2 + 4n^2 d_1^2}} d\phi' d\rho' d\rho \tag{3.50b}$$

C_1 is exactly equal to the static capacitance over the aperture when the first layer of the medium is infinitely thick, while C_2 is a correction term to account for the thickness of a finite layer. The sample thickness measurements for air and three types of well-known lossy liquids are compared with the results that are obtained from finite element computation, (3.38a) [5, 7] and (3.48) [10, 12] as revealed in Figure 3.7. The single integral of (3.38a) is solved by 32-order Gauss-Legendre integral method. However, (3.48) is solved by (5 × 5 × 7) order Gaussian triple integral routine. However, only one series term of finite thickness in (3.48) is considered in this comparison in order to simplify the computation. Therefore, the calculated values using (3.48) revealed are quite different from measured data and finite element simulation as indicated in Figure 3.7. In this work, it is observed that when the thickness of the samples exceeds 4 mm, the measured magnitude and phase of the reflection coefficient at 4 GHz is unchanged as shown in Figure 3.7. It can be concluded that the minimum significant thickness of the sample under test is assumed to be 4 mm for this study probe below 4 GHz. The sample under test can be considered infinite, as long as the sample is sufficient to cover the aperture probe, which the size of the sample equal to the diameter of the outer radius probe, $2b$ = 4.1 mm.

3.3 Semi-Empirical Closed-Form Modeling

3.3.1 Total Capacitance C_T in the Finite-Layer Medium

For empirical analysis, the dependence of each capacitance on the sample thickness d_1 as shown in Figure 3.8 is assumed to be exponential [13–15]:

$$C_1 = \varepsilon'_{r1} C_o \left[1 - \exp\left(-d_1/M\right)\right] \tag{3.51a}$$

$$C_2 = \varepsilon'_{r2} C_o \exp\left(-d_1/M\right) \tag{3.51b}$$

where M is an empirical parameter with the dimensions of length. The total terminal capacitance, C_T, can be written as:

$$C_T = C_f + C_1 + C_2 \tag{3.52}$$

3.3.2 Effective Complex Relative Permittivity, ε_{eff}

Substituting (3.12) and (3.13) into (3.14) yields

$$C_T = C_f + \varepsilon'_{r1} C_o + \left(\varepsilon'_{r2} - \varepsilon'_{r1}\right) C_o \exp\left(-d_1/M\right) \tag{3.53a}$$

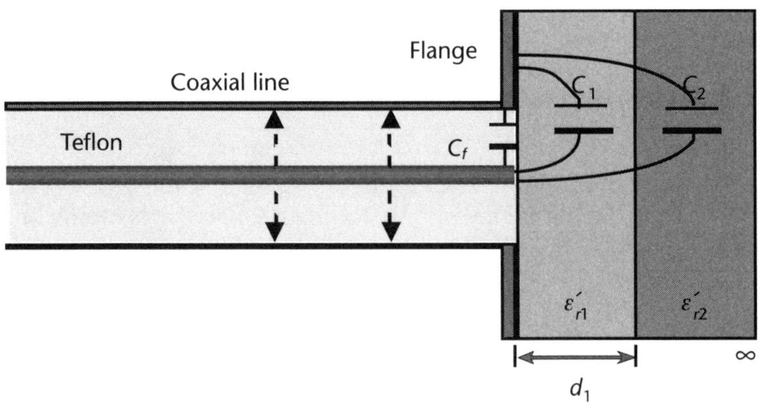

Figure 3.8 Configuration of parallel capacitance model.

$$= C_f + \varepsilon'_{\text{eff}} C_o \tag{3.53b}$$

where $\varepsilon'_{\text{eff}}$ is the effective dielectric constant of finite layer sample as:

$$\varepsilon'_{\text{eff}} = \varepsilon'_{r1} + \left(\varepsilon'_{r2} - \varepsilon'_{r1}\right)\exp\left(-d_1/M\right) \tag{3.54a}$$

However, (3.54a) is only valid for $\varepsilon'_{r1} > \varepsilon'_{r2}$ [14, 15]. For $\varepsilon'_{r1} < \varepsilon'_{r2}$, the effective dielectric constant, $\varepsilon'_{\text{eff}}$ can be expressed as [15]:

$$\frac{1}{\varepsilon'_{\text{eff}}} = \frac{1}{\varepsilon'_{r1}} + \left(\frac{1}{\varepsilon'_{r2}} - \frac{1}{\varepsilon'_{r1}}\right)\exp\left(-d_1/M\right) \tag{3.54b}$$

The formulation of actual complex relative permittivity ε_{eff} for a finite sample with one layer, d_1, which is backed by conducting metallic plate, can be expressed in simplified equation (3.54b). Since the dielectric constant of the metallic plate is assumed to be infinite ($\varepsilon'_{r2} \rightarrow \infty$),the $1/\varepsilon'_{r2}$ term in (3.54b) is neglected and subsequently yields:

$$\varepsilon_{\text{eff}} = \varepsilon'_{\text{eff}} - j\varepsilon''_{\text{eff}} = \frac{\varepsilon'_{r1} - j\varepsilon''_{r1}}{\left[1 - \exp\left(-d_1/M\right)\right]} \tag{3.55}$$

In this work, the empirical coefficient M is found to suit the study probe, which can be roughly represented by a single value as 0.0006 for a Teflon-filled coaxial probe with $2a$ = 1.3 mm and $2b$ = 4.1 mm. The comparison between (3.55) and measured permittivity data, which are obtained by using a commercial dielectric probe, is shown in Figure 3.9. The calculated values of (3.55) are represented by a solid line in Figure 3.9. The values of the actual dielectric constant, ε'_{r1}, and the actual loss factor, ε''_{r1}, in calculations of (3.55) are obtained from infinite medium measurements as tabulated in Table 3.1. Clearly, when the thickness of liquid sample has exceeded 4 mm, the effective permittivity is equal to the actual permittivity of the liquid sample, ($\varepsilon_{\text{eff}} = \varepsilon_{r1}$).

Table 3.1 Measured Values of ε'_{r1} and ε''_{r1} Used in the Calculations of (3.55)

	Actual relative permittivity, $\varepsilon'_{r1} - j\varepsilon''_{r1}$	
Sample	*1 GHz*	*5 GHz*
Propan-1-ol	6.68 – *j* 7.01	4.63 – *j* 2.20
Ethanol	12.72 – *j* 9.68	5.84 – *j* 3.89
Methanol	31.26 – *j* 9.25	12.13 – *j* 12.75

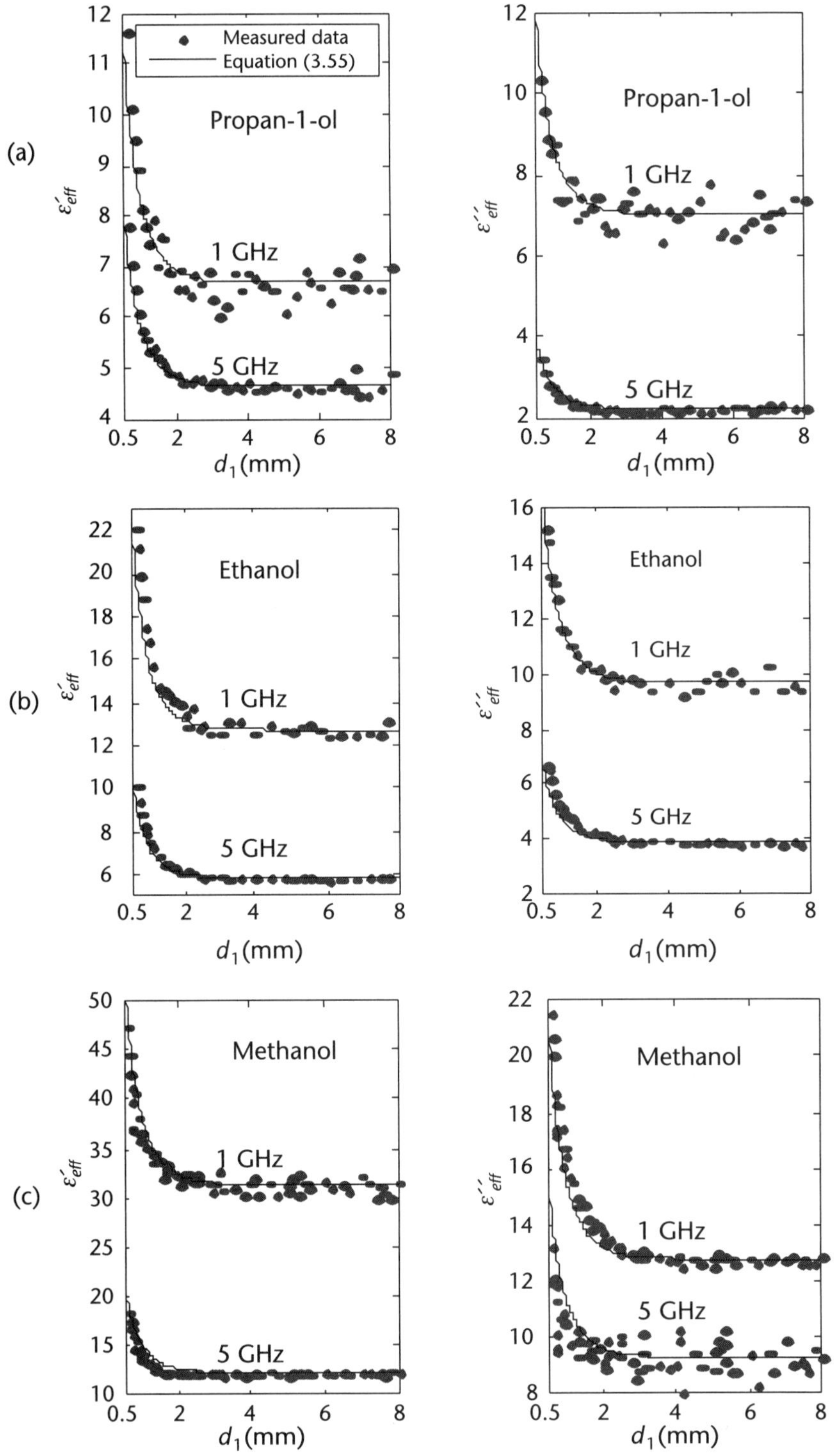

Figure 3.9 Variations in effective relative complex permittivity, ε_{eff}, with the thickness layer of (a) propan-1-ol, (b) ethanol, and (c) methanol, which are backed by a metallic plate. [The solid line is obtained from (3.55).]

3.3.3 Empirical Admittance Model of a Finite-Thickness Sample

From (3.55), the empirical normalized admittance, $\tilde{Y}$, for the finite-layer medium with thickness of d_1 and relative permittivity, $\varepsilon'_{r1} - \varepsilon''_{r1}$, which is backed by a metal plate, can be expressed as:

$$\tilde{Y} = \frac{j\omega\varepsilon_{\text{eff}}C_T}{Y_o} = \frac{j\left(\varepsilon'_{r1} - j\varepsilon''_{r1}\right)}{1 - \exp\left(-d_1/M\right)}\frac{\omega C_T}{Y_o} \tag{3.56}$$

where ω is the operating angular frequency (in rad/s). The total capacitance, C_T, in (3.56) can be calculated by using (2.8).

References

[1] Baker, J. J., et al., "Analysis of an Open-Ended Coaxial Probe with Lift-Off for Nondestructive Testing," *IEEE Trans. on Instrum. Meas.*, Vol. 43, 1994, pp. 1093–1099.

[2] Abramowitz, M., and I. A. Stegun, *Handbook of Mathematical Functions: With Formulas, Graphs, and Mathematical Table*, London, U.K.: Dover Publications, 1965.

[3] Bakhtiari, S., S. I. Ganchev, and R. Zoughi, "Analysis of Radiation from an Open-Ended Coaxial Line into Stratified Dielectrics," *IEEE Trans. on Microwave Theory Tech.*, Vol. 42, 1994, pp. 1261–1267.

[4] Li, L. L., et al., "Flanged Coaxial Microwave Probes for Measuring Thin Moisture Layers," *IEEE Trans. on Biomedical Engineering*, Vol. 39, 1992, pp. 49–57.

[5] Langhe, D. P., et al., "Measurements of Low Permittivity Materials based on a Spectral Domain Analysis for the Open Ended Coaxial Probe," *IEEE Trans. on Instrum. Meas.*, Vol. 41, 1993, pp. 639–645.

[6] Xu, Y. S., and R. G. Bosisio, "Nondestructive Measurements of the Resistivity of Thin Conductive Films and the Dielectric Constant of Thin Substrates Using an Open-Ended Coaxial Line," *IEE Proceeding-H*, Vol. 139, 2002, pp. 500–506.

[7] Galejs, J., *Antennas in Inhomogeneous Media*, Oxford, U.K.: Pergamon, 1969.

[8] Thomson, W. L. K., *Reprint of Papers on Electrostatics and Magnetism*, 2nd ed., London, U.K.: Macmillan, 1884.

[9] Anderson, L. S., G. B. Gajda, and S. S. Stuchly, "Analysis of an Open-Ended Coaxial Line Sensor in Layered Dielectric," *IEEE Trans. on Instrum. Meas.*, Vol. 35, 1986, pp. 13–18.

[10] Fan, S., K. Staebell, and D. Misra, "Static Analysis of an Open-Ended Coaxial Line Terminated by Layered Media," *IEEE Trans. on Instrum. Meas.*, Vol. 39, 1990, pp. 435–437.

[11] Silvester, P., "TEM Wave Properties of Microstrip Transmission Lines," *Proc. Inst. Elec. Eng.*, Vol. 115, 1968, pp. 43–48.

[12] Misra, D., "On the Measurement of the Complex Permittivity of Materials by an Open-Ended Coaxial Probe," *IEEE Microwave and Guided Wave Letters*, Vol. 5, 1995, pp. 161–163.

[13] Folgero, K., and T. Tjomsland, "Permittivity Measurement of Thin Liquid Layers Using Open Ended Coaxial Probes," *Meas. Sci. Technol.*, Vol. 7, 1996, pp. 1164–1173.

[14] Jiang, G. Q., et al., "Measurement of the Microwave Dielectric Constant for Low-Loss Samples with Finite Thickness Using Open-Ended Coaxial-Line Probes," *Rev. Sci. Instrum.*, Vol. 64, 1993, pp. 1622–1626.

[15] Alanen, E., T. Lahtinen, and J. Nuutinen, "Variational Formulation of Open-Ended Coaxial Line in Contact with Layered Biological Medium," *IEEE Trans. on Biomedical Engineering*, Vol. 45, 1998, pp. 1241–1248.

Selected Bibliography

Aburwein, A. A., A. A. B. Othman, and A. S. A. Kassim, "Input Admittance and Aperture Field of Flanged Coaxial Line Radiating into Layered Half-Space," *Int. J. Electronics*, Vol. 87, 2000, pp. 961–971.

Binns, K. J., and P. J. Lawrenson, *Analysis and Computation of Electric and Magnetic Field Problems*, Oxford, U.K.: Pergamon, 1973.

Chen, G., K. Li, and Z. Ji, "Bilayered Dielectric Measurement with an Open-Ended Coaxial Probe," *IEEE Trans. on Microwave Theory Tech.*, Vol. 42, 1994, pp. 966–971.

Ching-Lieh, L., and K. -M. Chen, "Complex Waves Excited by an Open-Ended Coaxial Probe in Layered Media," *Antennas and Propagation Society International Symposium*, 1992.

Ganchev, S. I., et al., "Calibration and Measurement of Dielectric Properties of Finite Thickness Composite Sheets with Open-Ended Coaxial Sensors," *IEEE Trans. on Instrum. Meas.*, Vol. 44, 1995, pp. 1023–1029.

Irzinski, E. P, "The Input Admittance of a TEM Excited Annular Slot Antenna," *IEEE Trans. Antennas Propagat.*, Vol. 23, 1975, pp. 829–834.

Lee, J. H., H. J. Eom, and K. H. Jun, "Reflection of a Coaxial Line Radiating into a Parallel Plate," *IEEE Microwave Guided Wave Letters*, Vol. 6, 1996, pp. 135–137.

Levine, H., and C. H. Papas, "Theory of the Circular Diffraction Antenna," *J. Appl. Phys.*, Vol. 22, 1951, pp. 29–43.

Li, C. L., and K. M. Chen, "Determination of Electromagnetic Properties of Materials Using Flanged Open-Ended Coaxial Probe-Full-Wave Analysis," *IEEE Trans. on Instrum. Meas.*, Vol. 44, 1995, pp. 19–27.

Martens, H. C. F., J. A. Reedijk, and H. B. Brom, "Measurement of the Complex Dielectric Constant Down to Helium Temperatures. I. Reflection Method from 1 MHz to 20 GHz Using an Open-Ended Coaxial Line," *Rev. Sci. Instrum.*, Vol. 71, 2000, pp. 473–477.

Noh, Y. C., and H. J. Eom, "Radiation from a Flanged Coaxial Line into a Dielectric Slab," *IEEE Trans. Microwave Theory Tech.*, Vol. 47, 1999, pp. 2158–2161.

Panariello, G., L.Verolino, and G. Vitolo, "Efficient and Accurate Full-Wave Analysis of the Open-Ended Coaxial Cable," *IEEE Trans. on Microwave Theory Tech.*, Vol. 49, 2001, pp. 1304–1309.

Qiu, Z., X. Li, and W. Jiang, "On Stability of Formulation of Open-Ended Coaxial Probe for Measurement of Electromagnetic Properties of Finite-Thickness Materials," *Journal of Electromagnetic Waves and Applications*, Vol. 23, 2009, pp. 501–511.

Swift, C. T. "Input Admittance of a Coaxial Transmission Line Opening onto a Flat, Dielectric Covered Ground Plane," *NASA Tech. Rep.* NASA TND-4158, 1967.

Trabelsi, S., A. Kraszewski, and S. Nelson, "Nondestructive Microwave Characterization for Determining the Bulk Density and Moisture Content of Shelled Corn," *Meas. Sci. Technol.*, Vol. 3, 1998, pp. 1548–1556.

Wu, M. Z., et al., "Determination of Microwave Complex Permittivity of Particulate Materials," *Meas. Sci. Technol.*, Vol. 3, 2001, pp. 1191–1202.

CHAPTER 4

Analytical Modeling of a Bare Monopole Driven from a Coaxial Line

In this chapter, the analysis of analytical modeling will begin with the finite dipole structure and continue to the finite monopole, which is driven from the coaxial line. In fact, the difference between the dipole and the monopole is that the monopole has half the radiation resistance as compared to the dipole. From the developmental history of the formulation, the analytical expression for the dipole and monopole radiators has been developed intensively compared to an open-ended coaxial radiator. A great emphasis is placed on three types of primary parameters for the dipole and monopole: (1) current I_z distribution on the dipole or monopole along a z-axis, (2) input impedance, Z_{in} (or input admittance, Y_{in}) at the feed point, and (3) z-component of the electric field, E_z.

The universal formulation of current distribution, I_z, on the monopole has been studied by researchers throughout the years. However, no findings show that there is a full analytical relationship between the current distribution, I_z, and an arbitrary dimension of the monopole. Previously, the distribution of the current I_z on the monopole had always been assumed to be in simple sinusoidal form. Subsequently, the current distributions were further investigated by several researchers and a lot of rigorous analytical current equations were developed in order to improve the reliability and accuracy of the current distribution along the length of the monopole. Although the current distribution I_z functions have been predicted, but the amplitude, I_o, of the function is normally an unknown value; it can be obtained by numerical methods. There are two types of governing integration, namely Pocklington's equation and Hallén's equation, which will always be considered to determine the unknown amplitude of the current iteratively. The calculated amplitude, I_o, that is applied to the integration by using the method of moments (MoM), which will be discussed in Chapter 6.

In this chapter, two analytical methods are reviewed, the induced electromagnetic field (EMF) method [1] and the image method (IM). The EMF method can be used to solve for the input impedance, Z_{in}, without involving the evaluation of the amplitude current, I_o. However, the combined method of EMF and IM is not only used for an input impedance solution, but also to show the correlation between the dipole and monopole. In addition, the author found that the input impedance, Z_{in}, derived from a combination of EMF and the IM methods, is the most reliable and simple quasi-static formulation for the thin and short monopole at present. Since most recent monopoles are driven from coaxial lines, which will produce a coupling fringing field at the gap between the monopole and the outer conductors at the coaxial line; hence, besides finding the actual current distribution, the open-end correction is another issue for the monopole at the driving point or the feed point. The fringing field will affect the accuracy of the calculated input impedance, Z_{in}, due to the usage of analytical quasi-static formulas. Thus, the fringing field effect must be taken into account in the theoretical calculation. Normally, the effects of fringing fields at the driving point are corrected for by the capacitance element circuits. In Section 4.4, various end-correction approaches are described in detail.

A closed-form equivalent circuit model is the simplest empirical technique that is able to relate the input impedance, Z_{in} (or input admittance, Y_{in}), of the monopole to its dimensions and the dielectric properties of the medium surrounding the monopole. However, the input impedance, Z_{in}, for longer monopoles (monopole length, $h \geq \lambda/4$ meter) is changed to irregular form, and with an increase in frequency will cause the simple equivalent circuit model to fail to handle this complicated shape. Thus, the simple equivalent circuit model is fitted for very short monopole or very low operating frequencies. For a long monopole that operates at a high frequency, a complex equivalent circuit is required. To be specific, more *RLC* elements are required in the circuit to match with the input impedance, Z_{in}, which is obtained from full-wave simulation or measurements.

4.1 Theoretical Background

4.1.1 Current Distribution on Dipole and Monopole

As previously mentioned, there is no analytical formula that is properly suited for the distribution current I_z of an arbitrary size of a dipole or monopole. Nevertheless, in the late 1920s, the equations for the distribution current I_z were proposed by assuming that I_z takes a convenient sinusoidal form such as:

$$I_z = \begin{cases} I_m \sin k(h - z'), & \text{for } 0 \leq z' \leq h \\ I_m \sin k(h + z'), & \text{for } -h \leq z' \leq 0 \end{cases} \tag{4.1}$$

where k (= ω/c) is the propagation constant of the medium surrounding the dipole or monopole and I_m is the maximum current amplitude and is expressed as:

$$I_m = \frac{I_o}{\sin(kh)} \tag{4.2}$$

where I_o is the driving point current amplitude at $z = 0$. However, when the dipole or monopole is relatively thick and long compared to the operation wavelength, λ, its current distribution is no longer sinusoidal. Thus, (4.1) is only valid when the monopole radius, a is less than $\lambda \times 10^{-4}$ meters and the monopole length h is less than $\lambda/4$ meters [2].

From (4.1), the current I_z at the end of the dipole (at $z = h$) is forced to be zero as shown in Figure 4.1, while the current is maximum at the driving point or the feed point (at $z = 0$). Several improved current distribution functions have been proposed

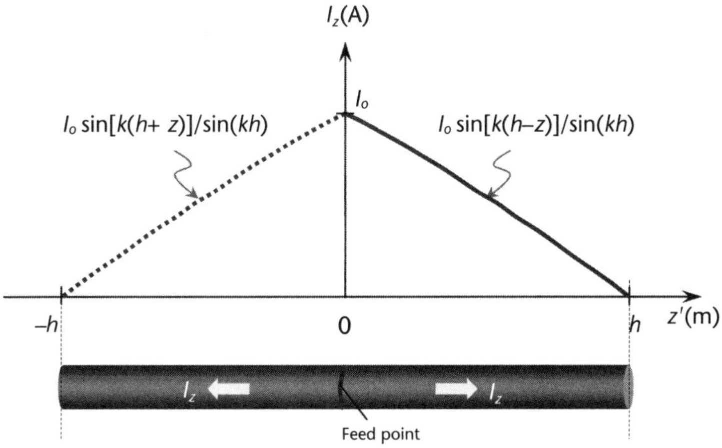

Figure 4.1 Sinusoidal current distribution along a dipole.

previously. The unknown coefficients in each function can be determined by the method of moment (MoM), which will be discussed in Chapter 6.

The following are various expressions of the current distribution I_z:

$$I_z = \sin k\left(h - |z|\right) + A\left[1 - \cos k\left(h - |z|\right)\right], \text{ for } h < \lambda \tag{4.3a}$$

where A is the unknown complex coefficient [3].

$$I_z = \sin k\left(h - |z|\right) + Bk\left(h - |z|\right)\cos k\left(h - |z|\right) \tag{4.3b}$$

where B is the unknown complex coefficients [4].

$$I_z = \sum_{n=1}^{N}\left\{I_n \cos\left[\frac{(2n-1)\pi z}{2h}\right] + I'_n \sin\left(\frac{n\pi z}{h}\right)\right\} \tag{4.3c}$$

where I_n and I'_n are the unknown expansion coefficients [5].

$$I_z = I_v\begin{bmatrix} \sin k\left(h - |z|\right) + T_U(\cos kz - \cos kh) \\ +T_D\left(\cos\frac{kz}{2} - \cos\frac{kh}{2}\right) \end{bmatrix}, \text{ for } kh \le 5\pi/4 \tag{4.3d}$$

where I_v, T_U, and T_D are unknown complex coefficients [6].

$$I_z = I_o\left(e^{jkz} + Re^{-jkz}\right) \tag{4.3e}$$

where I_o is the unknown amplitude of the incident current wave. R is the complex reflection coefficient at the open end of the line [7].

$$I_z = e^{-k|z|}\sum_{n=0}^{3} I_n\left|z^n\right|, \text{ for } h/\lambda \geq 0.625 \tag{4.3f}$$

where I_n is the unknown expansion coefficient [8].

$$I_z = \sum_{n=1}^{N} c_n \sin\left[\frac{n\pi}{2h}\left(h-|z|\right)\right] \tag{4.3g}$$

where c_n is the unknown expansion coefficient [9].

4.1.2 Pocklington's Equation

A thin conducting cylinder of length $2h$ (in meters) and radius a (in meters), located symmetrically along the z-axis, is shown in Figure 4.2(a). For cylindrically symmetrical cases, the nonvanishing components are H_ϕ, E_ρ, E_z, and A_z for the principal TEM mode as shown in Figure 4.2(b). From (1.28), we obtain

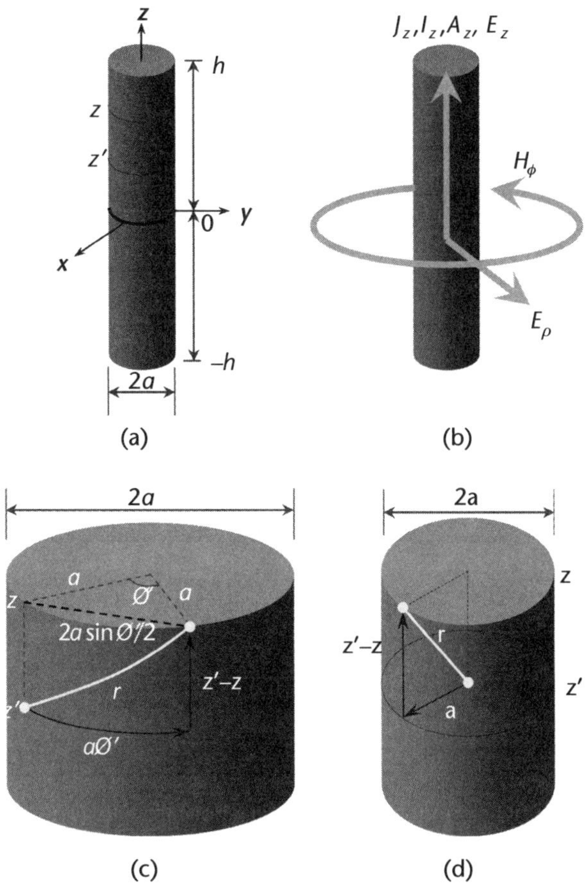

Figure 4.2 (a) Cylindrical coordinates. (b) TEM mode components for linear cylindrical conductor. (c) Distance between two points on the wire surface. (d) Distance between two points from the center of dipole.

$$\frac{1}{\rho}\frac{\partial(\rho H_\phi)}{\partial \rho} = j\omega\varepsilon E_z \tag{4.4a}$$

$$\frac{\partial H_\phi}{\partial z} = -j\omega\varepsilon E_\rho \tag{4.4b}$$

where $\varepsilon = \varepsilon_o\varepsilon_r$ is the permittivity of the medium. The magnetic field $\vec{H}$ can also be written in terms of the vector potential $\vec{A}$ as

$$\vec{H} = \frac{1}{\mu}\vec{\nabla} \times \vec{A} \tag{4.5}$$

As pointed out above, for linear wire, the component of vector potential $\vec{A}$ is only in the z-direction; thus, (4.5) becomes

$$H_\phi = -\frac{1}{\mu}\frac{\partial A_z}{\partial \rho}\hat{\phi} \tag{4.6}$$

Inserting (4.6) into (4.4a) and (4.4b) yields

$$\frac{1}{\rho}\frac{\partial}{\partial \rho}\left(\rho\frac{\partial A_z}{\partial \rho}\right) = j\omega\mu\varepsilon E_z \tag{4.7a}$$

$$\frac{\partial^2 A_z}{\partial z\,\partial \rho} = -j\omega\mu\varepsilon E_\rho \tag{4.7b}$$

The wave equation in a free space is written as

$$\vec{\nabla}^2\vec{A} + k^2\vec{A} = \frac{1}{\rho}\frac{\partial}{\partial \rho}\left(\rho\frac{\partial A_z}{\partial \rho}\right) + \frac{\partial^2 A_z}{\partial z^2} + k^2 A_z = 0 \tag{4.8}$$

Replacing (4.7a) into (4.8) gives

$$\frac{\partial^2 A_z}{\partial z^2} + k^2 A_z = -j\omega\varepsilon\mu E_z \tag{4.9}$$

In 1897, Pocklington [10] was the first to express (4.9) into an integro-differential form as:

$$\left(\frac{\partial^2}{\partial z^2} + k^2\right)\int_{-h}^{h}\int_{0}^{2\pi} I_z \frac{e^{-jkr}}{2\pi r} d\phi' dz' = -j4\pi\omega\varepsilon E_z \tag{4.10}$$

where

$$A_z = \frac{\mu}{4\pi}\int_{-h}^{h}\int_{0}^{2\pi} I_z \frac{e^{-jkr}}{2\pi r} d\phi' dz' \tag{4.11}$$

The I_z term is the equivalent filamentary line-source current located on the surface of the dipole. The e^{-jkr}/r term in (4.11) is the kernel of the integral equation with

$$r = \sqrt{\rho^2 + \rho'^2 - 2\rho\rho'\cos(\phi' - \phi) + (z - z')^2} \tag{4.12}$$

Since the current is only located on the surface of the cylindrical wire ($\rho' = \rho = a$), (4.12) gives [in Figure 4.2(c)]

$$r = \sqrt{4a^2 \sin^2(\phi'/2) + (z - z')^2} \tag{4.13}$$

For observations along the center of a very thin wire ($\rho' = 0$ and $\rho = a$), the distance r in (4.13) can be reduced to $r = \sqrt{a^2 + (z - z')^2}$ [in Figure 4.2(d)] and the (4.10) can be simplified as:

$$\left(\frac{\partial^2}{\partial z^2} + k^2\right)\int_{-h}^{h} I_z \frac{e^{-jkr}}{r} dz' = -j4\pi\omega\varepsilon E_z \tag{4.14}$$

Richmond et al. [5] solved the differential operator of (4.14) and gives it in a convenient form as:

$$\int_{-h}^{h} I_z \frac{e^{-jkr}}{r^5}\left[(1 + jkr)(2r^2 - 3a^2) + (kar)^2\right]dz' = -j4\pi\omega\varepsilon_o\varepsilon_r E_z \tag{4.15}$$

Referring to [11], (4.15) has been reexpressed in a more convenient form as:

$$E_z = \frac{j}{4\pi\omega\varepsilon_o\varepsilon_r}\int_{-h}^{h} I_z e^{-jkr}\left[\frac{2}{r^2}\left(jk + \frac{1}{r}\right) + \frac{k^2a^2}{r^3} - \frac{3a^2}{r^4}\left(jk + \frac{1}{r}\right)\right]dz' \tag{4.16}$$

The electric field, E_z, of the dipole can also be derived by using the image method, which is referring to the symmetry monopole configuration as shown in Figure 4.3(a). The monopole is equivalent to the half of the vertical dipole, while the bottom half of the dipole is replaced by a ground conducting plane as shown in Figure 4.3(b). The electromagnetic waves that are emitted from the monopole and subsequently being reflected from the ground plane becomes similar to the radiating wave from an image of the bottom half of the dipole. In addition, the feed voltage of the dipole is assumed to be twice the feed voltage for monopole. Thus, an actual monopole will have input impedance, which is only half of the dipole. If the current, I_z, term in (4.14) is replaced by (4.1), it yields

$$E_z = \frac{jI_o\eta_o}{4\pi\sqrt{\varepsilon_r}\sin(kh)}\left[\frac{e^{-jkr_1}}{r_1} + \frac{e^{-jkr_2}}{r_2} - 2\cos(kh)\frac{e^{-jkr_3}}{r_3}\right] \text{ (Dipole)} \tag{4.17}$$

where

$$r_1 = \sqrt{\rho^2 + (z - h)^2} \tag{4.18a}$$

$$r_2 = \sqrt{\rho^2 + (z + h)^2} \tag{4.18b}$$

$$r_3 = \sqrt{\rho^2 + z^2} \tag{4.18c}$$

where η_o (= $\sqrt{\mu_o}/\sqrt{\varepsilon_o}$) is the intrinsic impedance of free space. The model configuration of (4.17) is shown in Figure 4.4. Equation (4.17) is an alternative solution of

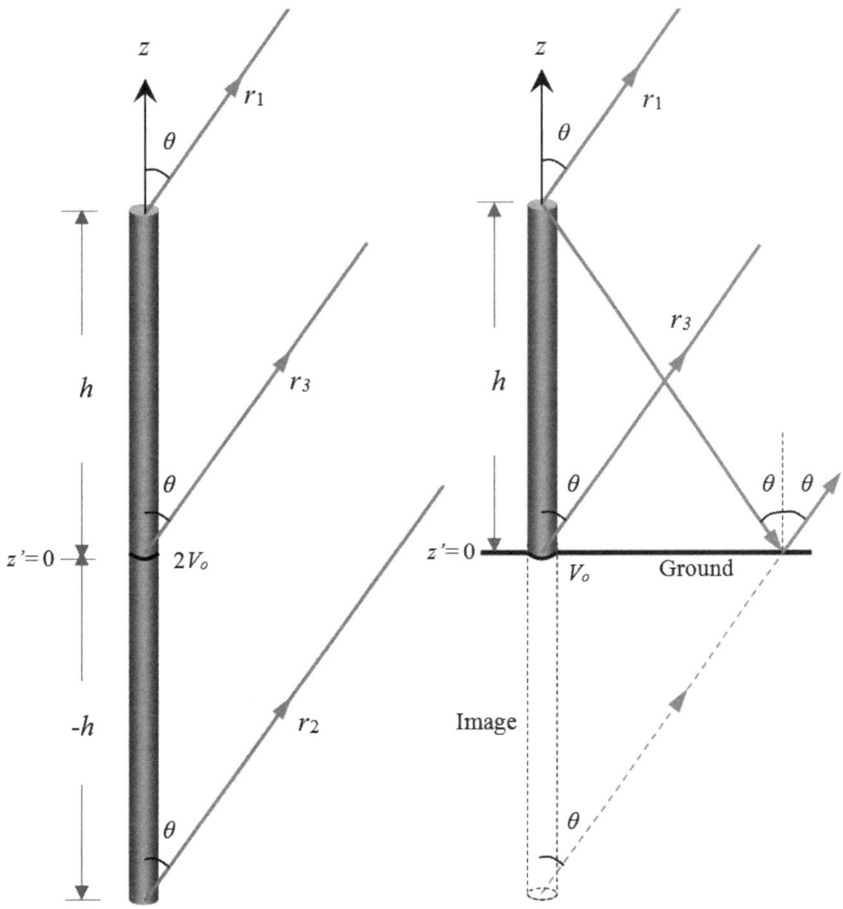

Figure 4.3 The correlation between the dipole model and the monopole model. (a) Mathematical model of the dipole. (b) Method of images for the monopole.

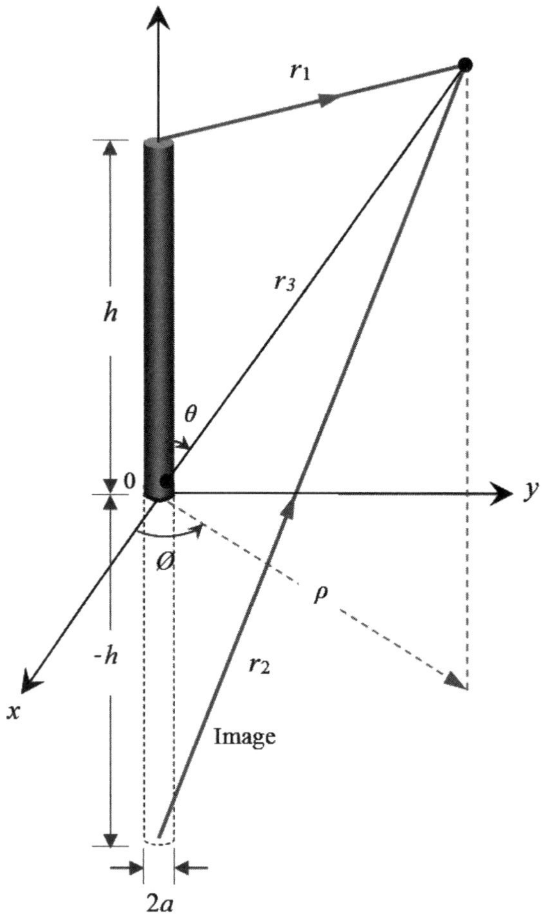

Note:

$r_1 = \sqrt{\rho^2 + (z-h)^2}$; $r_2 = \sqrt{\rho^2 + (z+h)^2}$; $r_3 = \sqrt{\rho^2 + z^2}$

Figure 4.4 Cylindrical coordinates for image theory used in evaluating the monopole radiation field.

E_z, but it is suitable only for very thin dipole; since (4.17) gives rise to $E_z \neq 0$, which is not satisfied with the boundary condition that $E_z = 0$ on the conducting surface at $\rho = a$. The derivation of (4.17) is described on the accompanying CD.

The electric field E_z can also be formulated by referring to coordinates as shown in Figure 4.5. Similarly, by substituting the sinusoidal current distribution of (4.1) into (4.14), the monopole driven from the ground is given as [2]:

$$E_z = \frac{jI_o\eta_o}{4\pi\sqrt{\varepsilon_r}\sin(kh)}\left\{\frac{e^{-jkr_1}}{r_1} - \frac{e^{-jkr_3}}{r_3}\cos(kh) - \sin(kh)(z)\left(\frac{1}{kr_3^2} + \frac{j}{r_3}\right)\frac{e^{-jkr_3}}{r_3}\right\} \quad \text{(Monopole)} \quad (4.19)$$

The derivation of (4.19) is described on the accompanying CD.

4.1.3 Hallén's Equation

Later, this integral equation for thin dipole was modified by Hallén in 1938 [12] as:

$$A_z = \frac{\mu}{4\pi}\int_{-h}^{h} I_z \frac{e^{-jkr}}{r} dz' = -j\sqrt{\varepsilon\mu}\left[\kappa\cos(kz) + \frac{V_o}{2}\sin\left(k|z|\right)\right] \tag{4.20}$$

where V_o is the applied voltage at a feed terminal of the dipole. κ is the constant coefficient that requires the current to vanish at the end of the monopole. Also, at the end of monopole (at $z' = h$), the ideal distributed current $I_z = 0$; thus, (4.20) can be rewritten as:

$$\kappa\cos(kh) + \frac{V_o}{2}\sin\left(k|h|\right) = 0 \tag{4.21}$$

From (4.21), the constant coefficient κ gives

$$\kappa = -\frac{V_o}{2}\frac{\sin\left(k|h|\right)}{\cos(kh)} \tag{4.22}$$

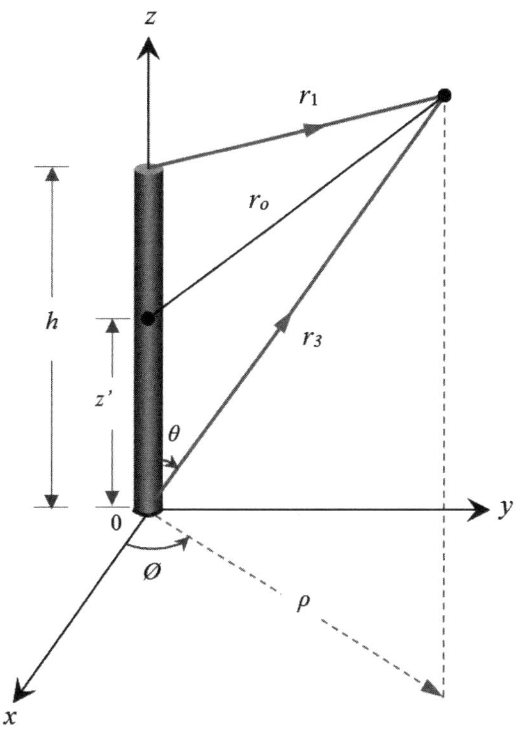

Note:

$r_1 = \sqrt{\rho^2 + (z-h)^2}$; $r_o = \sqrt{\rho^2 + (z-z')^2}$; $r_3 = \sqrt{\rho^2 + z'^2}$

Figure 4.5 Cylindrical coordinates used in evaluating the monopole radiation field.

Substituting (4.22) into (4.20) yields:

$$\frac{\mu}{4\pi}\int_{-h}^{h} I_z \frac{e^{-jkr}}{r} dz' = j\sqrt{\varepsilon\mu}\frac{V_o}{2\cos(kh)}\left[\sin\left(k|h|\right)\cos(kz) - \cos(kh)\sin\left(k|z|\right)\right]$$
$$= j\sqrt{\varepsilon\mu}\frac{V_o}{2\cos(kh)}\sin\left[k\left(h - |z|\right)\right] \quad (4.23)$$

Equation (4.23) will be applied to find the current distribution I_z by using the MoM routine, which is described in Chapter 6.

4.2 Induced Electromagnetic Field (EMF) Modeling

For dipole or monopole cases, the energy of the current is released through the extended conductor from an excitation point; thus, the inductive properties are normally dominant at the point $z = 0$. Hence, the analysis and prediction of the performance of the dipole or monopole are often expressed as the input impedance $Z_{in} = \mathfrak{Re}(Z_{in}) + j\mathfrak{Im}(Z_{in})$ instead of input admittance, $Y_{in} = G(0) + jB(0)$.

The induced electromagnetic field (EMF) method was introduced by Brillouin [1] to determine the input impedance Z_{in} at the driving point of linear monopole:

$$Z_{in} = \frac{V_o}{I_o} = -\frac{1}{|I_o|^2}\int_0^h I_z E_z dz' \quad (4.24)$$

where $V_o = \int_0^h E_z dz$. Substituting (4.1) and (4.17) into (4.24) yields [2]:

$$Z_{in} = \frac{j\eta_o}{4\pi\sqrt{\varepsilon_r}\sin^2(kh)}\int_0^h \left\{ \begin{array}{l} \sin\left[k(h-z)\right]\times \\ \left[\dfrac{e^{-jkr_1}}{r_1} + \dfrac{e^{-jkr_2}}{r_2} - 2\cos(kh)\dfrac{e^{-jkr_3}}{r_3}\right] \end{array} \right\} dz \quad \text{(Monopole)}$$
$$(4.25)$$

Similarly, if substituting (4.1) and (4.19) into (4.24), Z_{in} is given as [2]:

$$Z_{in} = \frac{j\eta_o}{4\pi\sqrt{\varepsilon_r}\sin^2(kh)}\int_0^h \left\{ \begin{array}{l} \sin\left[k(h-z)\right]\times \\ \left[\begin{array}{l} \dfrac{e^{-jkr_1}}{r_1} - \cos(kh)\dfrac{e^{-jkr_3}}{r_3} \\ -z\sin(kh)e^{-jkr_3}\left(\dfrac{jk}{r_3^2} + \dfrac{1}{r_3^3}\right) \end{array}\right] \end{array} \right\} dz \quad \text{(Monopole)} \quad (4.26)$$

Clearly, the unknown amplitude current I_o has vanished in input impedance Z_{in} formulations. Thus, integrations of (4.25) and (4.26) can be solved analytically as shown on the accompanying CD. The MATLAB calculation program of (4.25) and

(4.26) using a Simpson's rule is shown on the accompanying CD. A semi-empirical formulation of input impedance Z_{in} for the dipole with $2h$ of length was proposed in [13] as:

$$Z_{in} = R + jX - j120\left[\ln\left(\frac{2h}{a}\right) - 1\right]\cot(kh) \quad \text{(Dipole)} \tag{4.27a}$$

$$Z_{in} = \frac{1}{2}(R + jX) - j60\left[\ln\left(\frac{2h}{a}\right) - 1\right]\cot(kh) \quad \text{(Monopole)} \tag{4.27b}$$

The accuracy of (4.27) is limited in the range ($0 < kh \leq \pi/2$). The values of R and X are obtained by fitting with numerical simulation data. In [13], the R and X are expressed as a function of (kh) in third-order polynomials as:

$$R = -0.4787 + 7.3246(kh) + 0.3963(kh)^2 + 15.6131(kh)^3 \tag{4.28a}$$

$$X = -0.4456 + 17.0082(kh) + 8.6793(kh)^2 + 9.6031(kh)^3 \tag{4.28b}$$

It should be noted that the input impedance of the monopole is half of the value of (4.27). Figure 4.6 shows the calculated normalized resistance, $\Re e(Z_{in})/Z_o$, and the normalized reactance, $\Im m(Z_{in})/Z_o$, of a monopole for free space at room temperature using (4.25), (4.26), and (4.27). Z_o is the characteristic impedance in a coaxial line, which can be calculated as

$$Z_o = \frac{1}{Y_o} = \frac{60\ln(b/a)}{\sqrt{\varepsilon_c}} \tag{4.29}$$

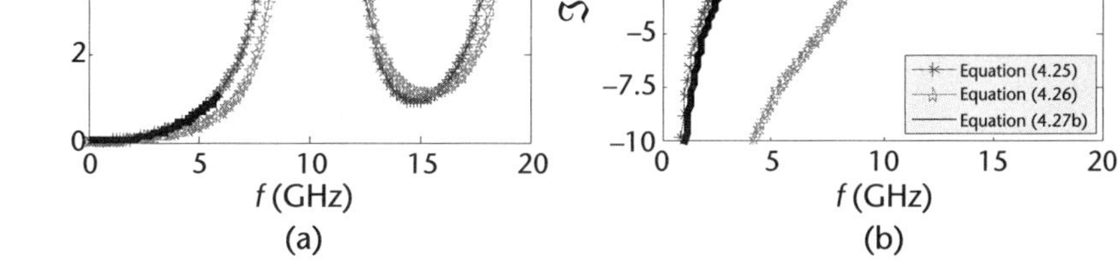

Figure 4.6 Comparison of calculated normalized resistance, $\Re e(Z_{in})/Z_o$, and normalized reactance, $\Im m(Z_{in})/Z_o$, of a monopole for air at 25°C using (4.25), (4.26), and (4.27) (a = 0.65 mm, b = 2.05 mm, and h = 14.32 mm).

where ε_c is a relative permittivity of a material filling the coaxial line (ε_c = 2.06 for Teflon). In Figure 4.6, the calculated $\Re e(Z_{in})/Z_o$ and $\Im m(Z_{in})/Z_o$ for (4.27) are only plotted in the range ($0 < kh \leq \pi/2$).

It is found that (4.25) and (4.26) give similar resistance values, $\Re e(Z_{in})/Z_o$, in the frequency range of 300 kHz to 20 GHz. However, both equations give less consistent values of reactance, $\Im m(Z_{in})/Z_o$, and this may be due to the fact that the field distribution near the source point was different between the two models. This issue will be discussed further in Section 4.3.

4.3 End Correction for a Monopole Driven from a Coaxial Line

There are two excitation source assumptions which are widely used at the driving point at $z = 0$ for the h length of the monopole. One is the delta-gap voltage model [14], and the other one is the magnetic frill model, which was proposed by Tsai [15] as shown in Figure 4.7(a). For the delta-gap model, the excitation source at driving point is assumed to have constant voltage V_o and its corresponding electric field, E_z^{exc}, becomes:

$$E_z^{exc} = \frac{V_o}{\delta z} \tag{4.30}$$

where δz (in meters) refers to the width of the gap between two connected monopoles and the voltage V_o (in volts) is usually set to unity. The delta-gap model is simple, but it produces less accurate results for input impedance and the fringing fields will not be taken into account. Nevertheless, it performs very well for computing radiation patterns. For instance, the excitation voltage in Hallén's equation is implemented based on the delta-gap concept.

The magnetic frill model, as shown in Figure 4.14(b), is more accurate for the real situation. For the magnetic frill model, the excitation field intensity, E_z^{exc}, is written as (1.29b). On the axis of the wire (ρ =0), we have $r = \sqrt{(z - z')^2 + \rho'^2}$ and (1.29b) can be reduced as:

$$\begin{aligned} E_z^{exc} &= \frac{V_o}{\pi \ln(b/a)} \int_a^b \int_0^\pi \cos\phi' \frac{\partial}{\partial \rho}\left(\frac{e^{-jkr}}{r}\right) d\phi' d\rho' \\ &= \frac{V_o}{\ln(b/a)}\left[\frac{e^{-jk\sqrt{(z-z')^2+a^2}}}{\sqrt{(z-z')^2+a^2}} - \frac{e^{-jk\sqrt{(z-z')^2+b^2}}}{\sqrt{(z-z')^2+b^2}}\right] \end{aligned} \tag{4.31}$$

Normally, the V_o in (4.31) is assumed to be equal to 1/2 V. Clearly, the calculated input impedance, Z_{in}, obtained from (4.30) cannot compare with the measurements using monopole, which is driven from a coaxial line as shown in Figure 4.8. In this case, (4.31) will provide a more accurate calculation result compared with (4.30). However, the solution using (4.31) still contains an approximation of small but unknown extent [16], which will be further studied in Chapter 6.

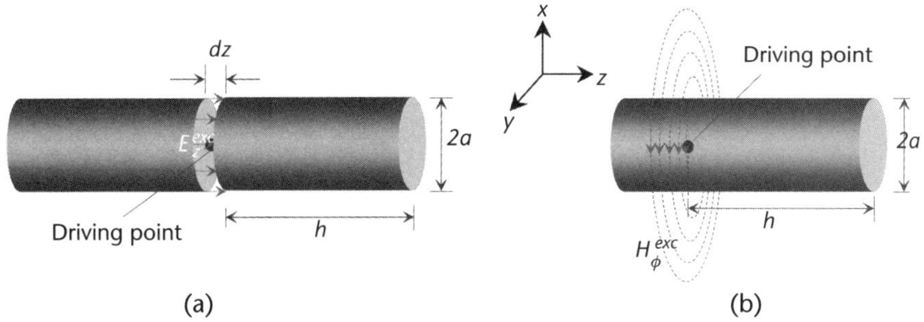

Figure 4.7 Monopole excitation models. (a) Delta-gap excitation model. (b) Magnetic frill excitation model.

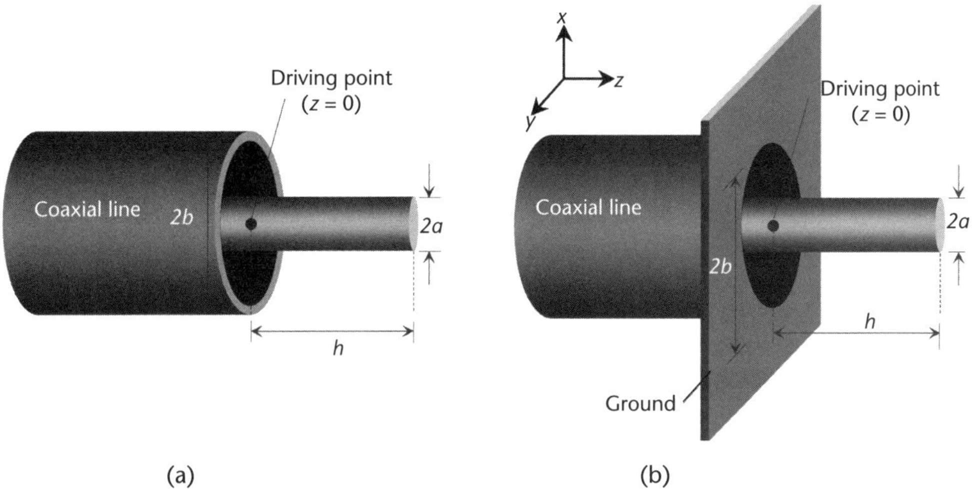

Figure 4.8 (a) Monopole fed by A coaxial transmission line with the inner radius *a* and the outer radius *b*. (b) Monopole driven from the coaxial line over the ground plane.

The input impedance Z_{in} obtained from measurements is strongly affected by the fringing field at the driving point (at the end of coaxial line). The reason is that the fringing field is caused by the coupling effect between the outer conductor and the monopole near the end of the coaxial line. Thus, the ideal input impedance needs an end correction term to correlate the theoretical solution to the experimental result.

Various solutions have been proposed to mitigate the end-effect problems by using capacitance approaches, but the applications are restricted to only up to the certain exposure length of the monopole in free space. Due to the capacitive properties of the fringing field, the expression of admittance Y_{in} is referred to in those problems, which is different from the earlier discussion of the use of impedance Z_{in}. The corrected apparent admittance Y_{app} for infinite monopole driven from a coaxial line was investigated by Chang [17] as:

$$Y_{\text{app}} = Y_{\text{in}} - j\omega C_T \tag{4.32}$$

where the end correction capacitance, C_T (in Farads), at the driving point ($z = 0$) is given as:

$$C_T \approx -4\varepsilon_o a \ln\left(\frac{4}{\pi}\right) \tag{4.33}$$

where ε_o is the permittivity of free space. The TEM input admittance, Y_{in}, in (4.32) is derived based on delta-gap voltage model, which can be found in [17]. Later, Williamson [16] reported that the Y_{app} for infinite monopole can be expressed as:

$$Y_{app} = Y_{in} - jB_c \tag{4.34}$$

where the end correction term, B_c (in Siemens), is given as:

$$B_c = \left[\left(4.4 \times 10^{-4}\right) + \left(7.2 \times 10^{-5}\right)\left(\frac{b}{a}\right)\right] ka, \text{ for } ka \le 0.1 \tag{4.35}$$

The Y_{in} in (4.34) is different from the Y_{in} in (4.32), which is derived based on the magnetic frill model as expressed in [18].

Recently, You et al. [19] proposed a semi-empirical capacitance correction for the analytical admittance formula over a wide range of length, h, radius, a, aspect ratio, b/a, and frequency, f, as well as the relative permittivity, ε_r, of the medium surrounding the monopole. The apparent admittance, Y_{app}, is written as [19]:

$$Y_{app} = Y_{in} - j\alpha\omega\varepsilon_r C \tag{4.36}$$

The Y_{in} in (4.36) is calculated by (4.25) as:

$$Y_{in} = \frac{1}{Z_{in}} = \frac{-j4\pi\sqrt{\varepsilon_r}\sin^2(kh)}{\eta_o \int_0^h \sin\left[k(h - z')\right] \times \left[\frac{e^{-jkr_1}}{r_1} + \frac{e^{-jkr_2}}{r_2} - 2\cos(kh)\frac{e^{-jkr_3}}{r_3}\right] dz'} \tag{4.37}$$

α is the weight parameter of the end correction, and C is the aperture capacitance between the inner cylindrical conductor and the ground plane at the open end of the coaxial line. The capacitance, C, as shown in Figure 4.9(b) can be calculated from [20]:

$$C = 2.48\varepsilon_o(b - a) \tag{4.38}$$

In this work, the weight parameter, α, is expressed in second-order polynomial form, which includes three variables (a, b, and h) as

$$\alpha = A_3 a^2 + A_2 a + A_1 \tag{4.39}$$

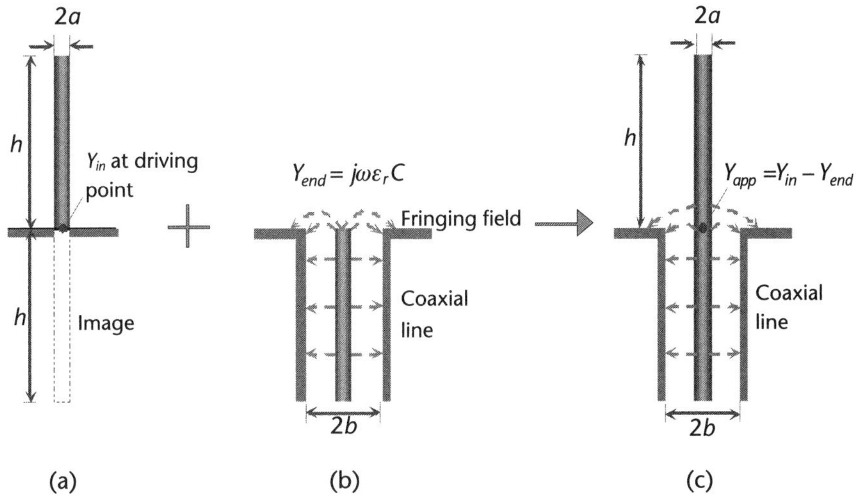

Figure 4.9 (a) The ideal input admittance, Y_{in}. (b) The aperture admittance, Y_{end}, of an open-ended coaxial line. (c) The apparent admittance, $Y_{app} = Y_{in} - \alpha Y_{end}$, for practical monopole.

where

$$A_3 = \sum_{n=0}^{4} B_n''' h^{-n},\ A_2 = \sum_{n=0}^{3} B_n'' h^{-n},\ A_1 = \sum_{n=0}^{2} B_n' h^{-n}$$

and

$$\begin{aligned}
B_0' &= \beta_4 R_t^3 + \beta_3 R_t^2 + \beta_2 R_t + \beta_1; & B_1' &= \beta_7 R_t^2 + \beta_6 R_t + \beta_5;\\
B_2' &= \beta_{10} R_t^2 + \beta_9 R_t + \beta_8; & B_0'' &= \beta_{13} R_t^2 + \beta_{12} R_t + \beta_{11};\\
B_1'' &= \beta_{16} R_t^2 + \beta_{15} R_t + \beta_{14}; & B_2'' &= \beta_{19} R_t^2 + \beta_{18} R_t + \beta_{17};\\
B_3'' &= \beta_{22} R_t^2 + \beta_{21} R_t + \beta_{20}; & B_0''' &= \beta_{25} R_t^2 + \beta_{24} R_t + \beta_{23};\\
B_1''' &= \beta_{28} R_t^2 + \beta_{27} R_t + \beta_{26}; & B_2''' &= \beta_{31} R_t^2 + \beta_{30} R_t + \beta_{29};\\
B_3''' &= \beta_{34} R_t^2 + \beta_{33} R_t + \beta_{32}; & B_4''' &= \beta_{37} R_t^2 + \beta_{36} R_t + \beta_{35}
\end{aligned}$$

where $R_t = b/a$ and the coefficients of β are the fitting values. Up to 37 fitting coefficients (β_1 to β_{37}) are chosen to provide a greater degree of freedom for the regression fitting. The values of β_1 to β_{37} are listed on the accompanying CD. The values of the weight parameter β are found by optimizing with up to 1,700 calculated admittance values obtained from finite-element method (FEM). Equation (4.36) is applicable for low-conductivity materials at low frequencies satisfying the conditions: ($2.3 \le b/a \le 4$), ($0.22 \le a \le 0.65$) mm, ($1 \le h \le 10$) mm, and $kh \le 3\pi/4$. For arbitrary values of size ratio, b/a, the weight parameter, α is close to unity when $h = (10 \times a)$ mm, whereas when $h \ge (10 \times b)$ mm, α becomes zero.

Figure 4.10 shows good agreement for the free space normalized admittance, $\tilde{Y}_{in} = G(0)/Y_o + jB(0)/Y_o$, results, which are obtained by using (4.36), (4.37), FEM,

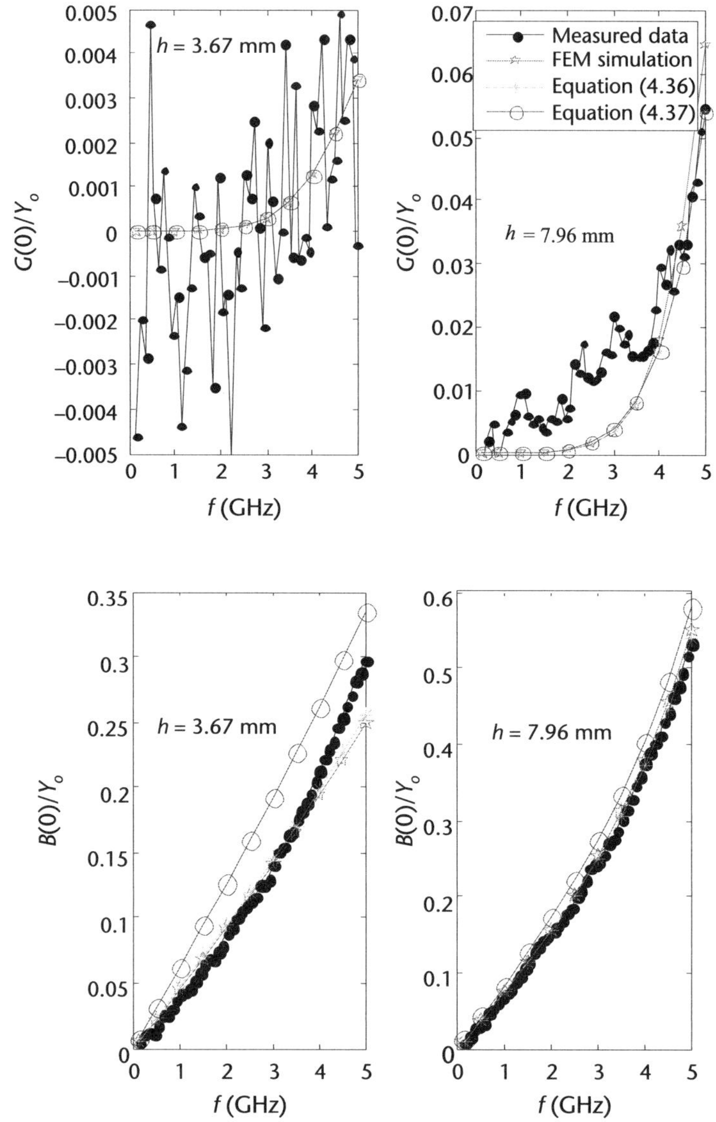

Figure 4.10 Variation in normalized conductance, $G(0)/Y_o$, and normalized susceptance, $B(0)/Y_o$, with frequency for h = 3.67 mm and h = 7.96 mm for a monopole in free space.

and measurement for the normalized conductance, $G(0)/Y_o$, and normalized susceptance, $B(0)/Y_o$, with frequencies for h = 3.67 mm and h = 7.96 mm. As expected, the effect of the finite gap fields can be clearly seen for the shorter monopole with h = 3.67 mm monopole when using (4.37). Additionally, Figure 4.10 clearly shows the limitation of (4.37) as well as FEM at higher frequencies where poor agreement with the measured normalized susceptance, $B(0)/Y_o$, can be observed at frequencies above 3 GHz.

The variation in normalized admittance, $\tilde{Y}_{in}$, with frequency for propan-1-ol, ethanol, methanol, and water using h = 3.67 mm monopole are shown in Figure 4.11. Again, (4.37) satisfactorily predicts the admittance values for all the liquids. However, good agreement between measured and calculated admittance for water

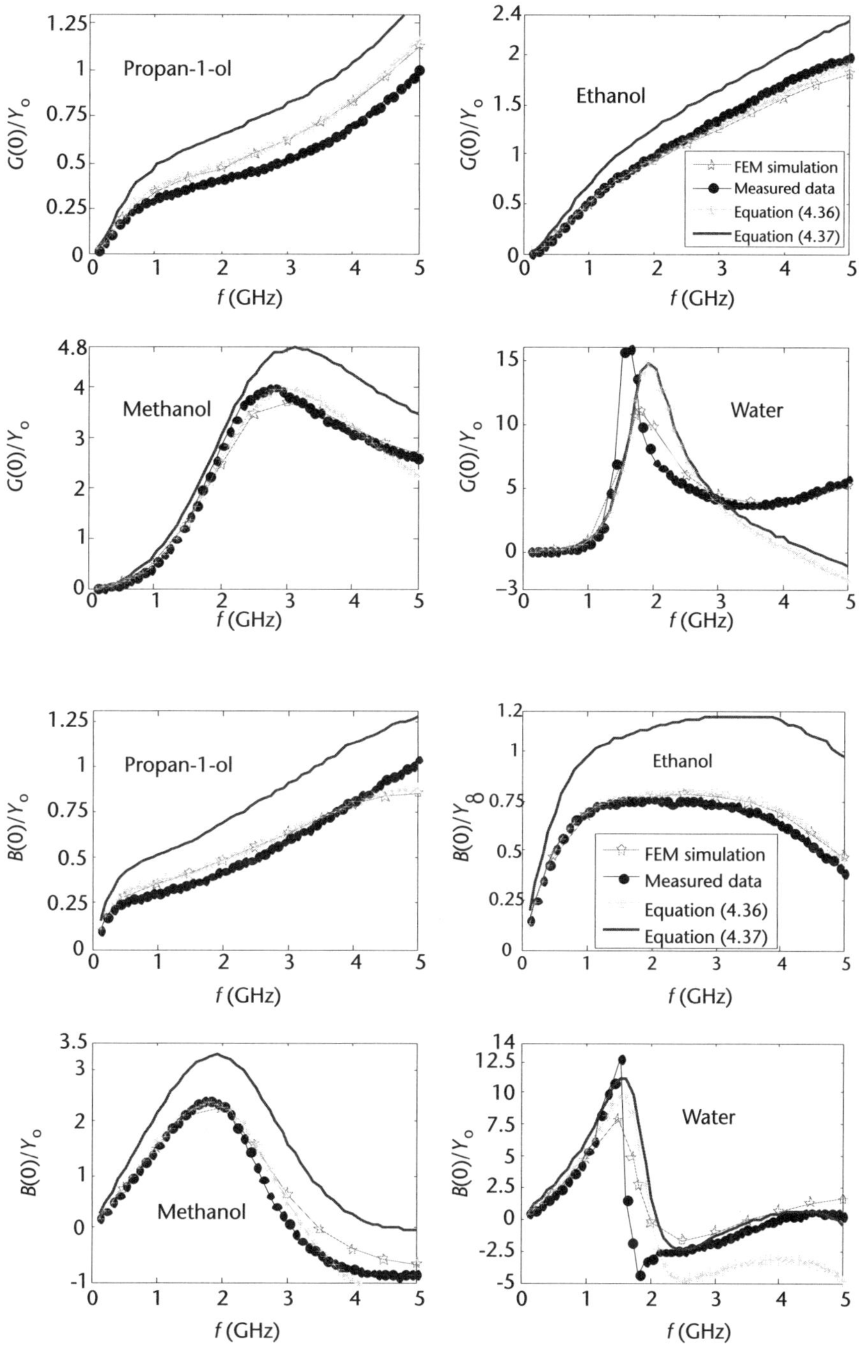

Figure 4.11 The normalized conductance, $G(0)/Y_o$, normalized susceptance, $B(0)/Y_o$, of monopole at $z = 0$ with $a = 0.65$ mm, $h = 3.67$ mm, and $b/a = 3.154$ for various liquids.

is detected at frequencies less than 2 GHz, due to the fact that water, being a lossy material, does not meet the criteria $kh \leq 3\pi/4$ at frequencies above 2 GHz. At higher frequencies, inductance, L, occurred due to ground plane effects and the resistance, R, near the ground plane becomes significant. To overcome this problem, we suggest that (4.37) must also include several lumped circuit correction elements (R, C, and L) in series and parallel combination.

4.4 Equivalent Lumped-Element Circuits Modeling

4.4.1 Equivalent Circuits Modeling (in Free Space)

As discussed in Chapter 2, the values of R, C, and L in the equivalent circuit model are found by matching to either a full-wave simulation or a measurement. From previous measurements and simulations, a general form of complex input impedance, Z_{in} for a dipole or monopole that varies with angular frequency, ω, can be represented by Figure 4.12(a).

Hamid and Hamid [21] modeled the input impedance, Z_{in} of dipole using a five-element equivalent circuit, as shown in the Figure 4.13. Consequently, the Z_{in} is expressed as:

$$\begin{aligned} Z_{\text{in}} &= \Re\text{e}\left(Z_{in}\right) + j\Im\text{m}\left(Z_{in}\right) \\ &= -j\frac{1}{\omega C_1} + j\omega L_2 + Z_3 \\ &= -j\frac{1}{\omega C_1} + j\omega L_2 + \frac{1}{\left(1/R_3\right) + \left(1/j\omega L_3\right) + j\omega C_3} \\ &= -j\frac{1}{\omega C_1} + j\omega L_2 + \frac{j\omega R_3 L_3}{j\omega L_3 + R_3 - \omega^2 R_3 L_3 C_3} \end{aligned} \tag{4.40}$$

where

$$\Re\text{e}\left(Z_{\text{in}}\right) = \Re\text{e}\left(Z_3\right) \tag{4.41a}$$

$$\Im\text{m}\left(Z_{\text{in}}\right) = \frac{-1}{\omega C_1} + \omega L_2 + \Im\text{m}\left(Z_3\right) \tag{4.41b}$$

and

$$\Re\text{e}\left(Z_3\right) = \frac{\omega^2 L_3^2 R_3}{\left(R_3 - \omega^2 C_3 L_3 R_3\right)^2 + \omega^2 L_3^2} \tag{4.42a}$$

$$\Im\text{m}\left(Z_3\right) = \frac{\omega R_3 L_3\left(R_3 - \omega^2 C_3 L_3 R_3\right)}{\left(R_3 - \omega^2 C_3 L_3 R_3\right)^2 + \omega^2 L_3^2} \tag{4.42b}$$

First, the single value of capacitance, C_1, is found by matching the reactance of the impedance, $\Im\text{m}(Z_{\text{in}})$, at the lower-frequency range ($0 < \omega \leq 0.2\omega_o$).

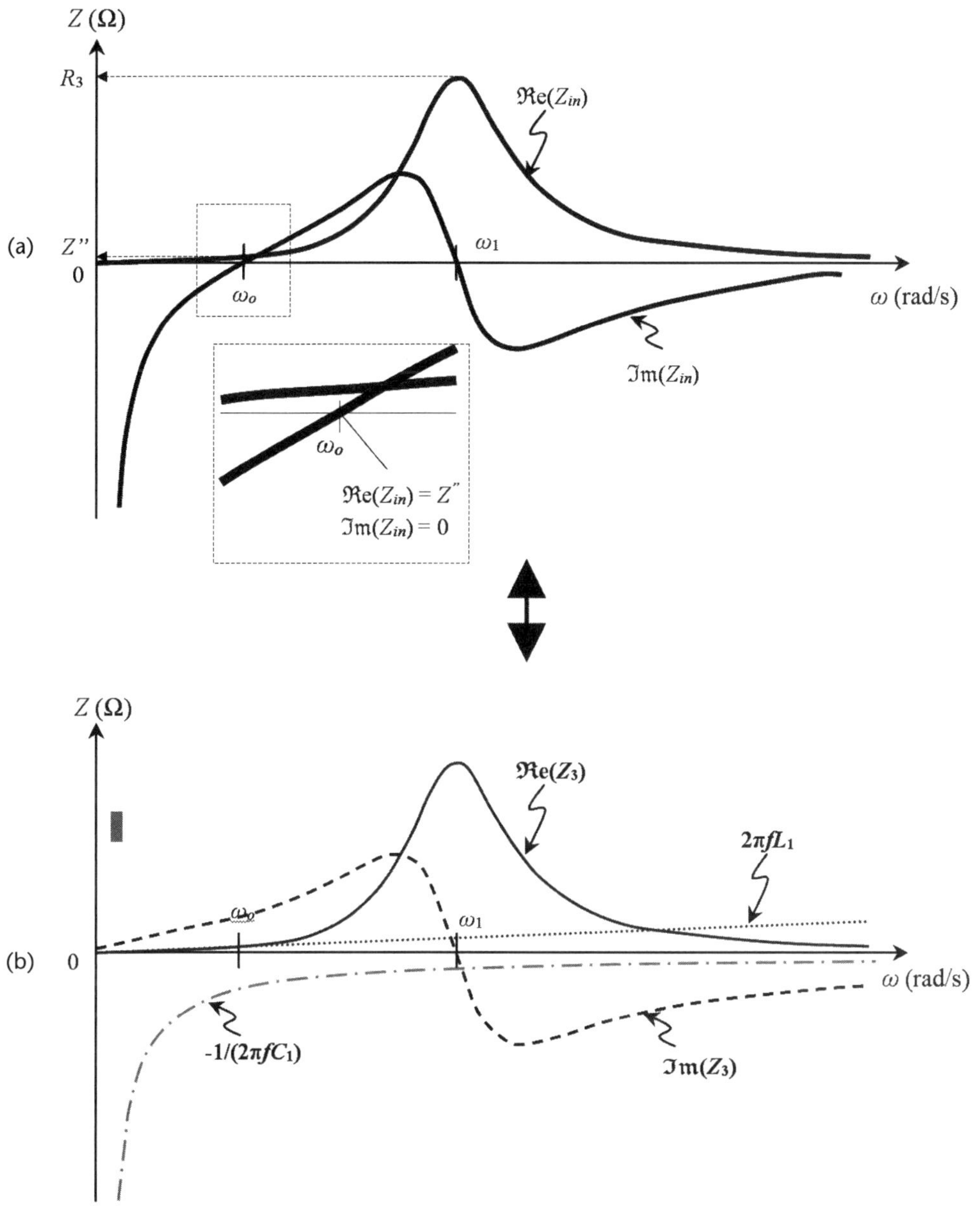

Figure 4.12 (a) Angular frequency response of input impedance, Z_{in} of dipole with length $2h$ in the range $0 \leq h/\lambda \leq 0.75$. (b) Components of input impedance for the five-element matching circuit.

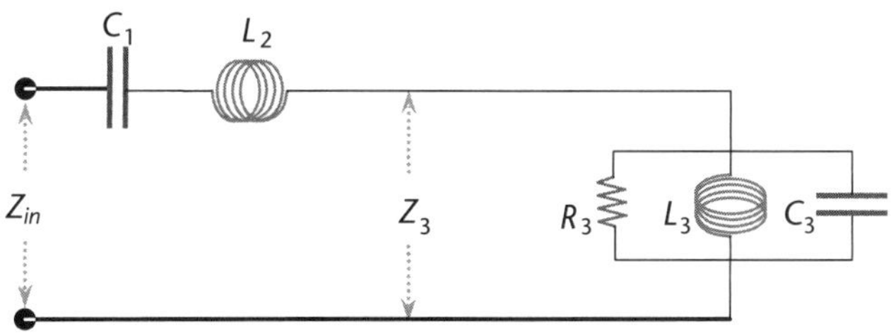

Figure 4.13 Five-element equivalent circuit.

$$C_1 \approx \frac{-1}{\omega \Im m\left(Z_{in}\right)}, \text{ for } (0 < \omega \leq 0.2\omega_o) \tag{4.43}$$

At lower frequencies, the capacitive properties dominate the total reactance, $\Im m(Z_{in})$ that is caused by the coupling fringing effect between the dipole conductor and the ground conductor at driving point.

At $h = \lambda/2$ (at the angular frequency, ω_1), the first anti-resonance occurs and causes a large and rapid change to Z_{in}, where $\Re e(Z_{in})$ shows a maximum value, while $\Im m(Z_{in})$ is equal to zero. This situation is equivalent to the cascaded $R_3L_3C_3$ parallel resonant circuit as shown in Figure 4.13. The series inductor L_2 can be used to make minor adjustments to matching with the $\Im m(Z_{in})$. In fact, L_2 is assumed to come from the self-inductance of the dipole conductor at the driving point. The value of L_2, which resonates with C_1 at the first anti-resonant angular frequency, ω_1, can be estimated using (4.41b):

$$0 = \frac{-1}{\omega_1 C_1} + \omega_1 L_2 + 0 \tag{4.44}$$

where $\Im m(Z_{in}) = \Im m(Z_3) = 0$ as shown in Figure 4.12. From (4.44), this gives:

$$L_2 = \frac{1}{\omega_1^2 C_1} \tag{4.45}$$

Similarly, by using (4.42b) at ω_1, the relationship between L_3 and C_3 at the first anti-resonant can be written as:

$$\begin{aligned} &\frac{\omega_1 R_3 L_3 \left(R_3 - \omega_1^2 C_3 L_3 R_3\right)}{\left(R_3 - \omega_1^2 C_3 L_3 R_3\right)^2 + \omega_1^2 L_3^2} = 0 \\ &\omega_1 R_3 L_3 \left(R_3 - \omega_1^2 C_3 L_3 R_3\right) = 0 \\ &L_3 = \frac{1}{\omega_1^2 C_3} \end{aligned} \tag{4.46}$$

For parallel resonant circuit, the value of R_3 is equal to the maximum value of $\Re e(Z_{in})$ [max $\Re e(Z_{in}) \approx \Re e(Z_3)$].

Before the first anti-resonance, there is a first resonance at an angular frequency of ω_o, when the condition of $h = \lambda/4$ has been met. At ω_o, the value of $\Re e(Z_{in})$ is approximately equal to characteristic impedance, $Z''_o \approx Z_o$ of the dipole, while the $\Im m(Z_{in})$ vanishes. Thus, (4.42a) at ω_o can be rewritten as:

$$\begin{aligned} &\frac{\omega_o^2 L_3^2 R_3}{\left(R_3 - \omega_o^2 C_3 L_3 R_3\right)^2 + \omega_o^2 L_3^2} = Z''_o \\ &\frac{\omega_o L_3 R_3}{\left(R_3 - \omega_o^2 C_3 L_3 R_3\right)^2 + \omega_o^2 L_3^2} = \frac{Z''_o}{\omega_o L_3} \end{aligned} \tag{4.47}$$

Substituting (4.47) into (4.41b), the $\mathfrak{Im}(Z_{in})$ vanishes at ω_o and yields

$$\begin{aligned} &\frac{-1}{\omega_o C_1} + \omega_o L_2 + \frac{Z_o''}{\omega_o L_3}\left(R_3 - \omega_o^2 C_3 L_3 R_3\right) = 0 \\ &\frac{Z_o''}{\omega_o L_3}\left(R_3 - \omega_o^2 C_3 R_3 L_3\right) = \frac{1}{\omega_o C_1} - \omega_o L_2 \end{aligned} \tag{4.48}$$

Substituting (4.46) into (4.48) yields

$$\begin{aligned} &\frac{Z_o'' C_3 R_3}{\omega_o}\left(\omega_1^2 - \omega_o^2\right) = \frac{1}{\omega_o C_1} - \omega_o L_2 \\ &C_3 = \frac{1 - \omega_o^2 L_2 C_1}{Z_o'' R_3 C_1\left(\omega_1^2 - \omega_o^2\right)} \end{aligned} \tag{4.49}$$

The steps to find the values of C_1, L_2, R_3, L_3, and C_3 in (4.40) have been summarized in Figure 4.14.

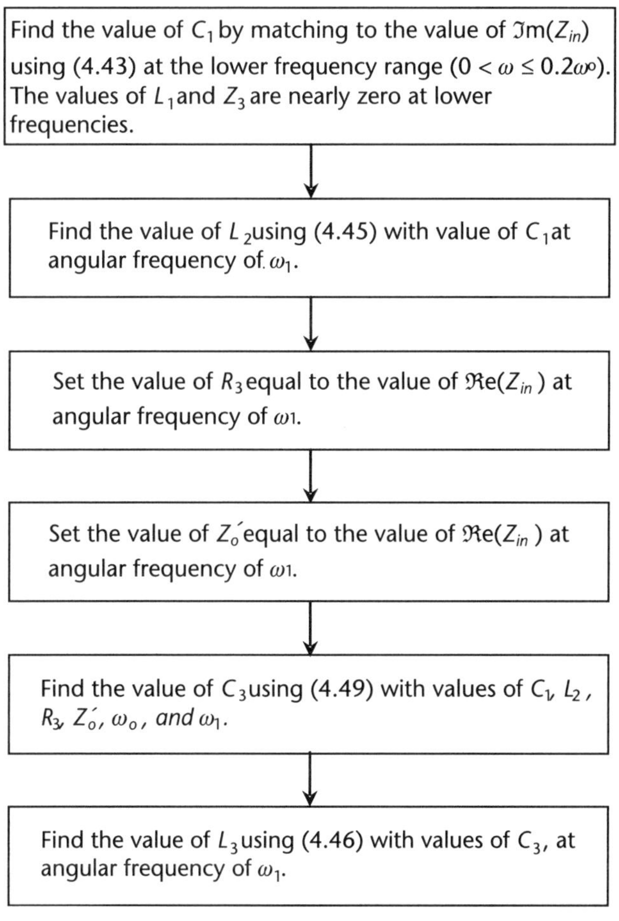

Figure 4.14 Flowchart of steps to find the values of C_1, L_2, R_3, L_3, and C_3 in (4.40).

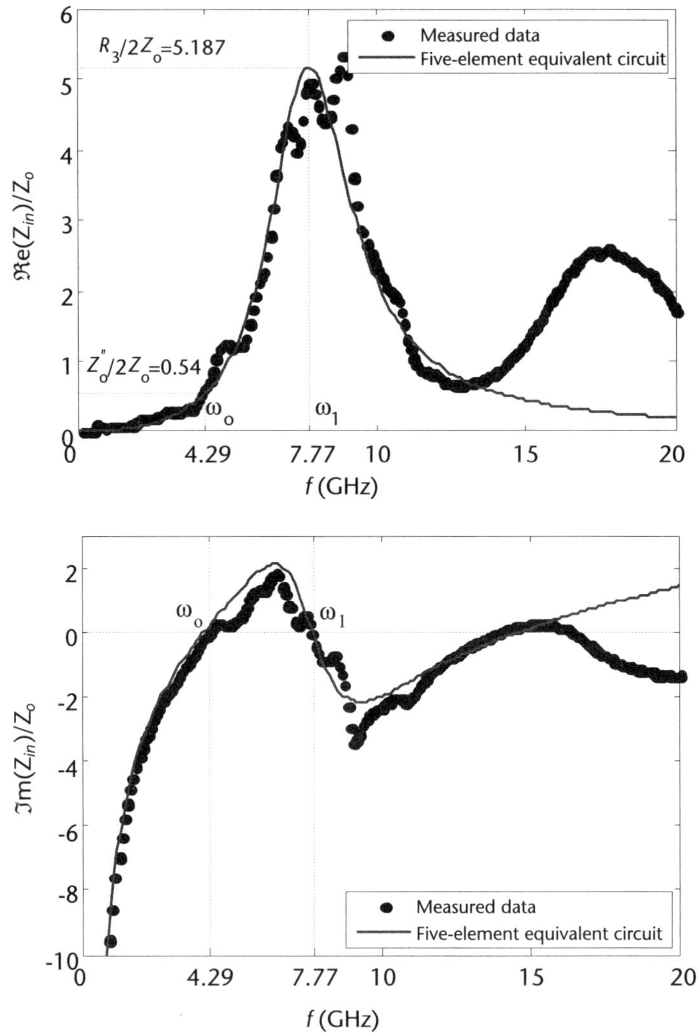

Figure 4.15 Variation in normalized resistance, $\Re e(Z_{in})$, and normalized reactance, $\Im m(Z_{in})$, with frequency for h = 14.32 mm on a monopole in free space.

For instance, the above method is tested using real measurement data from 300 kHz to 20 GHz. The input impedance, Z_{in}, is measured using h = 14.32 mm for monopole driven from a Teflon-filled coaxial line with a = 0.65 mm and b = 2.05 mm. In fact, the measured impedance is a value that has been normalized. Therefore, the lumped-element impedance should also be normalized with the characteristic impedance Z_o in (4.29), so that both values can be compared. From the measured normalized impedance, Z_{in}/Z_o, the first resonant angular frequency, ω_o, and anti-resonant angular frequency, ω_1, can be found, as well as the values of R_3 and Z''_o, as shown in Figure 4.15. In this work, the value of Z_o for a monopole driven from a coaxial line is equal to 48.2Ω. Based on the Figure 4.15, the five elements (C_1, L_2, R_3, C_3, and L_3) are given as:

$$C_1 = \frac{-1}{2\pi f \Im m\left(Z_{in}\right)} = 1.904 \times 10^{-13}\text{F, for } 0 < f \leq 1 \text{ GHz} \tag{4.50a}$$

$$L_2 = \frac{1}{\left[2\pi\left(7.77 \times 10^9\right)\right]^2 C_1} = 2.204 \times 10^{-9}\text{H} \tag{4.50b}$$

$$R_3 = 2Z_o(5.187) = 2(48.2)(5.187) = 500\Omega \tag{4.50c}$$

$$Z_o'' = 2Z_o(0.54) = 2(48.2)(0.54) = 52.06\Omega \tag{4.50d}$$

$$\begin{aligned} C_3 &= \frac{1 - \left[2\pi\left(4.29 \times 10^9\right)\right]^2 L_2 C_1}{(52.06) R_3 C_1 \left\{\left[2\pi\left(7.77 \times 10^9\right)\right]^2 - \left[2\pi\left(4.29 \times 10^9\right)\right]^2\right\}} \\ &= 8.467 \times 10^{-14}\text{F} \end{aligned} \tag{4.50e}$$

$$L_3 = \frac{1}{\left[2\pi\left(7.77 \times 10^9\right)\right]^2 C_3} = 4.956 \times 10^{-9}\text{H} \tag{4.50f}$$

It should be noted that the lumped-element modeling is typically referred to a dipole, thus, the input impedance of the monopole would correspond to half of the input impedance of the dipole. Finally, the modeling impedance line in Figure 4.15 is calculated by:

$$\begin{aligned} \frac{Z_{\text{in}}}{Z_o} &= \frac{1}{2Z_o}\left(-j\frac{1}{\omega C_1} + j\omega L_2 + \frac{j\omega R_3 L_3}{j\omega L_3 + R_3 - \omega^2 R_3 L_3 C_3}\right) \\ &= \frac{\Re\mathfrak{e}\left(Z_{\text{in}}\right)}{Z_o} + j\frac{\Im\mathfrak{m}\left(Z_{\text{in}}\right)}{Z_o} \end{aligned} \tag{4.51}$$

However, the accuracy of (4.51) is limited to the range ($0 < \omega \leq 1.5\omega_1$). Several equivalent circuit models for dipole and monopole have been proposed by a number of researchers, and are tabulated on the accompanying CD.

As mentioned in Chapter 2, the number of *CRL* elements in the equivalent circuit for impedance modeling depends on the frequency range of operation with a desired accuracy. In the case of modeling up to $h/\lambda = 1$ ($\omega \leq \omega_3$), for instance, in the frequency range which includes two red anti-resonances (shown in Figure 4.16), two sets of parallel circuits [(R_3, L_3, and C_3) and (R_4, L_4, and C_4)] are required in the equivalent circuit modeling as shown in Figure 4.17.

The approach for determining the *CRL* in Figure 4.17 is like the discussion above. In order to improve the accuracy of the fitting, the *CRL* can be adjusted by including the criteria in which the reactance, $\Im\mathfrak{m}(Z_{\text{in}})$ reaches a maximum value as it intersects with resistance, $\Re\mathfrak{e}(Z_{\text{in}})$ (that is to say, $\Im\mathfrak{m}(Z_{\text{in}})_{\text{Max}} = \Re\mathfrak{e}(Z_{\text{in}})$ at particular angular frequency). For instance, the same measurement data in Figure 4.15 is used in order to find C_4, R_4, and L_4. Normally, at the first anti-resonant frequency, the input impedance will show irregular form, thus, the regular rule should be modified in this work. From the measured normalized impedance, Z_{in}/Z_o, it is found

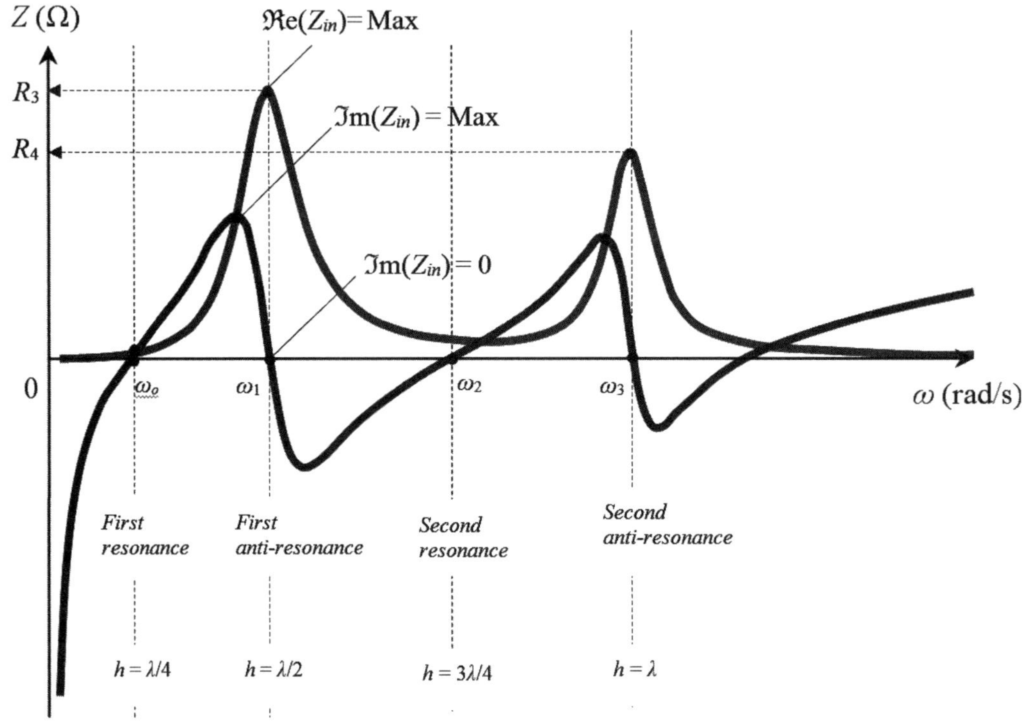

Figure 4.16 Angular frequency response of the input impedance Z_{in} of the dipole with a length $2h$ in the range $0 \le h/\lambda \le 1.5$.

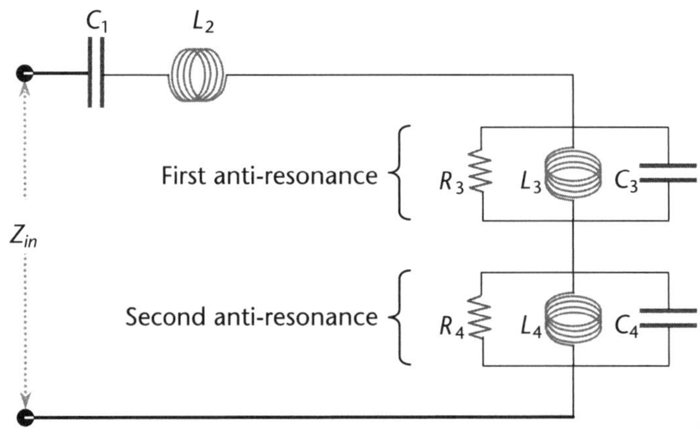

Figure 4.17 Broadband lumped-element equivalent circuit.

that the second anti-resonant frequency, f_3 of the real part, $\Re e(Z_{in})$ (f_3 = 17.5 GHz) is not consistent with the imaginary part, $\Im m(Z_{in})$ (f_3 = 16.35 GHz) as shown in Figure 4.18. Thus, the equivalent circuit modeling has to be done separately for the $\Re e(Z_{in})$ and the $\Im m(Z_{in})$ of the monopole.

$\Re e(Z_{in})/Z_o$ and $\Im m(Z_{in})/Z_o$ for the circuit in Figure 4.17 can be written as:

$$\frac{\Re\mathrm{e}(Z_{in})}{Z_o} = \frac{1}{2Z_o}\left\{\frac{\omega^2 L_3^2 R_3}{\omega^2 L_3 + \left(R_3 - \omega^2 C_3 L_3 R_3\right)^2} + \frac{\omega^2 L_{4\_\Re e}^2 R_4}{\omega^2 L_{4\_\Re e} + \left(R_4 - \omega^2 C_{4\_\Re e} L_{4\_\Re e} R_4\right)^2}\right\} \tag{4.52}$$

$$\frac{\Im\mathrm{m}(Z_{in})}{Z_o} = \frac{1}{2Z_o}\left\{\frac{-1}{\omega C_1} + \omega L_2 + \frac{\omega R_3^2 L_3 - \omega^3 R_3^2 L_3^2 C_3}{\omega^2 L_3 + \left(R_3 - \omega^2 C_3 L_3 R_3\right)^2} - \frac{\omega R_4^2 L_{4\_\Im m} - \omega^3 R_4^2 L_{4\_\Im m}^2 C_{4\_\Im m}}{\omega^2 L_{4\_\Im m} + \left(R_4 - \omega^2 C_{4\_\Im m} L_{4\_\Im m} R_4\right)^2}\right\} \tag{4.53}$$

where values of C_1, L_2, C_3, L_3, and R_3 in (4.52) and (4.53) are obtained from (4.50a) through (4.50f). However, the values of R_4, $L_{4\_\Re e}$, $L_{4\_\Im m}$, $C_{4\_\Re e}$, and $C_{4\_\Im m}$ can be found as:

$$R_4 = 2Z_o \Re\mathrm{e}(Z_{in}) = 2(48.2)(2.61) = 251.60\Omega \ \ (\text{at } \omega_3) \tag{4.54a}$$

$$Z_1'' = 2Z_o \Re\mathrm{e}(Z_{in}) = 2(48.2)(0.83) = 80.01\Omega \ \ (\text{at } \omega_2) \tag{4.54b}$$

$$C_{4\_\Re e} = \frac{1 - \omega_2^2 L_2 C_1}{Z_1'' R_4 C_1\left(\omega_3^2 - \omega_2^2\right)} = \frac{1 - \left[2\pi\left(13.95\times 10^9\right)\right]^2 L_2 C_1}{(80.01) R_4 C_1\left\{\left[2\pi\left(17.5\times 10^9\right)\right]^2 - \left[2\pi\left(13.95\times 10^9\right)\right]^2\right\}}$$
$$= -1.381\times 10^{-13}\mathrm{F} \tag{4.54c}$$

$$L_{4\_\Re e} = \frac{1}{\omega_3^2 C_{4\_\Re e}} = \frac{1}{\left[2\pi\left(17.5\times 10^9\right)\right]^2 C_{4\_\Re e}} = -6.093\times 10^{-10}\mathrm{H} \tag{4.54d}$$

$$C_{4\_\Im m} = \frac{1 - \left[2\pi\left(13.95\times 10^9\right)\right]^2 L_2 C_1}{(80.01) R_4 C_1\left\{\left[2\pi\left(16.35\times 10^9\right)\right]^2 - \left[2\pi\left(13.95\times 10^9\right)\right]^2\right\}} \tag{4.54e}$$
$$= -2.021\times 10^{-13}\mathrm{F}$$

$$L_{4\_\Im m} = \frac{1}{\left[2\pi\left(16.35\times 10^9\right)\right]^2 C_{4\_\Im m}} = -4.689\times 10^{-10}\mathrm{H} \tag{4.54f}$$

Finally, the calculated normalized impedances using (4.52) and (4.53) are plotted in Figure 4.18. In fact, the fitting between the models of (4.52) and (4.53) and the measurements can be improved by adjusting the values of Z_o'' or Z_1''.

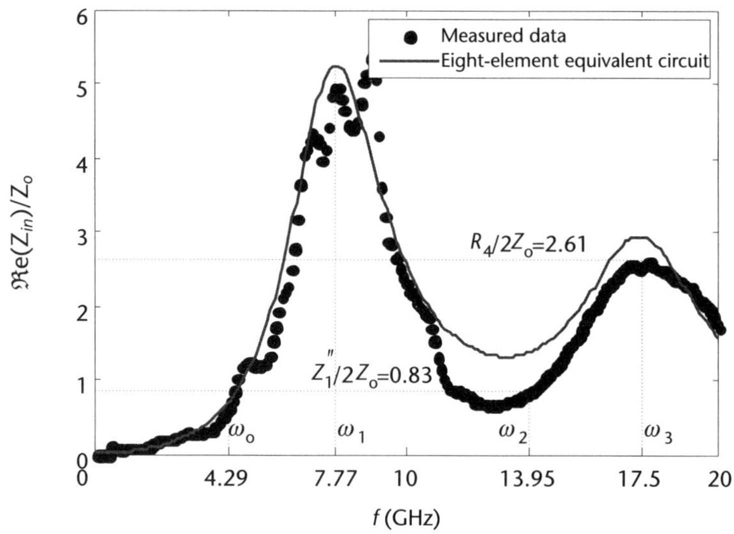

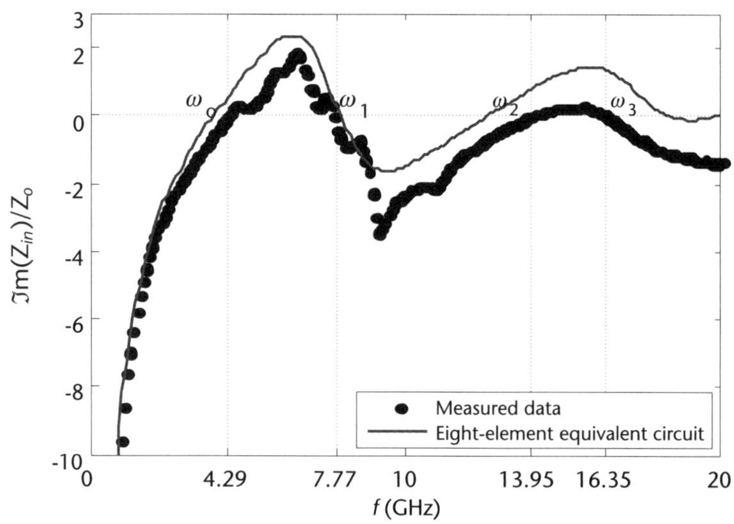

Figure 4.18 Variation in normalized resistance, $\Re e(Z_{in})$ and normalized reactance, $\Im m(Z_{in})$ with frequency for $h = 14.32$ mm for a monopole in free space.

4.4.2 Equivalent Circuits Modeling (in Dielectric Medium)

The complex relative permittivity, ε_r of a homogenous lossy medium can be written as:

$$\varepsilon_r = \varepsilon_r' - j\varepsilon_r'' = \varepsilon_r'\left(1 - j\frac{\varepsilon_r''}{\varepsilon_r'}\right) = \varepsilon_r'(1 - j\tan\delta) \qquad (4.55)$$

where the real part ε_r' is the relative dielectric constant, and the imaginary part ε_r'' is the relative dielectric loss factor of the lossy medium. The tan δ is the loss tangent of the lossy medium.

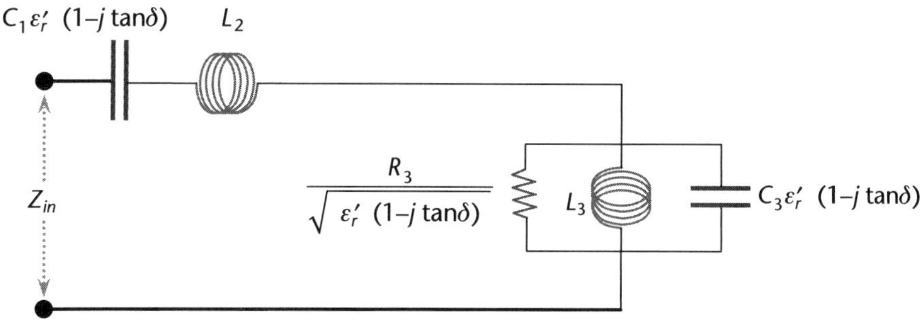

Figure 4.19 Five-element equivalent circuit of the dipole (or monopole) in a dielectric medium.

When the free space surrounding the dipole (or monopole) has been replaced by the lossy dielectric medium, the values of C_1, R_3, and C_3 in the equivalent circuit in Figure 4.19 will be changed, except for L_2 and L_3. In this case, Liao et al. [22, 23] proposed that the capacitance values of C_1 and C_3 obtained from the previous procedure (in Section 4.4.1) be multiplied by (4.55), while the resistance R_3 is divided by (4.55) as shown in Figure 4.19. In fact, the loss tangent (tan δ) part corresponds to the conductive loss in the capacitors (C_1 and C_3) and the reactance loss in the resistor (R_3). Thus, the five-element lumped circuit can be extended to eight-element lumped circuits as shown in Figure 4.19.

The new input impedance Z_{in} can be written as:

$$Z_{in} = \frac{R_1'}{1 + j\omega R_1' C_1 \varepsilon_r'} + j\omega L_2 - \frac{\omega^2 L_3 L_3' R_3' R_3''}{j\omega L_3' R_3' R_3'' + j\omega L_3 R_3' R_3'' - \omega^2 L_3 L_3' R_3'' - \omega^2 L_3 L_3' R_3' - j\omega^3 \varepsilon_r' C_3 L_3 L_3' R_3' R_3''} \tag{4.56}$$

The original capacitors C_1 and C_3 are shunted by parallel resistors of $R_1'$and R''_3, respectively, while the original resistor R_3 is shunted by the inductor L_1'. The expressions of R_1', R_3', R''_3, and L_1' in Figure 4.20 are given as:

$$R_1' = \frac{1}{\omega C_1 \varepsilon_r \tan\delta} \tag{4.57a}$$

$$R_3' = \frac{R_3}{\sqrt{\varepsilon_r'}\cosh\left[\frac{1}{2}\sinh^{-1}(\tan\delta)\right]} \tag{4.57b}$$

$$R_3'' = \frac{1}{\omega C_3 \varepsilon_r \tan\delta} \tag{4.57c}$$

$$L_3' = \frac{R_3}{\omega\sqrt{\varepsilon_r}\sinh\left[\frac{1}{2}\sinh^{-1}(\tan\delta)\right]} \tag{4.57d}$$

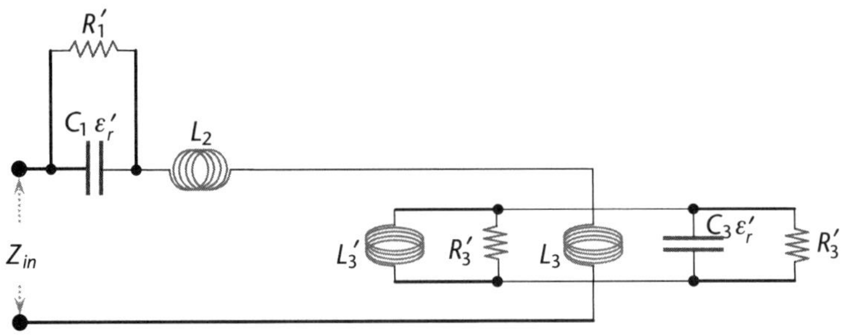

Figure 4.20 Eight-element equivalent circuit of the dipole (or monopole) in a lossy medium.

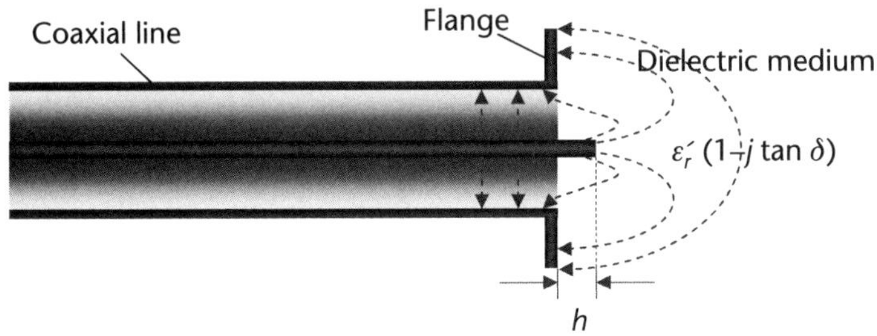

Figure 4.21 Short monopole driven from a coaxial line.

For a very short monopole ($h < \lambda/10$) in the lossy dielectric medium as shown Figure 4.21, the input impedance Z_{in} can be reduced as [24]:

$$Z_{\text{in}} = \Re\text{e}\left(Z_{\text{in}}\right) + j\Im\text{m}\left(X_{\text{in}}\right) = A\omega^2\sqrt{\varepsilon_r'(1-j\tan\delta)} + \frac{1}{jC\omega\varepsilon_r'(1-j\tan\delta)} \tag{4.58}$$

where A and C are constant values that are obtained by fitting with the simulation or measurement impedance of the probe in free space. However, (4.58) is only valid up to a few gigahertz of operation frequency (before first resonant frequency). The real part of resistance $\Re\text{e}(Z_{\text{in}})$ and the imaginary part of reactance $\Im\text{m}(Z_{\text{in}})$ are given as:

$$\Re\text{e}\left(Z_{\text{in}}\right) = \frac{\sin 2\delta}{2\varepsilon_r'\omega C} + A\sqrt{\varepsilon_r'}\omega^2\sqrt{\frac{\sec\delta + 1}{2}} \tag{4.59a}$$

$$\Im\text{m}\left(Z_{\text{in}}\right) = \frac{\cos^2\delta}{\varepsilon_r'\omega C} + A\sqrt{\varepsilon_r'}\omega^2\sqrt{\frac{\sec\delta - 1}{2}} \tag{4.59b}$$

4.5 Mutual Impedance

When two different lengths of monopoles or dipoles are placed at a distance D from one another, as shown in Figure 4.22, the combination of two electromagnetic fields occurs. Now the input impedance of the monopoles or dipoles requires the consideration of the self-radiation and mutual-radiation effects. The applications and advantages of the mutual-radiation effects will be discussed in Chapter 9.

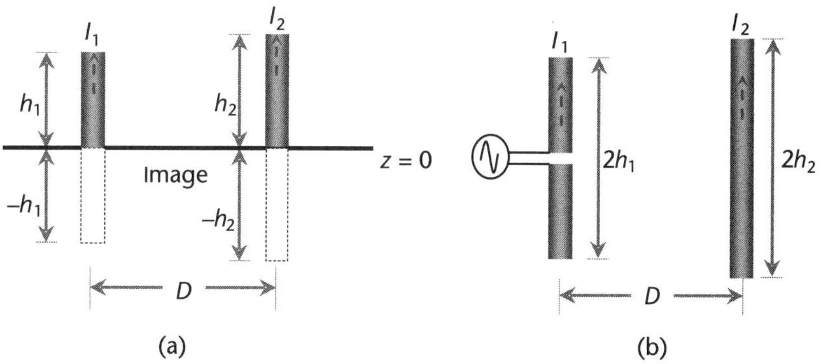

Figure 4.22 Arbitrary lengths of two parallel, mutually coupled (a) monopoles and (b) dipoles.

4.5.1 Two-Element Array

A coaxial radiator with unidirectional radiation pattern can be produced by optimizing the length, h and the distance spacing, D of the monopoles. Let us consider two parallel monopoles separated by a D distance as shown in Figure 4.23; one monopole is excited by a coaxial feed line (monopole 1) and the other acts as a parasitic element (monopole 2). By using Kirchhoff's law, the relationship between voltage, V, current, I, self-impedance, Z_{11} and Z_{22}, and mutual impedance, Z_{12} and Z_{21}, for the coupled parallel monopole can be written as [25]:

$$V_1 = I_1 Z_{11} + I_2 Z_{12} \tag{4.60a}$$

$$V_2 = I_1 Z_{21} + I_2 Z_{22} \tag{4.60b}$$

where I_1 and I_2 are the input currents for monopoles 1 and 2, respectively. In fact, I_2 is an induced current. The self-impedance, Z_{11}, at the excitation port in (4.60a) is equal to (4.25).

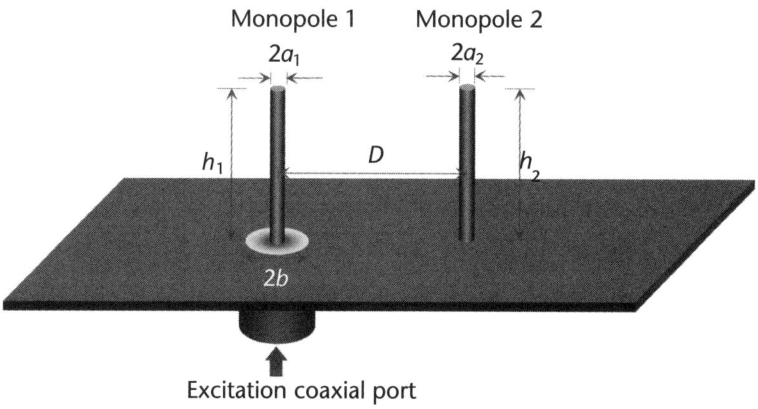

Figure 4.23 Coupled monopoles, one excited at a port and the other at a shorted parasitic element.

The self-impedance, Z_{11}, and mutual impedance, Z_{12}, contributed by monopole 2 are well expressed as [7]:

$$Z_{22} = \frac{j\eta_o}{4\pi\sqrt{\varepsilon_r}\sin^2(kh_2)} \times \int_0^{h_2}\left\{\sin\left[k(h_2 - z)\right]\left[\frac{e^{-jkR_1}}{R_1} + \frac{e^{-jkR_2}}{R_2} - 2\cos(kh_2)\frac{e^{-jkR_3}}{R_3}\right]\right\}dz \tag{4.61a}$$

$$Z_{12} = \frac{j\eta_o}{4\pi\sqrt{\varepsilon_r}\sin(kh_1)\sin(kh_2)} \times \int_0^{h_2}\left\{\sin\left[k(h_2 - z)\right]\left[\frac{e^{-jkR_1'}}{R_1'} + \frac{e^{-jkR_2'}}{R_2'} - 2\cos(kh_1)\frac{e^{-jkR_3'}}{R_3'}\right]\right\}dz \tag{4.61b}$$

where $R_1=\sqrt{a_2^2 + (z - h_2)^2}$, $R_2=\sqrt{a_2^2 + (z + h_2)^2}$, $R_3=\sqrt{a_2^2 + z^2}$, $R_1'=\sqrt{D^2 + (z - h_1)^2}$, $R_2' = \sqrt{D^2 + (z + h_1)^2}$, and $R_3' = \sqrt{D^2 + z^2}$. The input impedance, Z_{in}, of the monopole 1 can be calculated by dividing (4.60a) by the input current I_1 as [25]:

$$Z_{\text{in}} = \frac{V_1}{I_1} = Z_{11} + Z_{12}\frac{I_2}{I_1} \tag{4.62}$$

By using the relationship of (4.60a) and (4.60b), (4.62) can be rewritten as [25]:

$$Z_{\text{in}} = \frac{Z_{11}^2 - Z_{12}^2}{Z_{22} - Z_{21}(V_2/V_1)} \tag{4.63}$$

The value of V_2 in (4.60b) is equal to zero, since there is no source excited at monopole 2. Thus, (4.63) can be simplified as:

$$Z_{\text{in}} = Z_{11} - \frac{Z_{12}^2}{Z_{22}} \tag{4.64}$$

When the monopole 2 is shorter than monopole 1, it is called a director, and the radiation direction is forwarded from monopole 1 to monopole 2. However, when monopole 2 is longer than monopole 1, it is called a reflector, and the radiation direction from monopole 2 is passed through monopole 1. The optimum length of monopole 2, h_2, can be estimated using (4.64). For instance, the calculated value of Z_{11} using (4.25) for certain length of h_1 is given as:

$$Z_{11} = (76 + j14)\Omega \tag{4.65}$$

Thus, the optimum length h_2 and the spacing distance D of monopole 2 can be adjusted so that the calculated reactive component of $\mathfrak{Im}(-Z_{12}^2/Z_{22})$ using (4.61) becomes equal to -14Ω. Normally, the value of D only varies from 0.1λ to 0.4λ. If both parallel monopoles have the same height ($h_1 = h_2$), the self-impedance term of Z_{22} is equal to Z_{11}.

4.5.2 Arrays of Three or More Elements

When an array containing three elements, as shown in Figure 4.24, with one dipole is excited by RF source (dipole 2) and the other two elements act as a parasitic element (dipole 1 and dipole 3), the network relationship between voltage, V, current, I, and impedance, Z, is given as [25]:

$$V_1 = I_1 Z_{11} + I_2 Z_{12} + I_3 Z_{13} \tag{4.66a}$$

$$V_2 = I_1 Z_{21} + I_2 Z_{22} + I_3 Z_{23} \tag{4.66b}$$

$$V_3 = I_1 Z_{31} + I_2 Z_{32} + I_3 Z_{33} \tag{4.66c}$$

The values of V_1 and V_3 in (4.66b) and (4.66c) are equal to zero, since there is no source excited at dipole 1 and dipole 3. Hence, the simultaneous solution of (4.66a) and (4.66c) gives

$$\frac{I_1}{I_2} = \frac{Z_{13}Z_{23} - Z_{12}Z_{33}}{Z_{11}Z_{33} - Z_{13}^2} \quad \text{and} \quad \frac{I_3}{I_2} = \frac{Z_{13}Z_{12} - Z_{23}Z_{11}}{Z_{11}Z_{33} - Z_{13}^2} \tag{4.67}$$

Substitute (4.67) into (4.66b) and then divide (4.66b) by the input current I_2; finally, the input impedance, Z_{in}, referred to dipole 2 is given as [7]:

$$Z_{\text{in}} = \frac{V_2}{I_2} = Z_{22} + Z_{21}\frac{I_1}{I_2} + Z_{23}\frac{I_3}{I_2} = \frac{Z_{11}Z_{22}Z_{33} - Z_{23}^2 Z_{11} - Z_{13}^2 Z_{22} - Z_{12}^2 Z_{23} + 2Z_{12}Z_{13}Z_{23}}{Z_{11}Z_{33} - Z_{13}^2} \tag{4.68}$$

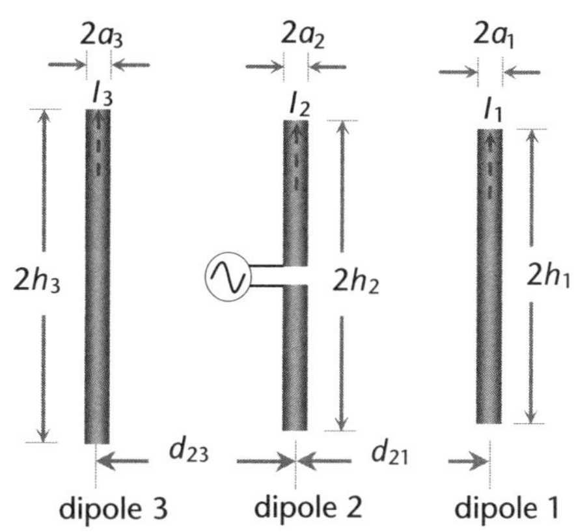

Figure 4.24 A linear array with three parallel dipoles.

The self-impedances and mutual impedances, Z_{ki}, in (4.68) can be calculated as:

$$Z_{ki} = \frac{j\eta_o}{4\pi\sqrt{\varepsilon_r}\sin(kh_i)\sin(kh_k)} \times \int_{-h_k}^{h_k} \left\{ \sin\left[k(h_k - z)\right]\left[\frac{e^{-jkR'_{ki}}}{R'_{ki}} + \frac{e^{-jkR''_{ki}}}{R''_{ki}} - 2\cos(kh_i)\frac{e^{-jkR'''_{ki}}}{R'''_{ki}}\right]\right\} dz \tag{4.69}$$

where subscripts $k = 1, 2, 3$ and $i = 1, 2, 3$. The auxiliary parameters in (4.69) are given as follows.

For $k \neq i$,

$$R'_{ki} = \sqrt{d_{ki}^2 + (z - h_k)^2},\ R''_{ki} = \sqrt{d_{ki}^2 + (z + h_k)^2},\ \text{and } R'''_{ki} = \sqrt{d_{ki}^2 + z^2}.$$

For $k = i$,

$$R'_{kk} = \sqrt{a_k^2 + (z - h_k)^2},\ R''_{kk} = \sqrt{a_k^2 + (z + h_k)^2},\ \text{and } R'''_{kk} = \sqrt{a_k^2 + z^2}.$$

References

[1] Brillouin, L., "Origin of Radiation Resistance," *Radioėlectricitė*, Vol. 3, 1922, pp. 147–152.

[2] Weiner, M. M., *Monopole Antennas*, New York: Marcel Dekker, 2003.

[3] Storer, J., "Variational Solution to the Problem of the Symmetrical Cylindrical Antenna," *Cruft Lab., Harvard University, Tech. Rep.*, Vol. 101, 1950.

[4] Tai, C. T., "A Variational Solution to the Problem of Cylindrical Antennas," Aircraft Radiation Systems Lab., Stanford Res. Inst., Stanford, Calif., Tech. Rep. 12, 1950.

[5] Richmond, J. H., "Digital Computer Solutions of the Rigorous Equations for Scattering Problems," *Proc. IEEE*, Vol. 53, 1965, pp. 796–804.

[6] King, R. W. P., and T. T. Wu, "The Electric Field Very Near a Cylindrical Driven Antenna," *Radio Science*, Vol. 1 (New Series), 1966, pp. 353–359.

[7] Collin, R. E., and F. J. Zucker, *Antenna Theory*, New York: McGraw-Hill, 1969.

[8] Scott, L. D., "Polynomial-Exponential-Product Theory for Antennas in Homogeneous Isotropic Media," Div. Eng. and Appl. Phys., Harvard Univeristy, Cambridge, MA, *NASA Sci. Rep. 7*, 1969.

[9] Neff, N. P., "A Simple Approximation to the Current on the Surface of an Isolated Thin Cylindrical Center-Fed Dipole," *IEEE Trans. on Antennas Propagat.*, Vol. 18, 1970, pp. 399–400.

[10] Pocklington, H. C., "Electrical Oscillations in Wire," *Cambridge Philos. Soc. Proc.*, Vol. 4, 1897, pp. 324–332.

[11] Thiele, G. A., "Wire Antennas," Ch. 2 in *Computer Techniques for Electromagnetics*, R. Mittra, (ed.), Oxford, U.K.: Pergamon, 1973.

[12] Hallén, E., "Theoretical Investigations into Transmitting and Receiving Antennae," *Nova Acta Regiae Sco. Sci. Upsaliensis*, Ser. IV, Vol. 4, No. 2, 1938, pp. 1–44.

[13] Tai, C. T., and S, A Long, "Dipoles and Monopoles," Ch. 4 in *Antenna Engineering Handbook*, J. L. Volakis, (ed.), New York: McGraw-Hill, 2007.

[14] Miller, E. K., "Admittance Dependence of the Infinite Cylindrical Antenna Upon Exciting Gap Thickness," *Radio Science*, Vol. 2, No. 12, 1967, pp. 1431–1435.

[15] Tsai, L. L., "A Numerical Solution for the Near and Far Fields of an Annular Ring of Magnetic Current," *IEEE Trans. on Antennas Propagat.*, Vol. 20, 1972, pp. 569–576.

[16] Williamson, A. G., "Admittance Correction Factor for Coaxially Excited Cylindrical Antennas," *IEEE Trans. on Antennas Propagat.*, Vol. 49, 2001, pp. 640–644.

[17] Chang, D. C., "Input Admittance and Complete Near-Field Distribution of an Annular Aperture Antenna Driven by a Coaxial Line," *IEEE Trans. on Antennas Propag.*, Vol. 18, 1970, pp. 610–617.

[18] Otto, D. V., "The Admittance of Cylindrical Antennas Driven from a Coaxial Line," *Radio Science*, Vol. 2, 1967, pp. 1031–1042.

[19] You, K. Y., et al., "Improved Formulation for Admittance of Thin and Short Monopole Driving from Coaxial Line into Dissipative Media," *IEEE Antennas and Wireless Propagation Letters.*, Vol. 8, 2009, pp. 1246–1249.

[20] Gajda, G., and S. S. Stuchly, "Numerical Analysis of Open-Ended Coaxial Line," *IEEE Trans. on Instrum. Meas.*, Vol. 32, 1983, pp. 380–384.

[21] Hamid, M., and R. Hamid, "Equivalent Circuit of Dipole Antenna of Arbitrary Length," *IEEE Trans. on Antennas Propagat.*, Vol. 45, 1997, pp. 1695–1696.

[22] Liao, Y., T. H. Hubing, and D. Su, "Equivalent Circuit for Dipole Antennas in a Lossy Medium," *IEEE Trans. Antennas Propagat.*, Vol. 60, 2012, pp. 3950–3953.

[23] Liao, Y., T. H. Hubing, and D. Su, "Equivalent Circuit with Frequency-Independent Lumped Elements for Coated Wire Antennas," *IEEE Trans. on Antennas Propagat.*, Vol. 60, 2012, pp. 5419–5423.

[24] Burdette, E. C., F. I. Cain, and J. Seals, "In Vivo Probe Measurement Technique at VHF Through Microwave Frequencies," *IEEE Trans. on Microwave Theory Tech.*, Vol. 28, 1980, pp. 414–427.

[25] Altshuler, E. E., "Self- and Mutual Impedances of Traveling-Wave Linear Antennas," *IEEE Trans. on Antennas Propagation*, Vol. 37, 1989, pp. 1312–1316.

Selected Bibliography

Cu, C. J., "Physical Limitation of Omnidirectional Antennas," *J. Appl. Phys.*, Vol. 19, 1948, pp. 1163–1175.

Duncan, R. H., "Theory of the Infinite Cylindrical Antenna Including the Feedpoint Singularity in Antenna Current," *J. Research N. B. S.*, Vol. 66D, No. 2, 1962.

King, R. W. P., "The Linear Antenna—Eighty Years of Progress," *Proceedings of the IEEE*, Vol. 55, No. 1, 1967, pp. 2–16.

Marcuvitz, N., *Waveguide Handbook*, Boston, MA: Boston Technical Publishers, 1964, pp. 208–209.

Misra, D. K., "In Situ Electric Probe Method for Determining the Dielectric Properties of Materials," *IEEE Trans. on Instrum. Meas.*, Vol. 37, 1988, pp. 157–160.

Rambabu, K., M. Ramesh, and A. T. Kalghatgi, "Broadband Equivalent Circuit of a Dipole Antenna," *IEE Proc. Microw. Antennas Propag.*, Vol. 146, 1999, pp. 391–393.

Streable, G. W., and L. W. Pearson, "A Numerical Study on Realizable Broad-Band and Equivalent Admittances for Dipole and Loop Antennas," *IEEE Trans. on Antennas Propagat.*, Vol. 29, 1981, pp. 707–717.

Tang, T. G., Q. M. Tieng, and M. W. Gunn, "Equivalent Circuit of a Dipole Antenna Using Frequency-Independent Lumped Elements," *IEEE Trans. on Antennas Propagat.*, Vol. 41, 1993, pp. 100–103.

CHAPTER 5

Analytical Modeling of a Dielectric-Coated Monopole Driven from the Coaxial Line

This chapter focuses on the analytical modeling of the low-loss dielectric-coated monopole that is immersed in a dissipative medium. The propagation of electromagnetic waves along dielectric-coated cylinder conductors were first investigated by Harms in 1907 [1]. Later, its electrical properties and analytical formulations, based on transmission line theory, had been studied extensively by researchers [2–4].

In practice, the coated monopole has several advantages compared to the bare monopole. One of the reasons is that the current on the bare monopole in a dissipative medium is rapidly attenuated compared to the coated monopole; to be specific, the bare monopole is inefficient for surface-wave transmission. The low-loss dielectric coating layer that is between the bare antenna and the dissipative medium can make the propagation of electromagnetic fields more concentrated along the monopole. In this case, the density of the electromagnetic field near the monopole or dipole conductor (within the dielectric coating layer) is increased, since the dielectric properties of dielectric coating are higher than air. In addition, the dielectric coating can avoid the contact between the bare conductor metal and the medium, and it can also prevent the leakage of electric charges from the conductor into the dissipative medium, especially for biological medium. Normally, the low-loss dielectric coating is the solid polyethylene or polytetrafluoroethylene (PTFE) or Teflon.

In this chapter, the transmission line theory model [2–10], the integral electric field model [11–14], and the equivalent lumped-element circuit model [15] are reviewed and analyzed. Among the three methods, the transmission line model is adapted easily to all circumstances, and it is able to analyze the coated monopole accurately. Generally, the transmission-line model requires the propagation constant of the external medium to be much larger than the propagation constant of the dielectric coating layer. Thus, the calculation of a coated monopole using the transmission line model is always implemented in the dissipative medium with higher dielectric properties than its insulation coat, such as soil, seawater, or biological tissue. In 2-D modeling, most of the analytical formulations are simplified and expressed either in Bessel's functions or Hankel's functions, since the Green's function term in the theoretical formulations can be expressed in those mathematical forms.

5.1 Theoretical Background

Generally, the analytical equations derived based on the dielectric-coated dipole are center-driven by the delta-gap voltage, $2V_o$. The $2h$ length of dipole has a radius a (region 1) that is coated with a radius b of the low-loss dielectric insulator (region 2), as shown in Figure 5.1. The insulator is surrounded by an infinite dissipative medium (region 3).

The same analytical methods for a coated dipole can also be applied to a coated monopole. The difference between both cases is that the feed voltage, $2V_o$ for the coated dipole is twice that of the feed voltage V_o for the coated monopole. Thus, the coated monopole will only have half of the electromagnetic field value compared to the coated dipole. The field components H_ϕ, E_ρ, and E_z always refer to the analytical formula derivation since those components are nonvanishing for the principal TEM mode.

From (1.28), the relationship between the H_ϕ, E_ρ, and E_z field components are given as:

$$\frac{1}{\rho}\frac{\partial\left(\rho H_\phi\right)}{\partial\rho} = j\omega\varepsilon E_z \tag{5.1a}$$

$$\frac{\partial H_\phi}{\partial z} = -j\omega\varepsilon E_\rho \tag{5.1b}$$

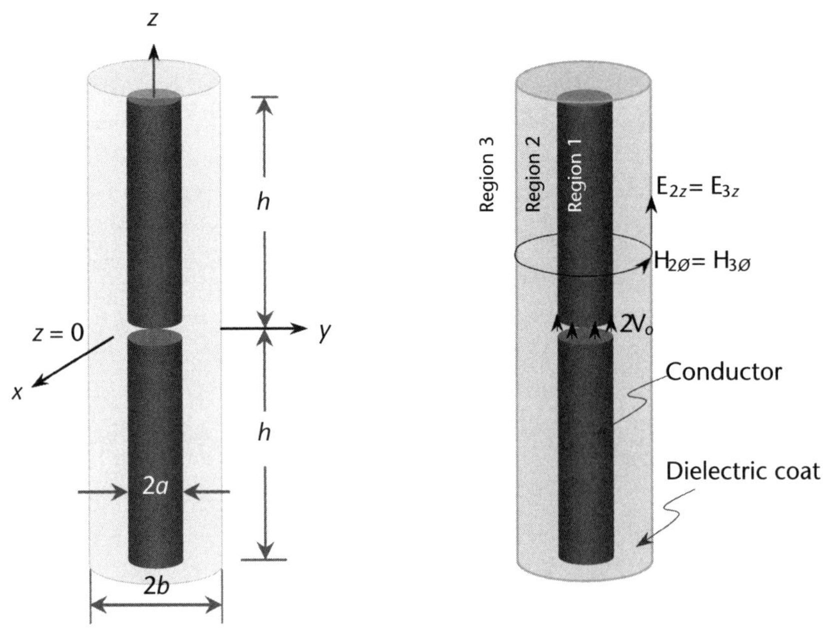

Figure 5.1 (a) Cylindrical coordinate. (b) TEM mode components for linear cylindrical conductor.

The dependence on the coordinate z of (5.1a) and (5.1b) can be removed by applying the Fourier transform with respect to the z as:

$$\tilde{E}_{\rho i} = \frac{-j\varsigma}{\xi_i^2}\frac{\partial \tilde{E}_{zi}}{\partial \rho},\ i = 1,2 \text{ and } 3 \tag{5.2a}$$

$$\tilde{H}_{\phi i} = \frac{-jk_i}{\omega \xi_i^2}\left(\frac{\partial \tilde{E}_{zi}}{\partial \rho}\right),\ i = 1,2 \text{ and } 3 \tag{5.2b}$$

$$\frac{1}{\rho}\frac{\partial}{\partial \rho}\left(\rho \frac{\partial}{\partial \rho}\tilde{E}_{zi}\right) + \xi_i^2 \tilde{E}_{zi} = 0 \tag{5.2c}$$

where

$$\xi_i = \sqrt{k_i^2 - \varsigma^2} \tag{5.3}$$

k_2 and k_3 are the propagation constants of the dielectric coating in region 2 and the dissipative medium in region 3, respectively. The subscript i represents the number of regions.

5.2 Transmission-Line Impedance Modeling

5.2.1 Propagation Constant k_L and Characteristic Impedance Z_c

The analytical solution uses transmission-line theory for a dielectric-coated monopole immersed in a dissipative medium that was first proposed by King and Smith [8]. The solutions of (5.2c) at $0 \le \rho \le \infty$ are given as:

$$\tilde{E}_{z1} = A_o J_o(\xi_2 \rho) \text{ for } 0 \le \rho \le a \tag{5.4a}$$

$$\tilde{E}_{z2} = B_o H_o^{(1)}(\xi_2 \rho) + C_o J_o(\xi_2 \rho) \qquad \text{for } a \le \rho \le b \tag{5.4b}$$

$$\tilde{E}_{z3} = D_o H_o^{(1)}(\xi_3 \rho) \qquad \text{for } b \le \rho \tag{5.4c}$$

and

$$\tilde{H}_{\phi 1} = \frac{jA_o k_2^2}{\omega \mu \xi_2} J_1(\xi_2 \rho) \qquad \text{for } 0 \le \rho \le a \tag{5.5a}$$

$$\tilde{H}_{\phi 2} = \frac{jk_2^2}{\omega \mu \xi_2}\left[B_o H_1^{(1)}(\xi_2 \rho) + C_o J_1(\xi_2 \rho)\right] \text{ for } a \le \rho \le b \tag{5.5b}$$

$$\tilde{H}_{\phi 3} = \frac{jD_o k_3^2}{\omega\mu\xi_3} H_1^{(1)}(\xi_3\rho) \qquad \text{for } b \le \rho \tag{5.5c}$$

A_o, B_o, C_o, and D_o in (5.4a) to (5.5c) are the unknown amplitudes of the electric field, E_z, and the magnetic field, H_ϕ. Moreover, $J_o(\xi_2\rho)$ is the zero-order Bessel function of the first kind, whereas $H_o^{(1)}(\xi_2\rho)$ is the zero-order Hankel function of the first kind. The terms A_o, B_o, C_o, and D_o can be found by referring to the boundary condition continuity at $\rho = a$ and $\rho = b$ for the tangential electric field, E_z, and the tangential magnetic field, H_ϕ, as:

$$\begin{gathered}\tilde{E}_{z1}\ (\text{at } \rho = a) = \tilde{E}_{z2}\ (\text{at } \rho = a) = -2V_o \\ B_o H_o^{(1)}(\xi_2 a) + C_o J_o(\xi_2 a) = -2V_o\end{gathered} \tag{5.6a}$$

$$\begin{gathered}\tilde{E}_{z2}\ (\text{at } \rho = b) = \tilde{E}_{z3}\ (\text{at } \rho = b) \\ B_o H_o^{(1)}(\xi_2 b) + C_o J_o(\xi_2 b) = D_o H_o^{(1)}(\xi_3 b)\end{gathered} \tag{5.6b}$$

$$\begin{gathered}\tilde{H}_{\phi 1}\ (\text{at } \rho = a) = \tilde{H}_{\phi 2}\ (\text{at } \rho = a) = \frac{I}{2\pi a} \\ \frac{jk_2^2}{\omega\mu\xi_2}\left[B_o H_1^{(1)}(\xi_2 a) + C_o J_1(\xi_2 a) - A_o J_1(\xi_2 a)\right] = \frac{I}{2\pi a}\end{gathered} \tag{5.6c}$$

$$\begin{gathered}\tilde{H}_{\phi 2}\ (\text{at } \rho = b) = \tilde{H}_{\phi 3}\ (\text{at } \rho = b) \\ \frac{k_2^2}{\xi_2}\left[B_o H_1^{(1)}(\xi_2 b) + C_o J_1(\xi_2 b)\right] = \frac{k_3^2}{\xi_3} D_o H_1^{(1)}(\xi_3 b)\end{gathered} \tag{5.6d}$$

It is well known that the electric field, $\tilde{E}_{z1}$, is equal to zero at the surface of a conductor (at $\rho = a$); thus, from (5.6a), the coefficient C_o can be expressed in term of B_o as:

$$C_o = -B_o \frac{H_o^{(1)}(\xi_2 a)}{J_o(\xi_2 a)} \tag{5.7}$$

Similarly, from (5.6b), the coefficient D_o can be expressed as:

$$D_o = \frac{B_o H_o^{(1)}(\xi_2 b) + C_o J_o(\xi_2 b)}{H_o^{(1)}(\xi_3 b)} = B_o\left[\frac{H_o^{(1)}(\xi_2 b)}{H_o^{(1)}(\xi_3 b)} - \frac{H_o^{(1)}(\xi_2 a) J_o(\xi_2 b)}{H_o^{(1)}(\xi_3 b) J_o(\xi_2 a)}\right] \tag{5.8}$$

Substituting (5.7) and (5.8) into (5.6d) yields

$$\begin{aligned}&\frac{k_2^2}{\xi_2}H_o^{(1)}(\xi_3 b)\left[H_1^{(1)}(\xi_2 b)J_o(\xi_2 a)-H_o^{(1)}(\xi_2 a)J_1(\xi_2 b)\right]\\&-\frac{k_3^2}{\xi_3}H_1^{(1)}(\xi_3 b)\left[H_o^{(1)}(\xi_2 b)J_o(\xi_2 a)-H_o^{(1)}(\xi_2 a)J_o(\xi_2 b)\right]=0\end{aligned}\tag{5.9}$$

The propagation constant, $k_L = \varsigma$ of a coated dipole in (5.9) takes the resistive losses and the radiation losses of the dissipative medium into account, which can be determined by solving (5.9) using a numerical method, such as the Newton-Raphson method [10].

In the case of a highly dissipative medium in region 3 [$(k_2/k_3)^2 < 0.06$] and small sizes of the radius conductor ($\xi_2 a \to 0$), the Hankel's functions and Bessel functions in (5.9) can be approximated as [3]:

$$J_o(\xi_2 a)\approx 1,\ J_1(\xi_2 a)\approx 0,\ H_1^{(1)}(\xi_2 a)\approx\frac{-2j}{\pi\xi_2 a},\ H_1^{(1)}(\xi_2 b)\approx\frac{-2j}{\pi\xi_2 b},$$

$$H_o^{(1)}(\xi_2 a)\approx 1+\frac{2j}{\pi}\left[\gamma+\ln\left(\frac{\xi_2 a}{2}\right)\right],\ H_o^{(1)}(\xi_2 b)\approx 1+\frac{2j}{\pi}\left[\gamma+\ln\left(\frac{\xi_2 b}{2}\right)\right]$$

$$\xi_2=\sqrt{k_2^2-k_L^2},\ \text{and}\ \xi_3=\sqrt{k_3^2-k_L^2}$$

where $\gamma = 0.577215$ is Euler's constant. Finally, the approximated equation (5.9) becomes:

$$\frac{k_2^2}{k_2^2-k_L^2}\left(\frac{-2j}{\pi b}\right)H_o^{(1)}(k_3 b)=\frac{2jk_3^2}{\pi\sqrt{k_3^2-k_L^2}}\ln\left(\frac{b}{a}\right)H_1^{(1)}(k_3 b)\tag{5.10}$$

Multiply the right hand side of (5.10) by $k_3/\sqrt{k_3^2-k_L^2}\approx 1$, in order to align the parameter k_L on both sides of (5.10), yielding,

$$\frac{k_2^2}{k_2^2-k_L^2}\left(\frac{-2j}{\pi b}\right)H_o^{(1)}(k_3 b)=\frac{2jk_3^3}{\pi\left(k_3^2-k_L^2\right)}\ln\left(\frac{b}{a}\right)H_1^{(1)}(k_3 b)$$

$$\frac{-k_2^2\left(k_3^2-k_L^2\right)}{k_3^3}\frac{H_o^{(1)}(k_3 b)}{b\ln(b/a)H_1^{(1)}(k_3 b)}=k_2^2-k_L^2\tag{5.11}$$

$$k_2^2\left(1+\frac{H_o^{(1)}(k_3 b)}{k_3 b\ln(b/a)H_1^{(1)}(k_3 b)}\right)=k_L^2\left(1+\frac{k_2^2}{k_3^3}\frac{H_o^{(1)}(k_3 b)}{b\ln(b/a)H_1^{(1)}(k_3 b)}\right)$$

From (5.11), the approximated formulation for k_L can be expressed as [4, 14, 16]:

$$k_L = k_2 \left[\frac{\ln\left(\frac{b}{a}\right) + \frac{H_o^{(1)}(k_3 b)}{k_3 b H_1^{(1)}(k_3 b)}}{\ln\left(\frac{b}{a}\right) + \frac{k_2^2}{k_3^3} \frac{H_o^{(1)}(k_3 b)}{b H_1^{(1)}(k_3 b)}} \right]^{1/2} \tag{5.12}$$

For the short dipole ($h < \lambda/4$) case, (5.10) can be further simplified by assuming $\xi_3 \approx k_3$, yielding:

$$\frac{k_2^2}{k_2^2 - k_L^2}\left(\frac{-2j}{\pi b}\right) H_o^{(1)}(k_3 b) = \frac{2jk_3}{\pi} \ln\left(\frac{b}{a}\right) H_1^{(1)}(k_3 b) \tag{5.13}$$

Thus, the formula of k_L is reduced as [3, 16]:

$$k_L = k_2 \left[\ln\left(\frac{b}{a}\right) + \frac{H_o^{(1)}(k_3 b)}{b k_3 H_1^{(1)}(k_3 b)} \right]^{1/2} \tag{5.14}$$

The ohmic losses in the coated dipole are neglected in (5.14).

5.2.2 Characteristic Impedance Z_C

From (1.35) and (1.36), the characteristic impedance, Z_C, of the dielectric-coated radiator in the dissipative medium can be expressed as:

$$Z_C = \sqrt{\frac{Z_L}{Y_L}} \tag{5.15}$$

and the propagation constant k_L can be written as:

$$jk_L = \sqrt{Z_L Y_L} \tag{5.16}$$

The characteristic impedance Z_C is treated as a lossy coaxial-transmission line impedance. Z_L and Y_L are the load impedance and admittance per unit length of the transmission line. The equivalent circuit of the transmission line for the dipole is shown in Figure 5.2.

Neglecting the resistive losses in the inner conductor (region 1), the series impedance Z_L per unit length in Figure 5.2 is given as:

$$Z_L = Z_1 + Z_2 = j\omega L + \frac{j\omega\mu_o}{2\pi k_3 b} \frac{H_o^{(1)}(k_3 b)}{H_1^{(1)}(k_3 b)} = \frac{j\omega\mu_o}{2\pi} \ln\left(\frac{b}{a}\right) + \frac{j\omega\mu_o}{2\pi k_3 b} \frac{H_o^{(1)}(k_3 b)}{H_1^{(1)}(k_3 b)} \tag{5.17}$$

And the shunt admittance, Y_L per unit length is written as:

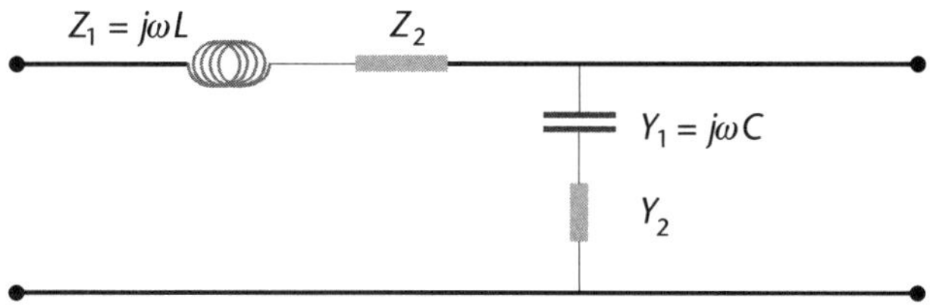

Figure 5.2 The lumped element equivalent circuit for a dielectric-coated monopole based on the transmission-line theory.

$$\begin{aligned}\frac{1}{Y_L} = \frac{1}{Y_1} + \frac{1}{Y_2} &= \frac{1}{j\omega C} + \frac{1}{j2\pi\omega\varepsilon_o\varepsilon_{r3}k_3b\dfrac{H_1^{(1)}(k_3b)}{H_o^{(1)}(k_3b)}} \\ &= \frac{\ln(b/a)}{j2\pi\omega\varepsilon_o\varepsilon_{r2}} + \frac{H_o^{(1)}(k_3b)}{j2\pi\omega\varepsilon_o\varepsilon_{r3}k_3bH_1^{(1)}(k_3b)}\end{aligned} \tag{5.18}$$

Thus,

$$Y_L = \frac{Y_1Y_2}{Y_1 + Y_2} = \frac{j2\pi\omega\varepsilon_o\varepsilon_{r2}\varepsilon_{r3}k_3b\dfrac{H_1^{(1)}(k_3b)}{H_o^{(1)}(k_3b)}}{\varepsilon_{r2} + \varepsilon_{r3}k_3b\ln\left(\dfrac{b}{a}\right)\dfrac{H_1^{(1)}(k_3b)}{H_o^{(1)}(k_3b)}} \tag{5.19}$$

The Z_1 (Ω/m) and Y_1 (S/m) terms are the external impedance and admittance associated with the cylinder dielectric coating with a radius of b (region 2). The Z_2 (Ω/m) and Y_2 (S/m) terms are the exterior impedance and admittance for the dissipative medium (region 3), respectively. The characteristic impedance Z_C can be formulated in terms of k_L by using the relationship of (5.17) and (5.19):

$$\begin{aligned}Z_C = \frac{\sqrt{Z_LY_L}}{Y_L} = \frac{jk_L}{Y_L} &= jk_L\left[\frac{\varepsilon_{r2} + \varepsilon_{r3}k_3b\ln\left(\dfrac{b}{a}\right)\dfrac{H_1^{(1)}(k_3b)}{H_o^{(1)}(k_3b)}}{j2\pi\omega\varepsilon_o\varepsilon_{r2}\varepsilon_{r3}k_3b\dfrac{H_1^{(1)}(k_3b)}{H_o^{(1)}(k_3b)}}\right] \\ &= \frac{\omega\mu_ok_L}{2\pi k_2^2}\left[\ln\left(\frac{b}{a}\right) + \frac{k_2^2H_o^{(1)}(k_3b)}{k_3^2k_3bH_1^{(1)}(k_3b)}\right]\end{aligned} \tag{5.20}$$

For the case of thick dielectric coatings ($b >> a$), (5.20) can be reduced as:

$$Z_C \approx \frac{\omega\mu_ok_L}{2\pi k_2^2}\ln\left(\frac{b}{a}\right) \tag{5.21}$$

The radiation effect contributed by Z_2 is ignored in (5.21). Substituting (5.17) and (5.19) into (5.16), the alternative expression for k_L is given as:

$$
\begin{aligned}
k_L &= -j\left[\frac{j\omega\mu_o}{2\pi}\ln\left(\frac{b}{a}\right) + \frac{j\omega\mu_o}{2\pi k_3 b}\frac{H_o^{(1)}(k_3 b)}{H_1^{(1)}(k_3 b)}\right]^{1/2}\left[\frac{j2\pi\omega\varepsilon_o\varepsilon_{r2}\varepsilon_{r3}k_3 b\dfrac{H_1^{(1)}(k_3 b)}{H_o^{(1)}(k_3 b)}}{\varepsilon_{r2} + \varepsilon_{r3}k_3 b\ln\left(\dfrac{b}{a}\right)\dfrac{H_1^{(1)}(k_3 b)}{H_o^{(1)}(k_3 b)}}\right]^{1/2} \\
&= k_2\sqrt{\varepsilon_{r3}}\left[\frac{k_3 bH_1^{(1)}(k_3 b)\ln(b/a) + H_o^{(1)}(k_3 b)}{\varepsilon_{r3}k_3 bH_1^{(1)}(k_3 b)\ln(b/a) + \varepsilon_{r2}H_o^{(1)}(k_3 b)}\right]^{1/2} \\
&= k_2 k_3\left[\frac{k_3 bH_1^{(1)}(k_3 b)\ln(b/a) + H_o^{(1)}(k_3 b)}{k_3^2 k_3 bH_1^{(1)}(k_3 b)\ln(b/a) + k_2^2 H_o^{(1)}(k_3 b)}\right]^{1/2}
\end{aligned}
\tag{5.22}
$$

For the dielectric coating with two layers (regions 2 and 3) as shown in Figure 5.3, the approximated formulation for k_L and the characteristic impedance Z_C are given as [6, 9, 13]:

$$
k_L = k_2\left[\frac{\ln\left(\dfrac{c}{a}\right)}{\ln\left(\dfrac{b}{a}\right) + \left(\dfrac{k_2}{k_3}\right)^2\ln\left(\dfrac{c}{b}\right)}\right]^{1/2}\left[\frac{\ln\left(\dfrac{c}{a}\right) + \dfrac{H_o^{(1)}(k_4 c)}{k_4 cH_1^{(1)}(k_4 c)}}{\ln\left(\dfrac{c}{a}\right) + \left(\dfrac{k_3}{k_4}\right)^2\dfrac{H_o^{(1)}(k_4 c)}{k_4 cH_1^{(1)}(k_4 c)}}\right]^{1/2}
\tag{5.23}
$$

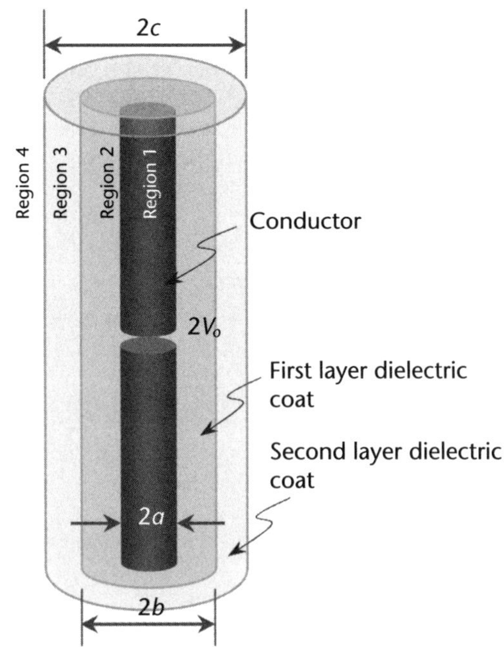

Figure 5.3 Two-layer, dielectric-coated dipole.

and

$$Z_c = \frac{\omega\mu_o k_L}{2\pi k_2^2}\left[\ln\left(\frac{b}{a}\right) + \left(\frac{k_2}{k_3}\right)^2 \ln\left(\frac{c}{b}\right) + \left(\frac{k_3}{k_4}\right)^2 \frac{H_o^{(1)}(k_4 c)}{k_4 c H_1^{(1)}(k_4 c)}\right] \quad (5.24)$$

where k_2 and k_4 are the wave propagation constants in regions 2 and 4, respectively. Symbol c in (5.23) and (5.24) is the radius of the second-layer dielectric coating.

5.2.3 Input Impedance Z_{in} at the Driving Point

By using Hallén's equation (4.23), the potential difference V_z between the upper end ($z = h$ and $z = -h$) and the driving point ($z = 0$) of the coated dipole can be written as:

$$V_z = \frac{\eta}{2\pi}\int_{-h}^{h} I_z \frac{e^{-jkr}}{r} dz' = \frac{jV_o}{\cos(k_L h)}\sin\left[k_L(h - |z|)\right] \quad (5.25)$$

From circuit theory Ohm's law ($V_z = I_z Z_c$), (5.25) becomes:

$$I_z = \frac{jV_o}{Z_c \cos(k_L h)}\sin\left[k_L(h - |z|)\right] \quad (5.26)$$

where Z_c is the characteristic impedance for the coated dipole transmission line. At the driving point ($z = 0$), $I_z = I_o$ and (5.26) can be rewritten as:

$$I_o = \frac{jV_o}{Z_c \cos(k_L h)}\sin(k_L h) \quad (5.27)$$

Thus, the input impedance, Z_{in} of the coated monopole at driving point $z = 0$ is given as:

$$Z_{in} = \frac{V_o}{I_o} = \frac{V_o Z_c \cos(k_L h)}{jV_o \sin(k_L h)} = -jZ_c \cot(k_L h) \quad (5.28)$$

In order to improve the calculated impedance, Z_{in} uses (5.28) to become closer to the measured results, and the fringing effects at the top end of the inner conductor must be considered. The effects can be corrected by replacing the actual length h of the inner conductor in (5.28) with the effective length, h'.

$$Z_{in} = -jZ_c \cot(k_L h') \quad (5.29)$$

You et al. [17] assumed that the fringing effects are inversely proportional to the square root of the operation. Thus, the effective length h' of the inner conductor is given as:

$$h' = h + \frac{\kappa}{\sqrt{f}} \tag{5.30}$$

κ is a coefficient value that can be adjusted in order to achieve a good match between the calculated and measured results. Figure 5.4 shows the calculated normalized resistance, $\mathfrak{Re}(Z_{in})/Z_o$, and reactance, $\mathfrak{Im}(Z_{in})/Z_o$, using (5.29) and the COMSOL simulation results and measurements for water (highly dissipative medium) at room temperature (25°C ± 1°C) over the frequency range from dc to 20 GHz. The results are measured using a Teflon-coated monopole driven from the coaxial line with a = 0.65 mm, b = 2.05 mm, and h = 13.92 mm as shown in Figure 5.5.

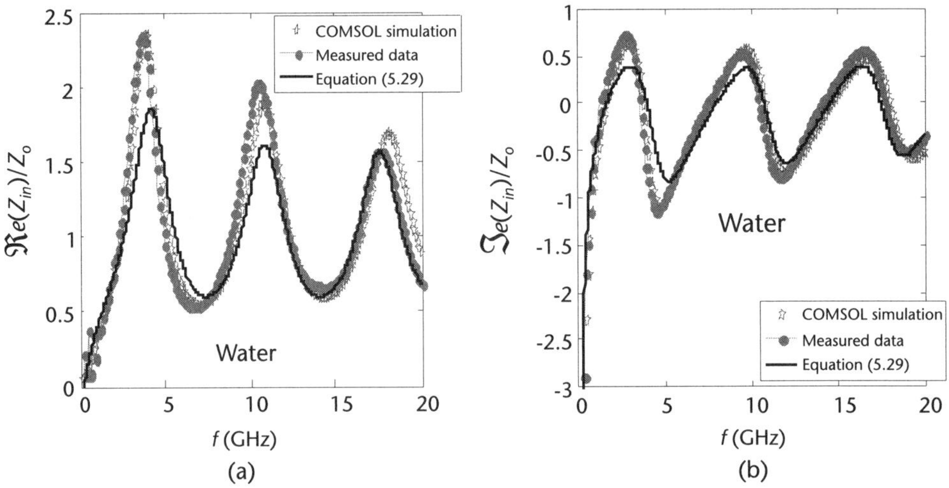

Figure 5.4 Comparison of (a) normalized resistance, $\mathfrak{Re}(Z_{in})/Z_o$, and (b) normalized reactance, $\mathfrak{Im}(Z_{in})/Z_o$, for water at 25°C by considering the size of the probe (ε_{r2} = 2.06) with a = 0.65 mm, b = 2.05 mm, and h = 13.92 mm.

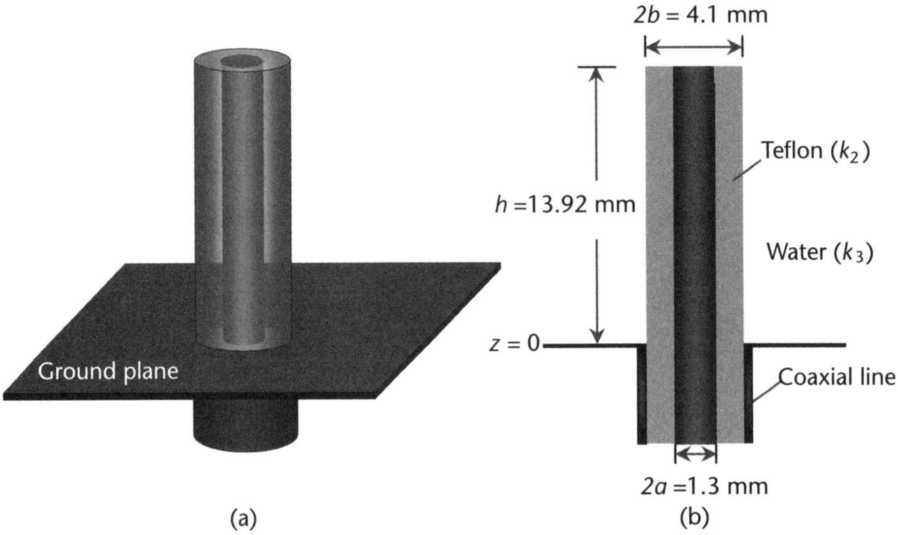

Figure 5.5 (a) Three-dimensional and (b) two-dimensional configurations of a coated monopole.

In this work, the κ term in (5.30) is found to match the measurements, which can be roughly represented by a single value as 550. It should be noted that when $k_L > k_2$, the calculated $\mathfrak{Re}(Z_{in})/Z_o$ using (5.29) will give a negative value due to ξ_2 becoming imaginary, since the case of $\xi_2 = \sqrt{k_2^2 - k_L^2}$ has previously been used in the formula derivation. Thus, the calculated values of $\mathfrak{Re}(Z_{in})/Z_o$ in Figure 5.4(a) have been modulated.

5.3 Integral Electric Field Modeling

5.3.1 Electromagnetic Fields in a Dielectric-Coated Layer (Region 2)

A rigorous integration of electric and magnetic scattering field for the dielectric-coated dipole was derived by King et al. [13]. In the derivation, the current flow I_z along the dipole was also assumed to be in the form of a simple sinusoidal distribution; thus, the electromagnetic fields ($H_{\phi 2}$, $E_{\rho 2}$, and E_{z2}) in the cylindrical dielectric-coated layer ($a \leq \rho \leq b$) are approximated by [11, 13]:

$$H_{\phi 2} \approx \frac{I_z}{2\pi\rho} = \frac{I_o}{2\pi\rho}\frac{\sin\left[k_L(h-|z|)\right]}{\sin(k_L h)}, \text{ for } -h \leq z \leq h \tag{5.31}$$

$$E_{\rho 2} \approx \frac{q_z}{2\pi\varepsilon_o\varepsilon_{r2}\rho} = \frac{jk_L I_o}{2\pi\omega\varepsilon_o\varepsilon_{r2}\rho}\frac{\cos\left[k_L(h-z)\right]}{\sin(k_L h)}, \text{ for } 0 \leq z \leq h \tag{5.32a}$$

$$= \frac{-jk_L I_o}{2\pi\omega\varepsilon_o\varepsilon_{r2}\rho}\frac{\cos\left[k_L(h+z)\right]}{\sin(k_L h)}, \text{ for } -h \leq z \leq 0 \tag{5.32b}$$

where the charge q_z (C/m) in (5.32) is derived using (5.26) as:

$$q_z = -\frac{j}{\omega}\frac{\partial I_z}{\partial z} = \frac{V_o k_L}{2\omega Z_c}\frac{\cos\left[k_L(h-z)\right]}{\cos(k_L h)}, \text{ for } 0 \leq z \leq h \tag{5.33a}$$

$$= -\frac{V_o k_L}{2\omega Z_c}\frac{\cos\left[k_L(h+z)\right]}{\cos(k_L h)}, \text{ for } -h \leq z \leq 0 \tag{5.33b}$$

while the E_{z2} field in the dielectric-coated layer can be found using (5.31) and (5.32) with the aid of Maxwell's equation in cylindrical coordinates, which is given as:

$$-\frac{\partial E_{z2}}{\partial \rho} + \frac{\partial E_{\rho 2}}{\partial z} = j\omega\mu_o H_{\phi 2} \qquad \text{(Maxwell's equation)} \tag{5.34}$$

Thus,

$$E_{z2} = \int_a^{\rho}\left(\frac{\partial E_{\rho 2}}{\partial z} - j\omega\mu_o H_{\phi 2}\right)\partial\rho = -\frac{j\omega\mu_o I_o}{2\pi}\ln\left(\frac{\rho}{a}\right)\left(1 - \frac{k_L^2}{k_2^2}\right)\frac{\sin\left[k_L\left(h - |z|\right)\right]}{\sin\left(k_L h\right)} \quad (5.35)$$

5.3.2 Electromagnetic Fields in a Dissipative Medium (Region 3)

The total electric field, $\vec{E}_3$, at any point in the dissipative medium ($b \le$ region 3) is given as [8, 13]:

$$\vec{E}_3 = \frac{\rho'}{4\pi}\int_{-h}^{h} dz'\int_{-\pi}^{\pi}\left[-j\omega\mu_o H_{\phi 3}\psi\hat{z}' - E_{z3}\left(\hat{\phi}\times\vec{\nabla}'\psi\right) + E_{\rho 3}\vec{\nabla}'\psi\right]d\phi' \quad (5.36)$$

where ψ is the Green's function and ∇' is the gradient operator with respect to the cylindrical coordinates (ρ', ϕ', z') at the driving point (source point):

$$\psi = \frac{e^{jk_3 r}}{r} \quad (5.37a)$$

$$\nabla' = \frac{\partial}{\partial\rho'}\hat{\rho}' + \frac{1}{\rho'}\frac{\partial}{\partial\phi'}\hat{\phi}' + \frac{\partial}{\partial z'}\hat{z}' \quad (5.37b)$$

$$r = \sqrt{\left(z - z'\right)^2 + \rho^2 + b^2 - 2\rho b\cos\phi'} \quad (5.37c)$$

The continuity of the transformed boundary conditions at $z = d_1$ for the tangential magnetic field, $\tilde{H}_{\phi i}$, and the tangential electric field, $\tilde{E}_{\rho i}$ is given as:

$$E_{z3}\ (\text{at}\ \rho = b) = E_{z2}\ (\text{at}\ \rho = b) \quad (5.38a)$$

$$E_{\rho 3}\ (\text{at}\ \rho = b) = \left(\varepsilon_{r2}/\varepsilon_{r3}\right)E_{\rho 2}\ (\text{at}\ \rho = b) \quad (5.38b)$$

$$H_{\phi 3}\ (\text{at}\ \rho = b) = H_{\phi 2}\ (\text{at}\ \rho = b) \quad (5.38c)$$

At $\rho = b$, (5.36) can be rewritten as:

$$\vec{E}_3 = \frac{b}{4\pi}\int_{-h}^{h} dz'\int_{-\pi}^{\pi}\left[j\omega\mu_o H_{\phi 2}\psi\hat{z}' - E_{z2}\left(\hat{\phi}\times\vec{\nabla}'\psi\right) + \left(\frac{\varepsilon_{r2}}{\varepsilon_{r3}}\right)E_{\rho 2}\vec{\nabla}'\psi\right]d\phi' \quad (5.39)$$

In (5.39), under the integral there are three terms corresponding to field components of $H_{\phi 2}$, $E_{\rho 2}$, and E_{z2} and those integrals with respect to ϕ' can be approximately solved as:

$$\int_{-\pi}^{\pi} j\omega\mu_o \psi \hat{z}'\, d\phi' = j2\pi\omega\mu_o \psi \hat{z}' \quad (5.40a)$$

$$\begin{aligned}\int_{-\pi}^{\pi} \hat{\phi}' \times \vec{\nabla}'\psi\, d\phi' &= -\int_{-\pi}^{\pi}\left[\left(\hat{\phi}' \times \hat{\rho}\right)\frac{\partial \psi}{\partial \rho} + \left(\hat{\phi}' \times \hat{z}\right)\frac{\partial \psi}{\partial z}\right] d\phi' \\ &= \pi b\left(\hat{\rho}\frac{\partial}{\partial z} - \hat{z}\frac{\partial}{\partial \rho}\right)\left(\frac{jk_3}{r} - \frac{1}{r^2}\right)\rho\psi\end{aligned} \quad (5.40b)$$

$$\int_{-\pi}^{\pi} \vec{\nabla}'\psi\, d\phi' = -2\left(\hat{\rho}\frac{\partial}{\partial \rho} + \hat{z}\frac{\partial}{\partial z}\right)\int_{0}^{\pi} \psi\, d\phi' = -2\pi\left(\hat{\rho}\frac{\partial}{\partial \rho} + \hat{z}\frac{\partial}{\partial z}\right)\psi \quad (5.40c)$$

and (5.37c) is simplified as:

$$r = \sqrt{\left(z - z'\right)^2 + \rho^2} \quad (5.41)$$

By using the approximated solutions of (5.40a) through (5.40c), (5.39) can easily be split into separate groups of field components by referring to the unit vector (ϕ', $\hat{\rho}$, and $\hat{z}$). Finally, it is found that the scattering vector electric field, $\vec{E}_3$, in region 3 is the summation of only two field components ($E_{\rho 3}$ and E_{z3}):

$$\vec{E}_3 = E_{\rho 3}\hat{\rho} + E_{z3}\hat{z} \quad (5.42)$$

where the $E_{\rho 3}$ and E_{z3} are given as:

$$E_{\rho 3} = \frac{b}{4\pi}\int_{-h}^{h}\left[-\pi b E_{z2}\left(\frac{jk_3}{r} - \frac{1}{r^2}\right)\rho\frac{\partial \psi}{\partial z} - 2\pi\left(\frac{\varepsilon_{r2}}{\varepsilon_{r3}}\right)E_{\rho 2}\frac{\partial \psi}{\partial \rho}\right]dz' \quad (5.43)$$

$$E_{z3} = \frac{b}{4\pi}\int_{-h}^{h}\begin{bmatrix} j2\pi\omega\mu_o H_{\phi 2}\psi + \pi b E_{z2}\dfrac{\partial \psi}{\partial \rho}\left(\dfrac{jk_3}{r} - \dfrac{1}{r^2}\right)\rho \\ -2\pi\left(\dfrac{\varepsilon_{r2}}{\varepsilon_{r3}}\right)E_{\rho 2}\dfrac{\partial \psi}{\partial z}\end{bmatrix}dz' \quad (5.44)$$

The solutions of differential terms in (5.43) and (5.44) are given as:

$$\frac{\partial \psi}{\partial \rho} = \left(\frac{jk_3}{r} - \frac{1}{r^2}\right)\rho\psi \quad \text{and} \quad \frac{\partial \psi}{\partial z} = \left(z - z'\right)\left(\frac{jk_3}{r} - \frac{1}{r^2}\right)\psi \quad (5.45)$$

Substituting (5.31), (5.32), and (5.35) into (5.43) and (5.44), the obvious mathematical representations for $E_{\rho 3}$ and E_{z3} are given as (5.46) and (5.47).

$$E_{\rho 3} = \frac{j\omega\mu_o I_o}{4\pi\sin(k_L h)}\left\{\begin{array}{l} \frac{b^2}{2}\ln\left(\frac{b}{a}\right)\left(1-\frac{k_L^2}{k_2^2}\right)\times \\ \int_{-h}^{h}\left\{\sin\left[k_L\left(h-|z'|\right)\right]\left(\frac{jk_3}{r}-\frac{1}{r^2}\right)^2\rho(z-z')\psi dz'\right\} \\ +\frac{\rho k_L}{k_3^2}\int_{-h}^{0}\cos\left[k_L(h+z')\right]\left(\frac{jk_3}{r}-\frac{1}{r^2}\right)\psi dz' \\ -\frac{\rho k_L}{k_3^2}\int_{0}^{h}\cos\left[k_L(h-z')\right]\left(\frac{jk_3}{r}-\frac{1}{r^2}\right)\psi dz' \end{array}\right\} \quad (5.46)$$

and

$$E_{z3} = \frac{j\omega\mu_o I_o}{4\pi\sin(k_L h)}\left\{\begin{array}{l} \int_{-h}^{h}\sin\left[k_L\left(h-|z|\right)\right]\psi\, dz' \\ -\frac{b^2}{2}\ln\left(\frac{b}{a}\right)\left(1-\frac{k_L^2}{k_2^2}\right)\int_{-h}^{h}\sin\left[k_L\left(h-|z'|\right)\right]\left(\frac{jk_3}{r}-\frac{1}{r^2}\right)^2\rho^2\psi dz' \\ +\frac{k_L}{k_3^2}\int_{-h}^{0}\cos\left[k_L(h+z')\right](z-z')\left(\frac{jk_3}{r}-\frac{1}{r^2}\right)\psi dz' \\ -\frac{k_L}{k_3^2}\int_{0}^{h}\cos\left[k_L(h-z')\right](z-z')\left(\frac{jk_3}{r}-\frac{1}{r^2}\right)\psi dz' \end{array}\right\}$$

(5.47)

Figures 5.6(a, b) show the calculated normalized electric field components, ($E_\rho/E_{\rho\max}$ and $E_z/E_{z\max}$) using (5.46) and (5.47), respectively, and the COMSOL

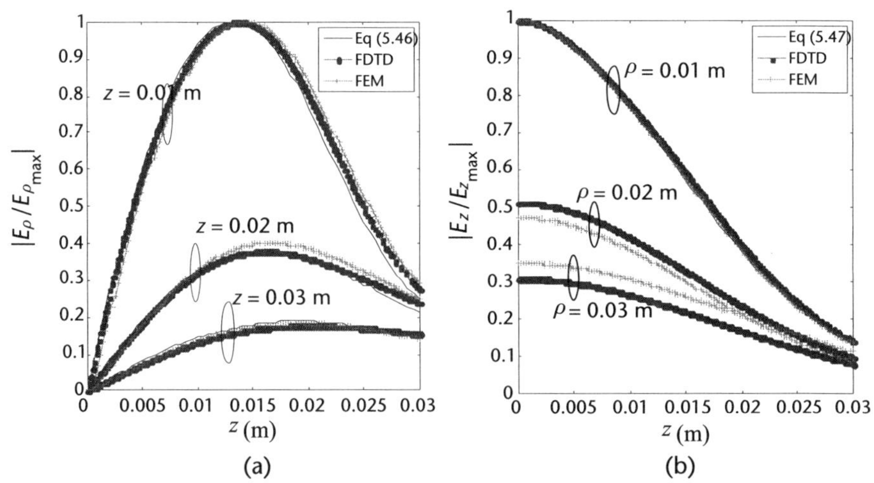

Figure 5.6 Comparison of normalized (a) $|E_z|$ and (b) $|E_\rho|$ at different cross-sections parallel to the ρ-axis (ρ = 0.01m, ρ = 0.02m, and ρ = 0.03m). (c) Dimensions and geometry of two-layer coated monopole.

simulation results and the finite-difference time-domain (FDTD) calculation for brain tissue (ε_{r3}=42.5, σ_3 = 0.88 S/m) at 915 MHz. Equations (5.46) and (5.47) are analyzed using a two-layer coated monopole driven from the coaxial line with a = 0.5 mm, b = 0.75 mm, c = 1.5 mm, and h = 21.1 mm as shown in Figure 5.6(c).

5.3.3 Near-Field and Far-Field Integral Formulations

For the near-field observation (observation point $P < b^2$), the exact solution of electric field is required. Therefore, Casey and Bansal [11] reexpressed (5.46) and (5.47) in more rigorous forms as:

$$E_{z3} = \frac{j\omega\mu_o I_o}{(2\pi)^2 \sin(k_L h)} \left\{ \begin{array}{l} -\int_{-h}^{h}\int_0^{\pi} \sin\left[k_L\left(h-|z'|\right)\right]\psi' d\phi' dz' \\ -b\ln\left(\frac{b}{a}\right)\left(1-\frac{k_L^2}{k_2^2}\right)\int_{-h}^{h}\int_0^{\pi}\left\{\begin{array}{l}\sin\left[k_L\left(h-|z'|\right)\right]\psi' \times \\ \left(\frac{jk_3}{r}+\frac{1}{r^2}\right)\left(b-\rho\cos\phi'\right)\end{array}\right\} d\phi' dz' \\ +\frac{k_L}{k_3^2}\int_{-h}^{0}\int_0^{\pi}\cos\left[k_L(h+z')\right](z-z')\left(\frac{jk_3}{r}+\frac{1}{r^2}\right)\psi' d\phi' dz' \\ -\frac{k_L}{k_3^2}\int_0^{h}\int_0^{\pi}\cos\left[k_L(h-z')\right](z-z')\left(\frac{jk_3}{r}+\frac{1}{r^2}\right)\psi' d\phi' dz' \end{array} \right\} \tag{5.48}$$

and

$$E_{\rho 3} = \frac{j\omega\mu_o I_o}{(2\pi)^2 \sin(k_L h)} \left\{ \begin{array}{l} -b\ln\left(\frac{b}{a}\right)\left(1-\frac{k_L^2}{k_2^2}\right)\int_{-h}^{h}\int_0^{\pi}\left\{\begin{array}{l}\sin\left[k_L\left(h-|z'|\right)\right] \times \\ \left(\frac{jk_3}{r}+\frac{1}{r^2}\right)\psi'(z-z')\cos\phi'\end{array}\right\} d\phi' dz' \\ -\frac{k_L}{k_3^2}\int_{-h}^{0}\int_0^{\pi}\cos\left[k_L(h+z')\right]\psi'\left(\frac{jk_3}{r}+\frac{1}{r^2}\right)\left(b\cos\phi'-\rho\right) d\phi' dz' \\ +\frac{k_L}{k_3^2}\int_0^{h}\int_0^{\pi}\cos\left[k_L(h-z')\right]\psi'\left(\frac{jk_3}{r}+\frac{1}{r^2}\right)\left(b\cos\phi'-\rho\right) d\phi' dz' \end{array} \right\} \tag{5.49}$$

where

$$\psi' = \frac{\exp\left(-jk_3\sqrt{(z-z')^2+(\rho-b)^2+4b\rho\sin^2(\phi'/2)}\right)}{\sqrt{(z-z')^2+(\rho-b)^2+4b\rho\sin^2(\phi'/2)}} \tag{5.50}$$

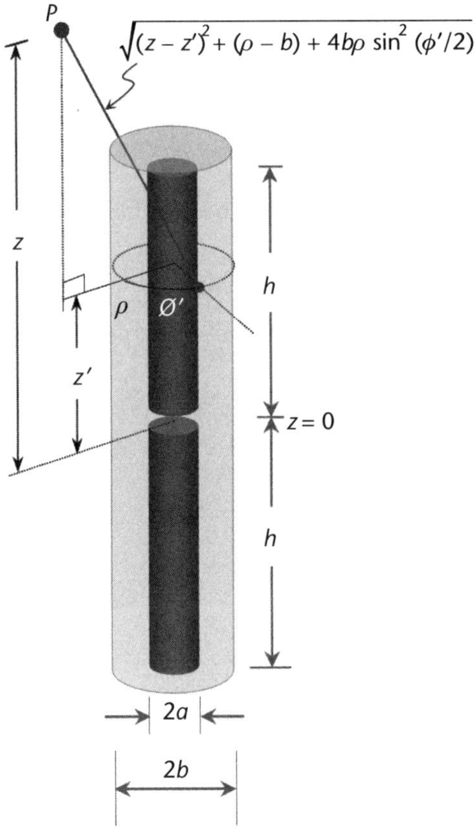

Figure 5.7 Cylindrical coordinates used in evaluating the near-field radiation of the dipole (observation point $P < b^2$).

The formula for electric field E_z refers to coordinates as shown in Figure 5.7. For the far-field observation (observation point $P \gg b^2$) with the thin dipole [$(k_4 b)^2 \ll 1$], (5.46) and (5.47) can be reduced as [13]:

$$E_{z3} = \frac{j\omega\mu_o I_o}{4\pi\sin(k_L h)}\left\{\begin{aligned}&\left(1-\frac{k_L^2}{k_2^2}\right)\int_0^h \sin[k_L(h-z')](\psi^+ + \psi^-)dz'\\&+\frac{k_L}{k_3^2}\left[\psi_o^+ + \psi_o^- - 2\psi_o\cos(k_L h)\right]\end{aligned}\right\} \tag{5.51}$$

$$E_{z3} = \frac{j\omega\mu_o I_o}{4\pi\sin(k_L h)}\frac{k_L\rho}{k_3^2}\int_0^h \cos[k_L(h-z')]\left\{\begin{aligned}&\left(\frac{jk_3}{r_1}-\frac{1}{r_1^2}\right)\psi^+\\&-\left(\frac{jk_3}{r_2}-\frac{1}{r_2^2}\right)\psi^-\end{aligned}\right\}dz' \tag{5.52}$$

where

$$\psi^+ = \frac{\exp\left(jk_3\sqrt{(z-z')^2+\rho^2}\right)}{\sqrt{(z-z')^2+\rho^2}}, \qquad \psi^- = \frac{\exp\left(jk_3\sqrt{(z+z')^2+\rho^2}\right)}{\sqrt{(z+z')^2+\rho^2}}$$

$$\psi_o^+ = \frac{\exp\left(jk_3\sqrt{(z-h)^2+\rho^2}\right)}{\sqrt{(z-h)^2+\rho^2}}, \qquad \psi_o^- = \frac{\exp\left(jk_3\sqrt{(z+h)^2+\rho^2}\right)}{\sqrt{(z+h)^2+\rho^2}}$$

$$\psi_o = \frac{\exp\left(jk_3\sqrt{z^2+\rho^2}\right)}{\sqrt{z^2+\rho^2}}$$

and

$$r_1 = \sqrt{(z-z')^2+\rho^2}, \qquad r_2 = \sqrt{(z+z')^2+\rho^2}$$

The model configuration of (5.51) and (5.52) are shown in Figure 5.8.

5.4 Equivalent Lumped-Element Circuit Modeling

Liao et al. [15] proposed that the equivalent circuit input impedance for a dielectric-coated dipole is a sum of two components. The first component is the impedance of the bare dipole, and the second component is the additional parallel inductance

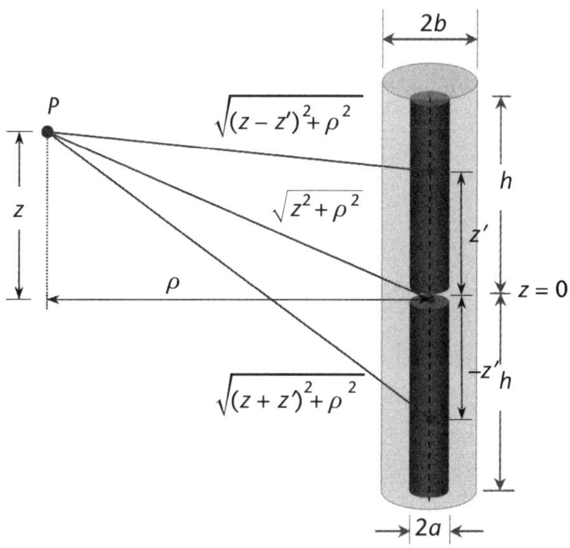

Figure 5.8 Cylindrical coordinates used in evaluating the far-field radiation of the dipole (observation point $P \gg b^2$).

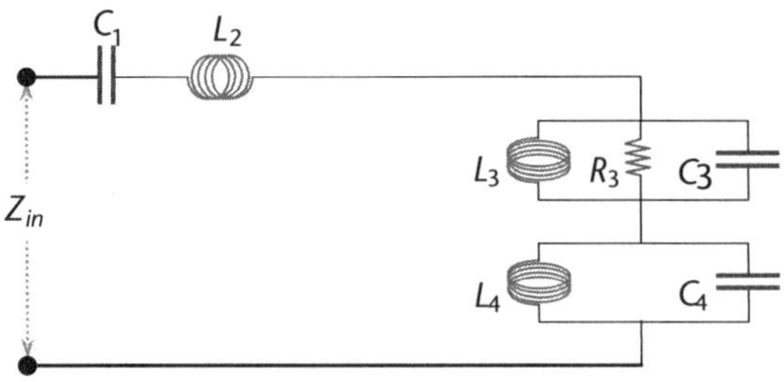

Figure 5.9 Seven-element equivalent circuit.

L_4 and capacitance C_4, which is contributed by the dielectric coating ($\mu_r = 1$) as shown in Figure 5.9.

$$Z_{in} = \underbrace{-j\frac{1}{\omega C_1} + j\omega L_2 + \frac{j\omega R_3 L_3}{j\omega L_3 + R_3 - \omega^2 R_3 L_3 C_3}}_{Base} + \underbrace{\frac{j\omega L_4}{1-\omega^2 C_4 L_4}}_{Coat} \tag{5.53}$$

where the empirical functions of C_1, L_2, C_3, L_3, and R_3 are expressed in [15] and can be found in Chapter 4.

The functions of L_4 and C_4 in (5.53) are derived using the combination of the mathematical works that have been reported in [18, 19] as:

$$L_4 \approx \frac{\mu_o h}{3\pi}\left(1-\frac{1}{\varepsilon_{r4}}\right)\ln\left(\frac{b}{a}\right) \tag{5.54}$$

$$C_4 \approx \frac{4h}{\mu_o \pi c^2}\frac{1}{\left(1-\frac{1}{\varepsilon_{r4}}\right)\ln\left(\frac{b}{a}\right)}\left\{\frac{1}{1-\left(1-\frac{1}{\varepsilon_{r4}}\right)\left[\ln\left(\frac{b}{a}\right)\Big/\left(\ln\left(\frac{h}{a}\right)-1\right)\right]}\right\} \tag{5.55}$$

5.5 Modified Dielectric-Coated Monopole

The coated monopole as shown in Figure 5.10 is widely used as the applicator in the interstitial microwave hyperthermia for treating deep-seated tumors and benign prostatic hyperplasia [6, 14, 16]. The dielectric coating (Teflon) is used as a hygienic protector when the monopole is in direct contact with the biological medium. Normally, the monopole array used in those treatments consists of four parallel monopoles, forming a square that surrounds a targeted tumor (in Figure 5.11), in order to obtain an optimum heating region covering the tumor. The calculation of the distribution of temperature and the further descriptions of this application will be discussed in Chapter 9.

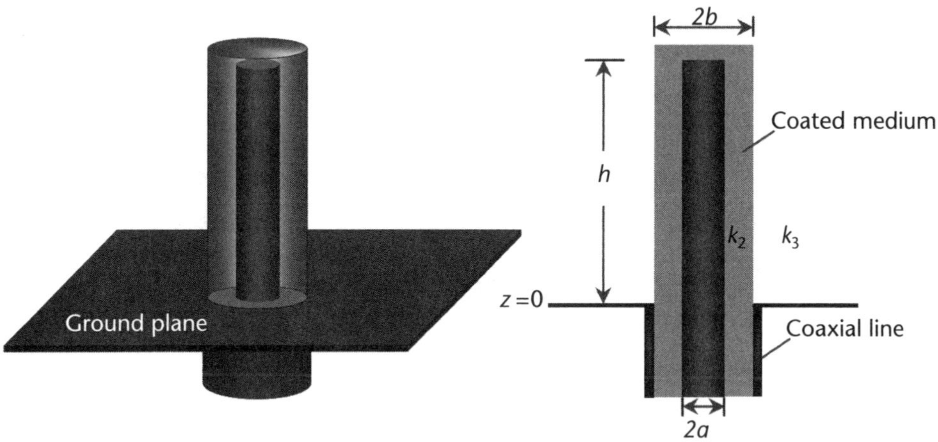

Figure 5.10 Typical dielectric-coated monopole driven from a coaxial line.

Over the years, many researchers [4, 6, 9, 10, 16] have modified the coated monopole into a multisection form as shown in Figure 5.12, so that the transmission of the RF power and the distribution of heat will be concentrated in the section along the monopole. Optimizing the heat distribution of the coated monopole not only will increase the heating efficiency, but will also reduce the number of elements in an array arrangement (in Figure 5.11), which will still lead to a desirable heating pattern and coverage region.

Z_{in} is the input impedance of the entire applicator at the driving point $z = 0$.

$$Z_{\text{in}} = jZ_{Ci} \coth\left(k_{Li} h_i + j\theta_i\right) \tag{5.56}$$

where θ_i is phase-shifted along the ith section.

$$\theta_i = \coth^{-1}\left(\frac{Z_i}{Z_{Ci}}\right) \tag{5.57}$$

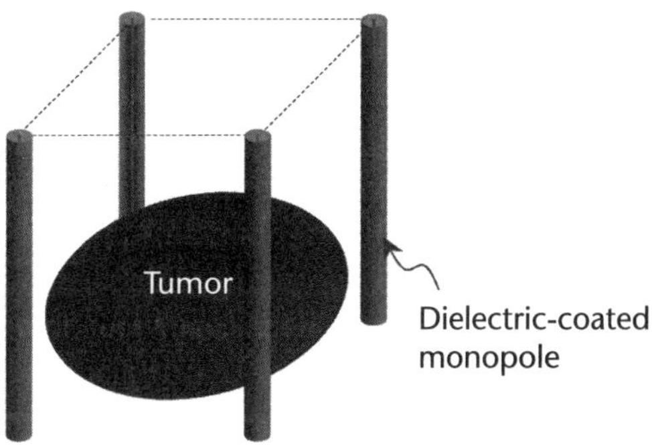

Figure 5.11 The array of four dielectric-coated monopoles surrounding a targeted tumor.

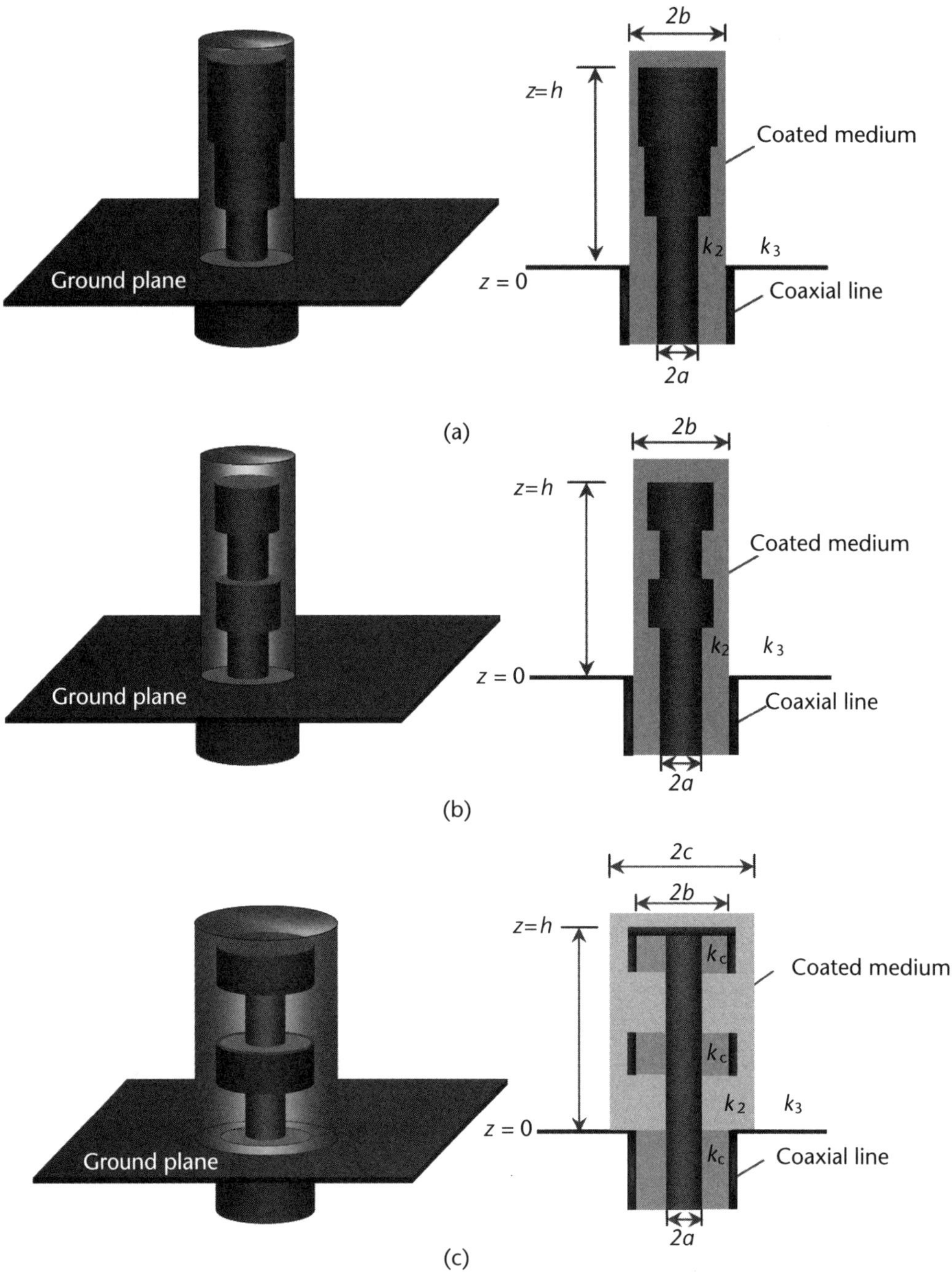

Figure 5.12 (a) Nonuniform multisection-coated monopole [6]. (b) Uniform multisection-coated monopole [16]. (c) Capacitive multisection-coated monopole [10].

Z_i, Z_{Ci}, K_{Li}, and h_i are the input impedance, characteristic impedance, propagation constant, and conductor length of the ith section, respectively. For instance, the input impedance, Z_{in} at the driving point $z = 0$ for the coated monopole at the three sections as shown in Figure 5.13 is observed.

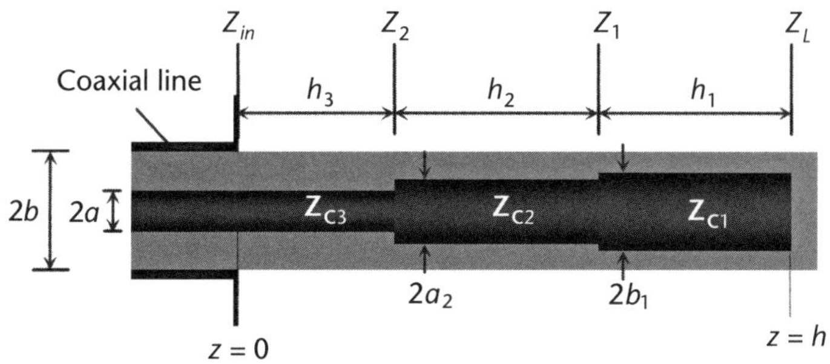

Figure 5.13 Multisection-coated monopole.

Starting from $z = h$ to $z = 0$, the impedances Z_i at the different reference planes are given by (5.58) through (5.60):

$$Z_1 = Z_{C1}\frac{Z_L + jZ_{C1}\tan(k_{L1}h_1)}{Z_{C1} + jZ_L\tan(k_{L1}h_1)} = jZ_{C1}\coth(k_{L1}h_1 + j\theta_1) \tag{5.58}$$

$$Z_2 = Z_{C2}\frac{Z_1 + jZ_{C2}\tan(k_{L2}h_2)}{Z_{C2} + jZ_1\tan(k_{L2}h_2)} = jZ_{C2}\coth(k_{L2}h_2 + j\theta_2) \tag{5.59}$$

$$Z_{\text{in}} = Z_{C3}\frac{Z_2 + jZ_{C3}\tan(k_{L3}h_3)}{Z_{C3} + jZ_2\tan(k_{L3}h_3)} = jZ_{C3}\coth(k_{L3}h_3 + j\theta_3) \tag{5.60}$$

where Z_1 and Z_2 are the input impedances at the first and second sections of the reference plane (at $z = h_3 + h_2$ and $z = h_3$). θ_1, θ_2, and θ_3 (rad) are the phase shifts along sections 1, 2, and 3, respectively.

$$\theta_1 = \coth^{-1}(Z_L/Z_{C1}) \tag{5.61a}$$

$$\theta_2 = \coth^{-1}(Z_1/Z_{C2}) \tag{5.61b}$$

$$\theta_3 = \coth^{-1}(Z_2/Z_{C3}) \tag{5.61c}$$

and Z_L is the open-circuited transmission-line impedance at the end of the coated monopole ($z = h$).

$$Z_L = \frac{-jZ_{C1}}{\tan(k_L h_1)} \tag{5.62}$$

The propagation constant k_L for each section in (5.58) through (5.60) is given as:

$$k_{L1} = k_2 k_3 \left[\frac{H_o^{(1)}(k_3 b) + k_3 b \ln(b/a_1) H_1^{(1)}(k_3 b)}{k_2^2 H_o^{(1)}(k_3 b) + k_2^3 b \ln(b/a_1) H_1^{(1)}(k_3 b)} \right]^{1/2} \tag{5.63a}$$

$$k_{L2} = k_2 k_3 \left[\frac{H_o^{(1)}(k_3 b) + k_3 b \ln(b/a_2) H_1^{(1)}(k_3 b)}{k_2^2 H_o^{(1)}(k_3 b) + k_2^3 b \ln(b/a_2) H_1^{(1)}(k_3 b)} \right]^{1/2} \tag{5.63b}$$

$$k_{L3} = k_2 k_3 \left[\frac{H_o^{(1)}(k_3 b) + k_3 b \ln(b/a) H_1^{(1)}(k_3 b)}{k_2^2 H_o^{(1)}(k_3 b) + k_2^3 b \ln(b/a) H_1^{(1)}(k_3 b)} \right]^{1/2} \tag{5.63c}$$

The characteristic impedance Z_C of each section is given as:

$$Z_{C1} = \frac{\omega \mu_o k_{L1}}{2\pi k_2^2} \left[\ln\left(\frac{b}{a_1}\right) + \left(\frac{k_2}{k_3}\right)^2 \frac{H_o^{(1)}(k_3 b)}{k_3 b H_1^{(1)}(k_3 b)} \right] \tag{5.64a}$$

$$Z_{C2} = \frac{\omega \mu_o k_{L2}}{2\pi k_2^2} \left[\ln\left(\frac{b}{a_2}\right) + \left(\frac{k_2}{k_3}\right)^2 \frac{H_o^{(1)}(k_3 b)}{k_3 b H_1^{(1)}(k_3 b)} \right] \tag{5.64b}$$

$$Z_{C3} = \frac{\omega \mu_o k_{L3}}{2\pi k_2^2} \left[\ln\left(\frac{b}{a}\right) + \left(\frac{k_2}{k_3}\right)^2 \frac{H_o^{(1)}(k_3 b)}{k_3 b H_1^{(1)}(k_3 b)} \right] \tag{5.64c}$$

The capacitive multisection-coated monopole as shown in Figure 5.14 is one of the applicators that is widely used in hyperthermia treatments, since it is more stable and less sensitive to biological medium (dissipate medium) compared to the

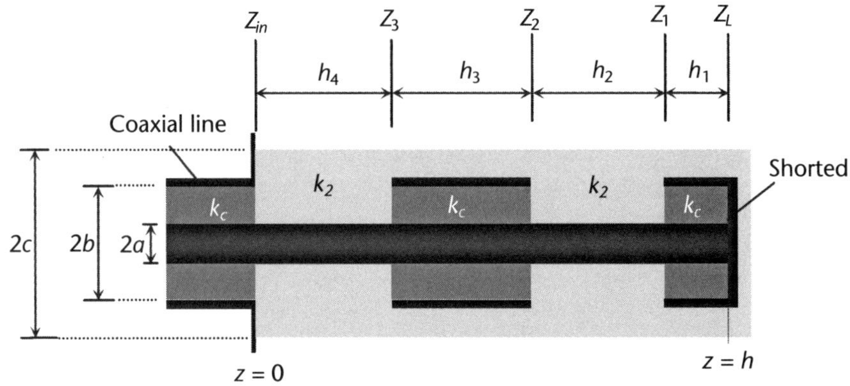

Figure 5.14 Cross-section view of the multisection capacitive monopole terminated by shorted coaxial line.

previous structure. Similarly, the impedances Z_i for each reference plane from $z = h$ to $z = 0$ are given by:

$$Z_1 = jZ_o \tan(k_c h_1) \tag{5.65}$$

$$Z_2 = Z_C \frac{Z_1 + jZ_C \tan(k_L h_2)}{Z_C + jZ_1 \tan(k_L h_2)} = jZ_C \coth(k_L h_2 + j\theta_2) \tag{5.66}$$

$$Z_3 = Z_o \frac{Z_2 + jZ_o \tan(k_c h_3)}{Z_o + jZ_2 \tan(k_c h_3)} = jZ_C \coth(k_L h_3 + j\theta_3) \tag{5.67}$$

$$Z_{in} = Z_C \frac{Z_3 + jZ_C \tan(k_L h_4)}{Z_C + jZ_3 \tan(k_L h_4)} = jZ_C \coth(k_L h_4 + j\theta_4) \tag{5.68}$$

The characteristic impedance Z_o and the propagation constant k_c of the coaxial line are given as:

$$Z_o = \frac{\ln(b/a)}{2\pi} \sqrt{\frac{\mu_o}{\varepsilon_o \varepsilon_c}} \tag{5.69}$$

$$k_c = k_o \sqrt{\varepsilon_c} \tag{5.70}$$

where k_o is the propagation constant in free space and ε_c is the relative permittivity of the coaxial-filled medium (Teflon: $\varepsilon_c = 2.05$). k_L in (5.66) and (5.68) is expressed as:

$$k_L = k_2 k_3 \left[\frac{H_o^{(1)}(k_3 c) + k_3 c \ln(c/a) H_1^{(1)}(k_3 c)}{k_2^2 H_o^{(1)}(k_3 c) + k_2^3 c \ln(c/a) H_1^{(1)}(k_3 c)} \right]^{1/2} \tag{5.71}$$

The characteristic impedance Z_C of the gap section is given as:

$$Z_C = \frac{\omega \mu_o k_L}{2\pi k_2^2} \left[\ln\left(\frac{c}{a}\right) + \left(\frac{k_2}{k_3}\right)^2 \frac{H_o^{(1)}(k_3 c)}{k_3 b H_1^{(1)}(k_3 c)} \right] \tag{5.72}$$

References

[1] Harms, F., "Electromagnetische Wellen an Einem Draht mit lsolierender Zylindrischer Hiille," *Ann. Phys.*, Vol. 23, 1907, pp. 44–60.

[2] King, R. W. P., "Theory of the Terminated Insulated Antenna in a Conducting Medium," *IEEE Trans. on Antennas Propagat.*, Vol. 15, 1964, pp. 305–318.

[3] Lee, K. M., "Insulated Linear Antenna," Ch. 11 in *Research Topics in Electromagnetic Wave Theory*, J. A. Kong, (ed.), New York: John Wiley & Sons, 1967.

[4] Prasad, S., and R. W. P. King, "Experimental Study of Embedded Insulated Antennas," *IEEE Trans. on Antennas Propagat.*, Vol. 30, 1982, pp. 1013–1017.

[5] Hertel, T. W., and G. S. Smith, "The Insulated Linear Antenna-Revisited," *IEEE Trans. on Antennas Propagat.*, Vol. 48, 2000, pp. 914–920.

[6] Iskander, M. F., and A. M. Tumeh, "Design Optimization of Interstitial Antennas," *IEEE Trans. on Biomedical Engineering*, Vol. 36, 1989, pp. 238–246.

[7] King, R. W. P., and T. T. Wu, "The Electric Field Very Near a Cylindrical Driven Antenna," *Radio Science*, Vol. 1 (New Series), 1966, pp. 353–359.

[8] King, R. W. P., and G. S. Smith, *Antennas in Matter: Fundamentals, Theory, and Applications*, Cambridge, MA: MIT Press, 1981.

[9] Su, D. W. F., and L. K. Wu, "Input Impedance Characteristics of Coaxial Slot Antennas for Interstitial Microwave Hyperthermia," *IEEE Trans. on Microwave Theory Tech.*, Vol. 47, 1999, pp. 302–307.

[10] Wang, P., et al., "Improved Calculation of Reflection Coefficient for Coaxial Antennas with Feed Gap Effect," *IEEE Trans. on Antennas Propagat.*, Vol. 57, 2009, pp. 559–563.

[11] Casey, J. P., and R. Bansal, "The Near Field of an Insulated Dipole in a Dissipative Dielectric Medium," *IEEE Trans. Microwave Theory Tech.*, Vol. 34, 1986, pp. 459–463.

[12] Clibbon, K. L., and A. McCowen, "Efficient Computation of SAR Distributions from Interstitial Microwave Antenna Arrays," *IEEE Trans. on Microwave Theory Tech.*, Vol. 42, 1994, pp. 595–600.

[13] King, R. W. P., B. S. Trembly, and J. W. Strohbehn, "The Electromagnetic Field of an Insulated Antenna in a Conducting or Dielectric Medium," *IEEE Trans. on Microwave Theory Tech.*, Vol. 31, 1983, pp. 574–583.

[14] Trembly, B. S., "The Effects of Driving Frequency and Antenna Length on Power Deposition Within a Microwave Antenna Array Used for Hyperthermia," *IEEE Trans. on Biomedical Engineering*, Vol. 32, 1985, pp. 152–157.

[15] Liao, Y., T. H. Hubing, and D. L. Su, "Equivalent Circuit with Frequency-Independent Lumped Element for Coated Wire Antennas," *IEEE Trans. on Antennas Propagat.*, Vol. 60, 2012, pp. 5419–5423.

[16] Debicki, P. S., and M. A. Astrahan, "Calculating Input Impedance of Electrically Small Insulated Antennas for Microwave Hyperthermia," *IEEE Trans. on Microwave Theory Tech.*, Vol. 41, 1993, pp. 357–360.

[17] You, K. Y., et al., "Improved Formulation for Admittance of Thin and Short Monopole Driving from Coaxial Line into Dissipative Media," *IEEE Antennas and Wireless Propagation Letters*, Vol. 8, 2009, pp. 1246–1249.

[18] Moore, J., and M. A. West, "Simplified Analysis of Coated Wire Antennas and Scatters," *IEE Proc. Microwaves Antennas Propag.*, Vol. 142, 1995, pp. 14–18.

[19] Jaisson, D., "Simple Model for the Input Impedance of a Wire Monopole Radiator with a Dielectric Coat," *IET Microwaves Antennas Propag.*, Vol. 2, 2008, pp. 316–323.

Selected Bibliography

Brillouin, L., "Origin of Radiation Resistance," *Radioélectricité*, Vol. 3, 2000, pp. 147–152.

Duncan, R. H., "Theory of the Infinite Cylindrical Antenna Including the Feedpoint Singularity in Antenna Current," *J. Research N. B. S.*, Vol. 66D, No. 2, 1962.

Hallén, E., "Theoretical Investigations into Transmitting and Receiving Antennae," *Nova Acta Regiae Sco. Sci. Upsaliensis*, Ser. IV, Vol. 4, No. 2, 1938, pp. 1–44.

Pocklington, H. C., "Electrical Oscillations in Wire," *Cambridge Philos. Soc. Proc.*, Vol. 4, 1897, pp. 324–332.

Richmond, J. H., "Digital Computer Solutions of the Rigorous Equations for Scattering Problems," *Proc. IEEE*, Vol. 53, 1965, pp. 796–804.

Scott, L. D., "Polynomial-Exponential-Product Theory for Antennas in Homogeneous Isotropic Media," *Div. Eng. and Appl. Phys., Harvard Univ., Cambridge, Mass., NASA Sci. Rep. 7*, 1969.

Storer, J., "Variational Solution to the Problem of the Symmetrical Cylindrical Antenna," *Cruft Lab., Harvard Univ., Tech. Rep.*, Vol. 101, 1950.

Tai, C. T., "A Variational Solution to the Problem of Cylindrical Antennas," Aircraft Radiation Systems Lab., Stanford Res. Inst., Stanford, Calif., Tech. Rep. 12, 1950.

Tsai, L. L., "A Numerical Solution for the Near and Far Fields of an Annular Ring of Magnetic Current," *IEEE Trans. on Antennas Propagat.*, Vol. 20, 1972, pp. 569–576.

Weiner, M. M., et al., *Monopole Elements on Circular Ground Plane*, Norwood, MA: Artech House, 1987.

CHAPTER 6

Numerical Modeling: Method of Moments

In this chapter, the reviews on the analysis of the method of moments (MoM) for open-ended coaxial probes, dipoles and monopoles will be introduced. The MoM was first proposed by Harrington [1] and was used to solve the Fredholm integral equation of the first kind for cylindrical wire problems. There is a close relationship between the analytical and numerical MoM modeling, due to the fact that MoM cannot be used if there is no established analytical integral equation that serves as a governing function in particular problems. Moreover, the accuracy of the MoM method is dependent on the rigors of the integral equation. Fortunately, several rigorous quasi-static integrals have been derived, such as Sommerfeld and Nevels' admittance integral formulations for open-ended coaxial probe [2] and Pocklington and Hallén's integral formulations for dipoles or monopoles [3, 4]. The difference between the analytical and numerical modeling is the complexity of the solutions. For instance, the difference in the voltages between the coaxial gap $a < \rho < b$ at the aperture surface $z = 0$ is assumed to be 1V (or 0.5V) for the analytical solution, as discussed in Chapter 1. However, in the numerical solution, the difference in the voltages is an unknown parameter. The MoM routines may divide the aperture surface into small element areas, and the unknown differences in voltages of each area are solved individually.

MoM is able to solve electrostatic (scalar) and electrodynamic (vector) problems. The fundamental steps of both MoM solutions are summarized later. However, examples of MoM solutions in this chapter are focused on the electrodynamics cases, since the relevant information on the scalar MoM can be found in many textbooks and references. The application of scalar MoM to the open-ended coaxial probe was first studied by Gajda and Stuchly [5] to solve for the surface charge density, σ on the aperture surface of the coaxial probe using a scattering electrostatic potential formulation (Poisson's equation). The final results were interpreted as capacitance, C, on the aperture surface of the probe. Later, the vector voltage, V, on the aperture surface of the coaxial probe was computed using vector MoM routine by Nevels et al. [2].

6.1 Summary of MoM Computation Steps for Electrostatic Cases

6.1.1 Electrostatic Cases

The integral form of the Poisson's equation with an unknown charge density, $\sigma(r')$ is discretized into an algebraic equation. The $G(r|r')$ term in the integral formulation is the Green's function (the kernel of the integral) as:

$$\underbrace{\int \sigma(r')G(r\,|\,r')\,dr' = V(r)}_{\text{Poisson's equation}} \Rightarrow \underbrace{\sum_{n=1}^{N} \sigma_n G_{mn} = V_m}_{\text{Algebraic equation}} \tag{6.1}$$

r and r' are the observation point and source point, respectively. The σ_n in the algebraic equation is the unknown discrete charge density. G_{mn} is the discrete Green's function, and V_m is the known discrete voltage.

6.1.2 Open-Ended Coaxial Probe

For instance, the Poisson's equation for open-ended coaxial probe is expressed as:

$$\int_0^{2\pi}\int_a^b \frac{\sigma_s}{4\pi\varepsilon_o\sqrt{\rho^2 + \rho'^2 - 2\rho\rho'\cos\phi'}}\rho'\,d\rho'\,d\phi' = V \tag{6.2}$$

where σ_s is the charge density on the surface of the aperture probe. V in (6.2) is the difference voltage between the inner and outer conductors on the surface of the aperture. The radius of inner and outer conductors of the probe are represented by a and b, respectively. Equation (6.2) can be converted into the linear algebraic equation as:

$$\sum_{n=1}^{N} \sigma_n G_{mn} = V_m \tag{6.3}$$

where

$$G_{mn} = \sum_{m=1}^{M}\int_0^{2\pi}\int_{\rho_n-\Delta\rho/2}^{\rho_n+\Delta\rho/2} \frac{1}{4\pi\varepsilon_o\sqrt{\rho_m^2 + \rho'^2 - 2\rho_m\rho'\cos\phi'}}\rho'\,d\rho'\,d\phi' \tag{6.4}$$

The closed-form expression of (6.4) can be written as:

$$G_{mn} = \frac{2}{4\pi\varepsilon_o}\sum_{m=1}^{M}\int_0^{\pi}\left[\begin{array}{l}\sqrt{\rho_m^2 + \rho'^2 - 2\rho_m\rho'\cos\phi'} \\ +\rho_m\cos\phi'\ln\left(\rho' - \rho_m\cos\phi' + \sqrt{\rho_m^2 + \rho'^2 - 2\rho_m\rho'\cos\phi'}\right)\end{array}\right]_{\rho_n-\frac{\Delta\rho}{2}}^{\rho_n+\frac{\Delta\rho}{2}} d\phi' \tag{6.5}$$

where the match point ρ_m in (6.5) is defined as:

$$\rho_m = a + \left(\frac{b-a}{N}\right)(m - 0.5), \qquad m = 1,2,\ldots,M$$

and the value of V_m is fixed as unity:

$$V_m = \begin{cases} 1\text{V} & \text{for } a \le \rho \le b \\ 0 & \text{other} \end{cases} \tag{6.6}$$

6.1.3 Dipole

However, for a dipole, Poisson's equation is given similarly as:

$$\int_0^h \frac{\sigma_l}{4\pi\varepsilon_o\sqrt{a^2 + (z - z')^2}}\, dz' = V \tag{6.7}$$

where σ_l is the line charge density along the dipole. V in (6.7) is the excitation voltage on the dipole. The length of the dipole is represented by h. Equation (6.7) can be converted into the linear algebraic equation as:

$$\sum_{n=1}^{N} \sigma_n G_{mn} = V_m \tag{6.8}$$

where

$$G_{mn} = \int_{z_n+\Delta z/2}^{z_n+\Delta z/2} \frac{1}{4\pi\varepsilon_o\sqrt{a^2 + \left(z_m - z'\right)^2}}\, dz' \tag{6.9}$$

The closed-form expression of (6.9) can be written as:

$$G_{mn} = \begin{cases} \dfrac{2\ln}{4\pi\varepsilon_o}\left(\dfrac{(\Delta z/2) + \sqrt{a^2 + (\Delta z/2)^2}}{a}\right) & m = n \\ \dfrac{1}{4\pi\varepsilon_o}\ln\left(\dfrac{z_m + (\Delta z/2) - z_n + \sqrt{a^2 + \left(z_m - z_n + \Delta z/2\right)^2}}{z_m - (\Delta z/2) - z_n + \sqrt{a^2 + \left(z_m - z_n - \Delta z/2\right)^2}}\right) & m \ne n \end{cases} \tag{6.10}$$

where the match point z_m in (6.10) is defined as:

$$z_m = \frac{h}{N}(m - 0.5), \qquad m = 1,2,\ldots,M$$

and the value of V_m is fixed as a unity:

$$V_m = \begin{cases} 1\text{V} & \text{for } 0 \le z \le h \\ 0 & \text{other} \end{cases} \tag{6.11}$$

The set of algebraic equations is expressed in the form of an $M \times N$ matrix ($M = N$), and the unknown discrete charge density terms, σ_n are solved using the Gaussian elimination method:

$$[\sigma_n] = [G_{mn}]^{-1}[V_m]$$

$$\begin{bmatrix} \sigma_1 \\ \sigma_2 \\ \vdots \\ \sigma_N \end{bmatrix} = \begin{bmatrix} G_{11} & G_{12} & \cdots & G_{1N} \\ G_{21} & G_{22} & \cdots & G_{2N} \\ \vdots & \vdots & \ddots & \vdots \\ G_{M1} & G_{M2} & \cdots & G_{MN} \end{bmatrix}^{-1} \begin{bmatrix} V_1 \\ V_2 \\ \vdots \\ V_M \end{bmatrix} \tag{6.12}$$

The values of σ_n, which have been obtained from MoM, can be used to find other parameters, such as capacitance C.

For instance, the fringing field capacitance C at the aperture coaxial probe can be computed from the charge density σ_n as:

$$C = \frac{1}{V_o} \sum_{n=1}^{N} \int_0^{2\pi} \int_{\rho_n - \Delta\rho/2}^{\rho_n + \Delta\rho/2} \sigma_n \rho' d\rho' d\phi' \tag{6.13}$$

where the voltage $V_o = 1$.

6.2 Summary of MoM Computation Steps for Electrodynamics Cases

6.2.1 Electrodynamics Cases

The procedure for applying MoM to solve vector scattering problems basically involves four main steps. The first step is to derive the appropriate governing integral equation, which is the most difficult step. However, in the past hundred years, there have been several rigorous quasi-static integrals derived for the dipole, monopole, and open-ended coaxial probe.

In MoM solutions of a vector-scattering field associated with the dipole or monopole, both Pocklington and Hallén integral equations are widely used. However, the Sommerfeld and Nevels's admittance integral equations are used to solve the aperture scattering problem for coaxial probe.

6.2.2 Open-Ended Coaxial Probe: Governing Equations

With regard to Sommerfeld's integral equation (without TM_{0n} modes):

$$V\left\{\frac{\ln(b/a)}{\eta_c} + \int_0^{\infty} \frac{1}{\sqrt{k_c^2 - k^2}} \left[\int_a^b \int_a^b J_1(k_c\rho) J_1(k_c\rho')\rho' d\rho' d\rho\right] k_c dk_c\right\} = \frac{2A_o}{\eta_c} \ln\left(\frac{b}{a}\right) \tag{6.14}$$

With regard to Nevel's integral equation (without TM_{0n} modes):

$$V\left\{\frac{1}{\eta_c \ln(b/a)}\int_a^b \frac{1}{\rho'}d\rho' + \frac{j\omega\varepsilon_o\varepsilon_r\rho}{2\pi}\int_a^b\int_{-\pi}^{\pi}\frac{\exp(-jkr)}{r}\cos\phi'\rho' d\phi' d\rho'\right\} = \frac{2V_o}{\eta_c \ln(b/a)} \tag{6.15}$$

where the r in (6.15) is given as:

$$r = \sqrt{\rho^2 + \rho'^2 - 2\rho\rho'\cos\phi'}$$

and J_1 in (6.14) is the Bessel function of the first order. a is the outer radius of inner conductor (in meters), and b is the inner radius of outer conductor (in meters). The terms ε_o ($\approx 8.854187817 \times 10^{-12}$ F/m) and μ_o ($\approx 4\pi \times 10^{-7}$ H/m) are the permittivity and permeability of free space, respectively. η_c, k_c, and ε_c are the intrinsic impedance, the propagation constant of the coaxial line, and the dielectric constant of the material (Teflon) that fills the coaxial line, respectively. In addition, k and ε_r are the propagation constant and relative permittivity of the exterior medium, respectively.

6.2.3 Dipole: Governing Equations

With regard to Pocklington's integro-differential equation (reduced kernel type):

$$\left(\frac{\partial^2}{\partial z^2} + k^2\right)\int_{-h}^{h} I_z \frac{e^{-jk\sqrt{a^2+(z-z')^2}}}{\sqrt{a^2 + (z - z')^2}}dz' = -j4\pi\omega\varepsilon_o\varepsilon_r E_z \tag{6.16}$$

With regard to Hallén's equation (reduced kernel type):

$$\frac{1}{4\pi}\int_{-h}^{h} I_z \frac{e^{-jk\sqrt{a^2+(z-z')^2}}}{\sqrt{a^2 + (z - z')^2}}dz' = \xi_o\cos(kz) - \frac{jV_o}{2\eta}\sin\left(k|z|\right) \tag{6.17}$$

where I_z in (6.16) and (6.17) is the distribution current on the dipole with length $2h$ and radius a. In addition, E_z and V_o are the excitation electric field and voltage, respectively. ξ_o in (6.17) is the constant coefficient that requires the current to vanish at the end of the dipole. η is the intrinsic impedance while k and ε_r are the propagation constant and relative permittivity of the medium surrounding the dipole, respectively.

The quasi-static integral equation (impedance integral or admittance integral formulation) is discretized into an algebraic equation with the aid of Ohm's law ($IZ = V$ or $VY = I$) and Kirchhoff's law as:

$$\sum_{n=1}^{N} I_n Z_{mn} = V_m \tag{6.18a}$$

or

$$\sum_{n=1}^{N} V_n Y_{mn} = I_m \tag{6.18b}$$

where I_n and V_n are the unknown current and the unknown voltage, respectively. Z_{mn} and Y_{mn} are the discrete of impedance and admittance integral formulations, and I_m and V_m are the known discrete voltage and current, respectively.

The algebraic equation (6.18a) [or (6.18b)] for every observation point (m = 1, 2, … , M) are found. The set of algebraic equation is expressed in an $M \times N$ matrix form ($M = N$):

$$\left[I_n\right] = \left[Z_{mn}\right]^{-1}\left[V_m\right] \tag{6.19a}$$

or

$$\left[V_n\right] = \left[Y_{mn}\right]^{-1}\left[I_m\right] \tag{6.19b}$$

The unknown voltage V_n or the unknown current I_n can be solved using the Gaussian elimination method.

The values of I_n or V_n, which have been obtained from MoM, can be used to find other parameters, such as H field, E field, input impedance Z_{in}, or input admittance Y_{in}.

6.2.4 Open-Ended Coaxial Probe: Final Interpretation

The total potential difference V between the inner and outer conductors is:

$$V = \sum_{n=1}^{N} V_n \tag{6.20}$$

The radial electric field E_ρ on the aperture probe is:

$$E_\rho = \frac{V}{\rho} \tag{6.21}$$

The total capacitance C on the aperture probe is:

$$C = \frac{\varepsilon_o \varepsilon_r}{V_o} \int_a^b \int_0^{2\pi} E_\rho \rho \, d\phi \, d\rho \tag{6.22}$$

The azimuth magnetic field H_ϕ on the aperture probe is:

$$H_\phi = \frac{j\omega\varepsilon_o\varepsilon_r}{\pi}\int_a^b\int_0^\pi E_\rho\frac{\exp(-jkr)}{r}\cos\phi'\rho' d\phi' d\rho' \tag{6.23}$$

The normalized input admittance $\tilde{Y}_{\text{in}}$ on the aperture probe is:

$$\tilde{Y}_{\text{in}} = \frac{2V_o}{\int_a^b E_\rho d\rho'} - 1 \tag{6.24}$$

The values of E_ρ, C, H_ϕ, and $\tilde{Y}_{\text{in}}$ can be determined by replacing (6.20) into (6.21), (6.22), (6.23), and (6.24), respectively.

6.2.5 Dipole: Final Interpretation

The total current I_z along the dipole is:

$$I_z = \sum_{n=1}^{N} I_n \tag{6.25}$$

The elevation electric field E_θ (far field) from the dipole is:

$$\begin{aligned} E_\theta &\approx jk\eta\sin\theta\int_{-h}^{h} I_z\frac{e^{-jkr}}{4\pi r}dz' \\ &\approx jk\eta\sin\theta\frac{e^{-jk\rho}}{4\pi\rho}\int_{-h}^{h} I_z e^{jkz'\cos\theta}dz' \end{aligned} \tag{6.26}$$

The azimuth magnetic field H_ϕ (far field) from the dipole is:

$$H_\phi = \frac{E_\theta}{\eta} \tag{6.27}$$

The values of E_θ and H_ϕ can be determined by replacing (6.25) into (6.26) and (6.27), respectively.

6.3 Mathematical Background of MoM

6.3.1 Point-Matching Method

In order to understand the foundation of MoM thoroughly, an example of the application of MoM of Pocklington's equation with the delta-gap excitation model $V_o/\delta z$ is carried out. As mentioned in Chapter 4, Pocklington's equation is used to model the scattering wave along the dipole with a length of $2h$ and a diameter of $2a$ as shown in Figure 6.1. The current distribution I_z in (6.16) is treated as a filament current line on the z-axis, and the scattering field is computed on the dipole surface.

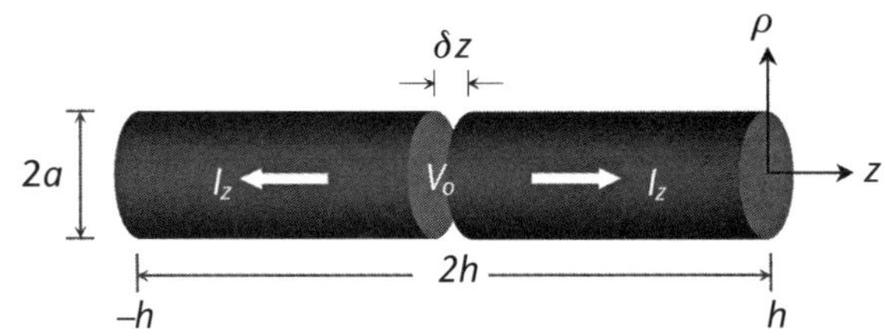

Figure 6.1 Center feed dipole with length 2*h* and radius *a*.

In the MoM solution, the current distribution I_z along the center of the dipole is assumed to be an unknown constant value that needs to be determined. First, the surface of the dipole is divided into N segments and the analysis of the scattering fields on each segment is carried out. Indirectly, the total current distribution I_z along the dipole is the sum of the discrete current at each segment.

Accordingly, I_z is discretized in terms of a finite series of known functions $F_n(z)$ (basis functions) with unknown constant current amplitudes I_n at each segment as:

$$I_z = \sum_{n=1}^{N} I_n F_n(z), \qquad n = 1,2,3,\ldots,N \tag{6.28}$$

The subscript n represents the number of segments along the dipole. The purpose of multiplying the amplitudes of I_n by the basis functions $F_n(z)$ is to improve the shape of the current so that it becomes similar to the actual current distribution, and this is further reviewed later.

The dipole in Figure 6.2 is divided into nine segments and the excited voltage V_o at the center of the dipole is assumed to be 1V. By incorporating (6.28) into (6.16), the discretization form for the left side of (6.16) can be written as:

$$\sum_{n=1}^{9} I_n \int_{z_n-\Delta z/2}^{z_n+\Delta z/2} F_n(z') \left(\frac{\partial^2}{\partial z^2} + k^2 \right) \frac{e^{-jk\sqrt{a^2+(z_m-z')^2}}}{\sqrt{a^2+(z_m-z')^2}} dz' \tag{6.29}$$

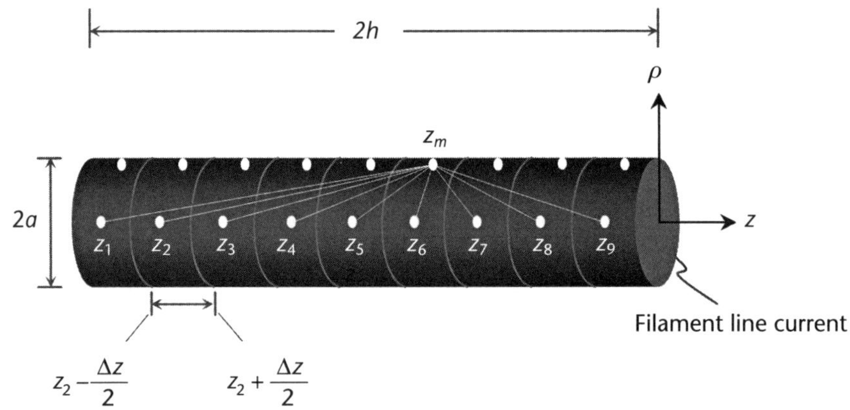

Figure 6.2 Dipole segmentation and its equivalent filament line current.

as the right side of (6.14) is reduced as:

$$\frac{-j4\pi\omega\varepsilon}{\delta z} \tag{6.30}$$

The integrations with respect to z' in (6.29) can be solved by Simpson's rule or the Gauss-Legendre quadrature rule. Lastly, the values of I_n at each segment area ($z_n - \Delta z/2 \leq z \leq z_n + \Delta z/2$) along the dipole are attempted to be optimized in order to match (6.29) with (6.30) at the single discrete point of z_m (the observation point), as shown in Figure 6.2. Thus, this kind of MoM is also known as the point-matching (collocation) method.

However, the matching between (6.29) and (6.30) for the mth point located at z_m can only produce a single linear equation, which is not sufficient to determine the unknown current I_n at each segment. To overcome this problem, the matching of (6.29) and (6.30) at every observation point of z_m (m = 1, 2, ..., 9) is matching to be computed. For instance, the interaction between each segment current I_n with observation point z_1 is shown in Figure 6.3.

Let the integral part of (6.29) be represented by Z_{mn} as:

$$Z_{mn} = \int_{z_n-\Delta z/2}^{z_n+\Delta z/2} F_n(z')\left(\frac{\partial^2}{\partial z^2} + k^2\right)\frac{e^{-jk\sqrt{a^2+(z_m-z')^2}}}{\sqrt{a^2+(z_m-z')^2}}dz' \tag{6.31}$$

Thus, the linear equation that based on the first observation point z_1 on the surface of the dipole can be written as:

$$I_1Z_{11} + I_2Z_{12} + I_3Z_{13} + \cdots + I_9Z_{19} = \frac{-j4\pi\omega\varepsilon_o\varepsilon_r}{\delta z} \tag{6.32a}$$

as the coupling for all segments of the current I_n, which is satisfied at the second observation point z_2, is shown in Figure 6.4.

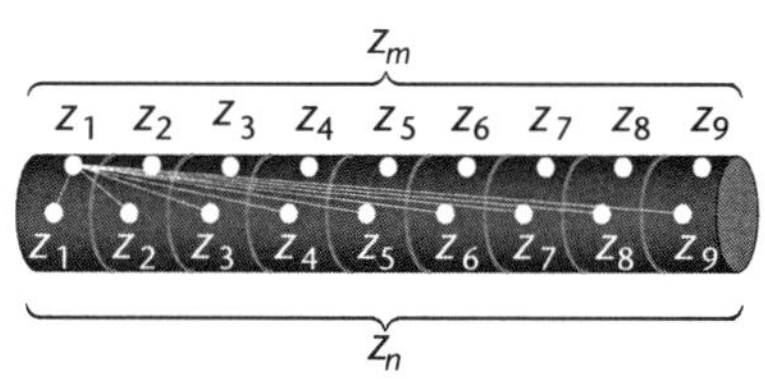

Figure 6.3 The matching of the scattering field at the first point z_1 on the surface of the dipole.

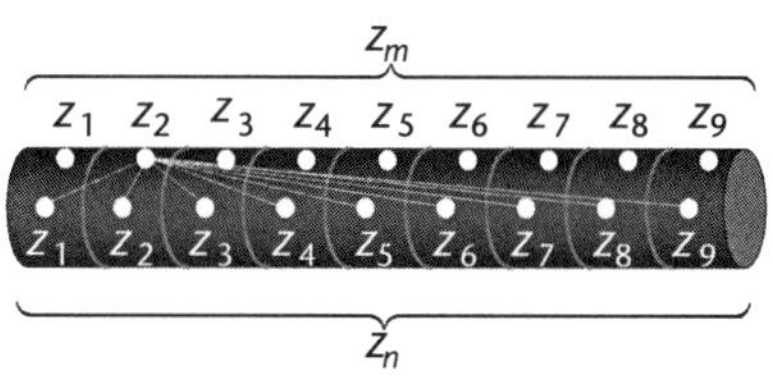

Figure 6.4 The matching of the scattering field at the second point z_2 on the surface of the dipole.

Based on Figure 6.4, the linear equation is given as:

$$I_1Z_{21} + I_2Z_{22} + I_3Z_{23} + \cdots + I_9Z_{29} = \frac{-j4\pi\omega\varepsilon_o\varepsilon_r}{\delta z} \tag{6.32b}$$

Subsequently, the linear equations for other observation points ($z_3, z_4, z_5, z_6, \ldots$) are computed. Figure 6.5 shows the last matching point at z_9.

The last linear equation to be satisfied at z_9 is expressed as:

$$I_1Z_{91} + I_2Z_{92} + I_3Z_{93} + \cdots + I_9Z_{99} = \frac{-j4\pi\omega\varepsilon_o\varepsilon_r}{\delta z} \tag{6.32c}$$

Finally, nine sets of linear equations are produced and corrected as:

$$\begin{aligned}
I_1Z_{11} + I_2Z_{12} + I_3Z_{13} + \cdots + I_9Z_{19} &= \frac{-j4\pi\omega\varepsilon_o\varepsilon_r}{\delta z} \\
I_1Z_{21} + I_2Z_{22} + I_3Z_{23} + \cdots + I_9Z_{29} &= \frac{-j4\pi\omega\varepsilon_o\varepsilon_r}{\delta z} \\
\vdots \qquad \vdots \qquad \vdots \qquad\quad \vdots \quad &\qquad\quad \vdots \\
I_1Z_{91} + I_2Z_{92} + I_3Z_{93} + \cdots + I_9Z_{99} &= \frac{-j4\pi\omega\varepsilon_o\varepsilon_r}{\delta z}
\end{aligned} \tag{6.33}$$

Equation (6.33) can be rewritten in a 9×9 matrix form as:

$$\begin{bmatrix} Z_{11} & Z_{12} & Z_{13} & \cdots & Z_{19} \\ Z_{21} & Z_{22} & Z_{23} & \cdots & Z_{29} \\ \vdots & \vdots & \vdots & \ddots & \vdots \\ Z_{91} & Z_{92} & Z_{93} & \cdots & Z_{99} \end{bmatrix} \begin{bmatrix} I_1 \\ I_2 \\ \vdots \\ I_9 \end{bmatrix} = \begin{bmatrix} \frac{-j4\pi\omega\varepsilon_o\varepsilon_r}{\delta z} \\ \frac{-j4\pi\omega\varepsilon_o\varepsilon_r}{\delta z} \\ \vdots \\ \frac{-j4\pi\omega\varepsilon_o\varepsilon_r}{\delta z} \end{bmatrix} \tag{6.34}$$

The unknown vector currents $[I_n]$ on the dipole are determined by inverting the admittance matrix $[Z_{mn}]$ and multiplying with by $(-j4\pi\omega\varepsilon_o\varepsilon_r/\delta z)$.

$$\left[I_n\right] = \left[Z_{mn}\right]^{-1}\left[\frac{-j4\pi\omega\varepsilon_o\varepsilon_r}{\delta z}\right] \tag{6.35}$$

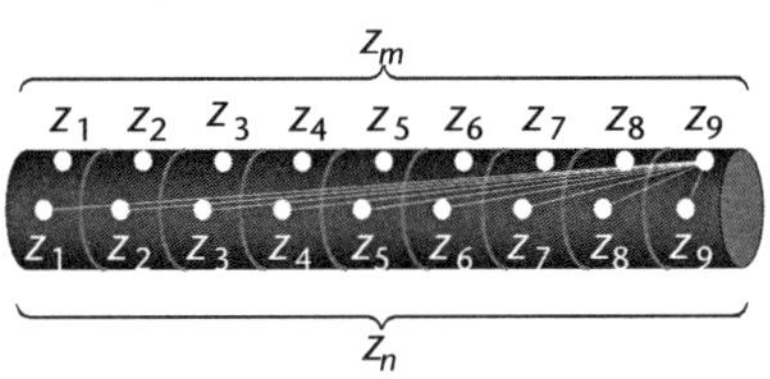

Figure 6.5 The matching of the scattering field at the last point z_9 on the surface of the dipole.

Finally, (6.35) can be solved by the Gaussian elimination method. In the mid-1970s, there were various forms of the basis functions, $F_n(z)$, which were intensively studied to solve the problem in the current distribution I_z on the dipole (or monopole), to reduce the derivation of the solution and to improve the convergence speed of the solution [6–11]. There are three types of basis functions $F_n(z)$ that are commonly used in MoM: the piecewise uniform functions, the piecewise linear functions, and the piecewise sinusoidal functions.

As well as the three kinds of basis functions, some higher-order functions, such as Fourier, Maclaurin, Chebyshev, Hermite, and Legendre polynomial functions, are also used to represent the series expansion for the current distribution on the dipole by Richmond [12]. Those higher orders of the basis functions can achieve a specified accuracy in MoM computation with fewer numbers N of current unknowns as compared to the linear basis functions. However, the usage of higher-order functions will increase the complexity in MoM formulation. The intensive discussion on the higher order of the basis functions can be found in Richmond [12], Mittra [13], and Moore and Pizer [14]. Besides the linear and higher-order functions, some interpolated trigonometric basis functions have been used by MoM-based public software or commercial software, such as the Numerical Electromagnetics Code (NEC). Normally, the basis functions involving three different coefficient terms (for constant, sine, and cosine) as follows. The total current, I_n on segment n is defined as [15]:

$$I_n = A_n + B_n \sin k\left(z - z_n\right) + C_n \cos k\left(z - z_n\right)$$

where z is the distance within a segment $(z_n - \Delta z/2) \le z \le (z_n + \Delta z/2)$ along the dipole and z_n is the distance point at the center of the segment.

A_n, B_n, and C_n are determined by the requirement in which the current and its derivative have to be continuous along the dipole, except at the both ends of the dipole and the fed point. This kind of trigonometric basis function is developed based on the analytical current distribution functions, as described in Section 4.1.1.

6.4 Summary of Three Kinds of Basis Functions $F_n(z)$

6.4.1 Piecewise Uniform Functions (Rectangular Pulse Functions)

The piecewise uniform functions $F_n(z)$ as shown in Figure 6.6 are defined as:

$$F_n(z) = \begin{cases} 1 & \text{for} \left(z_n - \dfrac{\Delta z}{2}\right) \le z \le \left(z_n + \dfrac{\Delta z}{2}\right) \\ 0 & \text{other} \end{cases} \tag{6.36}$$

The number of pulses segments N must be odd numbers. For instance, the $2h$ length of the dipole has been divided into $N = 9$ segments as shown in Figure 6.6. The geometry of the functions $F_n(z)$ for each segment is an orthogonal rectangle pulse and they do not overlap with each other. Normally, all segments will have equal lengths of Δz.

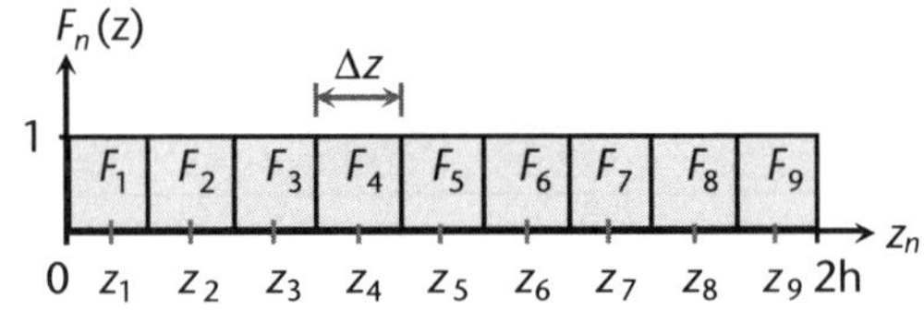

Figure 6.6 Piecewise uniform function, $F_n(z)$ versus segment length, Z_n.

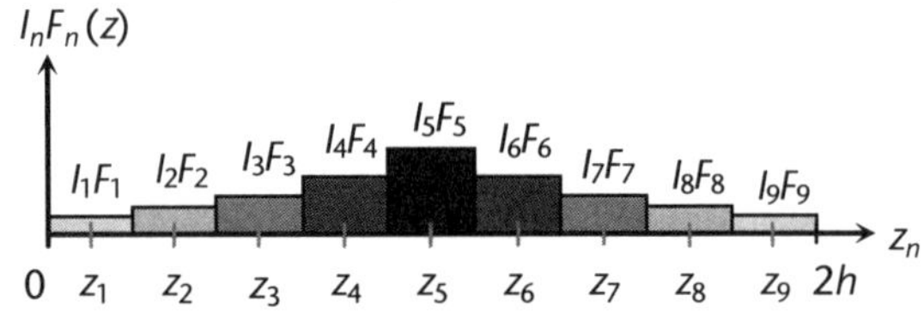

Figure 6.7 Current amplitudes in rectangle pulse form.

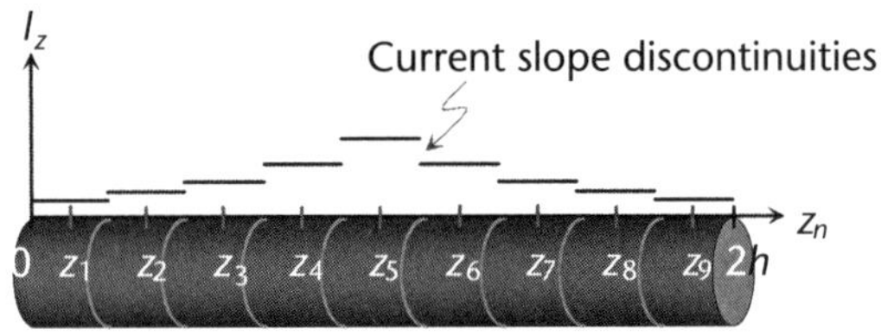

Figure 6.8 Discrete current distribution along the dipole.

The different amplitudes for each rectangle pulse, as shown in Figure 6.7, are based on the constant values of I_n, which have been determined by MoM.

The sum of the current from all segments $I_z = \Sigma I_n F_n$ is shown in Figure 6.8. However, this kind of basis function is simple, and the slope discontinuities between the pulses can cause discontinuities in the current distribution solution I_z.

6.4.2 Piecewise Linear Functions (Linear Triangular Functions)

The piecewise linear functions $F_n(z)$ as shown in Figure 6.9 are defined as:

$$F_n(z) = \begin{cases} \dfrac{z - z_{n-1}}{\Delta z} & \text{for } z_{n-1} \le z \le z_n \\ \dfrac{z_{n+1} - z}{\Delta z} & \text{for } z_n \le z \le z_{n+1} \\ 0 & \text{other} \end{cases} \tag{6.37}$$

The number of pulse segments N must be an odd number (for example, $N = 9$ points, $N - 1 = 8$ segments, $N - 2 = 7$ piecewise linear functions). The geometry of the functions $F_n(z)$ is a triangular pulse and the functions overlap with each other by half a triangle as shown in Figure 6.9. Normally, all segments will have equal lengths of Δz.

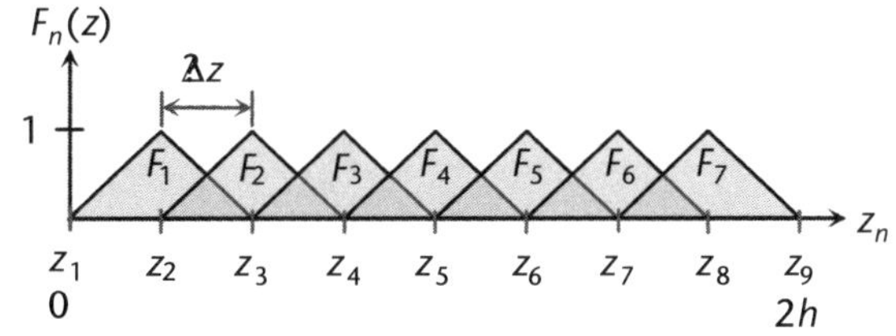

Figure 6.9 Piecewise linear function, $F_n(z)$, versus segment length, Z_n.

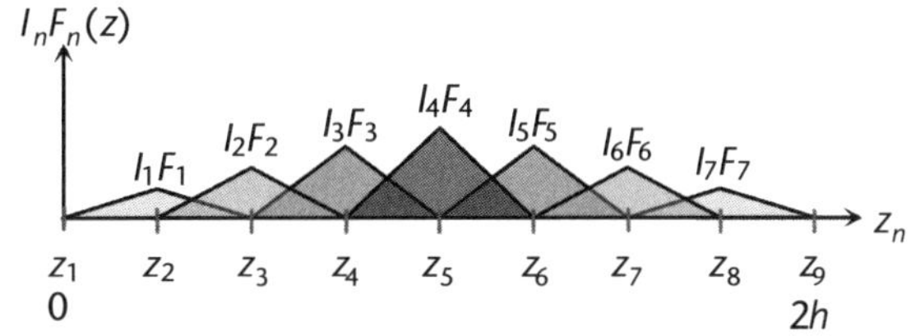

Figure 6.10 Current amplitudes in triangle pulse form.

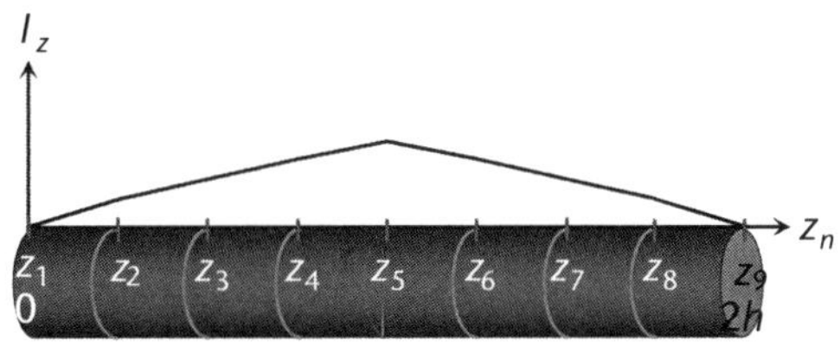

Figure 6.11 Current distribution along the dipole.

The maximum amplitudes of each triangle pulse, as shown in Figure 6.10, are determined by the constant values of I_n, which have been determined by MoM.

The sum of the current from all segments $I_z = \Sigma I_n F_n$ is shown in Figure 6.11. The piecewise linear functions allow the current distribution I_z to be continuous at each boundary segment. In addition, it enforces the current distribution I_z to be zero at the end of the dipole at $z_n = 0$ and $z_n = 2h$.

6.4.3 Piecewise Sinusoidal Functions

The piecewise sinusoidal functions $F_n(z)$, as shown in Figure 6.12, are defined as:

$$F_n(z) = \begin{cases} \dfrac{\sin\left[k\left(z - z_{n-1}\right)\right]}{\sin(k\Delta z)} & \text{for } z_{n-1} \leq z \leq z_n \\ \dfrac{\sin\left[k\left(z_{n+1} - z\right)\right]}{\sin(k\Delta z)} & \text{for } z_n \leq z \leq z_{n+1} \\ 0 & \text{other} \end{cases} \tag{6.38}$$

The piecewise sinusoidal functions are similar to piecewise linear functions. This kind of basis functions was first proposed by Richmond [10].

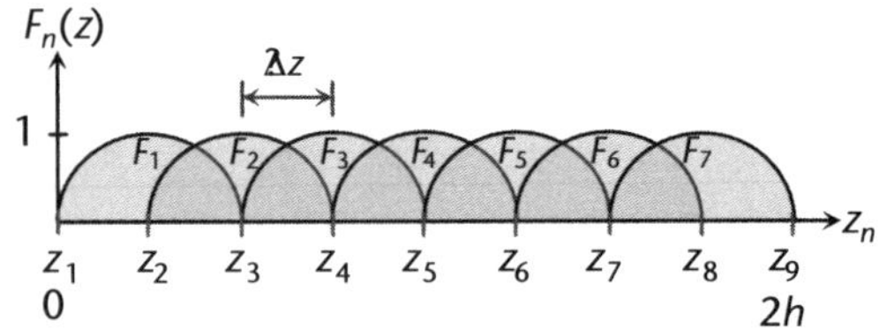

Figure 6.12 Piecewise sinusoidal function, $F_n(z)$, versus segment length, Z_n.

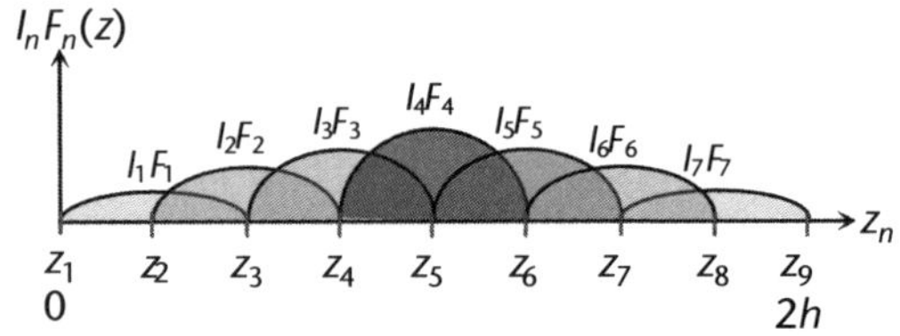

Figure 6.13 Current amplitudes in sinusoidal pulse form.

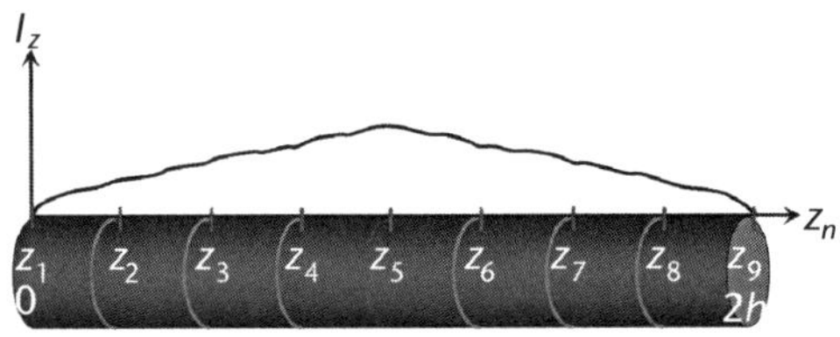

Figure 6.14 Current distribution along the dipole.

The maximum amplitudes of each sinusoidal pulse, as shown in Figure 6.13, are determined by the constant values of I_n, which have been determined by the MoM.

The sum of the current from all segments $I_z = \Sigma I_n F_n$ is shown in Figure 6.14. The piecewise sinusoidal functions allow the current distribution I_z to be continuous at each boundary segment. In addition, this basis function enforces the current distribution I_z to be zero at the end of the dipole at $z_n = 0$ and $z_n = 2h$.

6.4.4 Galerkin's Method in Basis Functions

As discussed earlier, by using the point-matching method, the matching between the right and left sides of (6.16) is entirely concentrated at each observation point z_m.

The point-matching method is

$$\sum_{n=1}^{9} I_n \int_{z_n-\Delta z/2}^{z_n+\Delta z/2} F_n(z') K_{mn} dz' \quad \underset{\text{At point } z_m}{\overset{\text{Matching}}{\Leftrightarrow}} \quad \frac{-j4\pi\omega\varepsilon_o\varepsilon_r}{\delta z}$$

where

$$K_{mn} = \left(\frac{\partial^2}{\partial z^2} + k^2\right) \frac{e^{-jk\sqrt{a^2+(z_m-z')^2}}}{\sqrt{a^2+(z_m-z')^2}}$$

Galerkin's method is

$$\sum_{n=1}^{9} I_n \int_{z_m-\Delta z/2}^{z_m+\Delta z/2} F_m(z) \int_{z_n-\Delta z/2}^{z_n+\Delta z/2} F_n(z') K_n dz' dz \quad \underset{\text{Over segment}}{\overset{\text{Matching}}{\Leftrightarrow}} \quad \frac{-j4\pi\omega\varepsilon_o\varepsilon_r}{\delta z} \int_{z_m-\Delta z/2}^{z_m+\Delta z/2} F_m(z) dz$$

where

$$K_n = \left(\frac{\partial^2}{\partial z^2} + k^2 \right) \frac{e^{-jk\sqrt{a^2+(z-z')^2}}}{\sqrt{a^2 + (z-z')^2}}$$

Clearly, they will have matching discontinuities at a place between two observation points. The derivation of the solution due to the above-mentioned issue can be improved by integrating both sides of (6.16) with a weighting function $F_m(z)$ over the observation segment ($z_m - \Delta z/2 \leq z \leq z_m + \Delta z/2$). Hence, the left and right sides of (6.16) are matched together in an average sense over the entire observation segment ($z_m - \Delta z/2 \leq z \leq z_m + \Delta z/2$) instead of at a discrete point z_m at the center segment. This MoM is known as Galekin's method when the basis functions $F_n(z')$ and weighting functions $F_m(z)$ are the same. The detailed mathematical calculation for the example is further described in Section 6.5.

6.5 Example of Galerkin's MoM Computation

6.5.1 Galerkin's Method with Piecewise Sinusoidal Functions

For instance, the basis functions $F_n(z)$ and weighting functions $F_m(z)$ use the same piecewise sinusoidal functions:

$$F_n(z') = \begin{cases} \dfrac{\sin\left[k\left(z' - z_{n-1}\right)\right]}{\sin(k\Delta z)} & \text{for } z_{n-1} \leq z \leq z_n \\ \dfrac{\sin\left[k\left(z_{n+1} - z'\right)\right]}{\sin(k\Delta z)} & \text{for } z_n \leq z \leq z_{n+1} \\ 0 & \text{other} \end{cases} \tag{6.39}$$

$$F_m(z) = \begin{cases} \dfrac{\sin\left[k\left(z - z_{m-1}\right)\right]}{\sin(k\Delta z)} & \text{for } z_{m-1} \leq z \leq z_m \\ \dfrac{\sin\left[k\left(z_{m+1} - z'\right)\right]}{\sin(k\Delta z)} & \text{for } z_m \leq z \leq z_{m+1} \\ 0 & \text{other} \end{cases} \tag{6.40}$$

First, let the integral with respect for z' for the left side of Pocklington's equation be represented by K_{mn} as:

$$K_{mn} = \int_{z_{n-1}}^{z_{n+1}} F_n(z')\left(\frac{\partial^2}{\partial z^2} + k^2\right)\frac{e^{-jk\sqrt{a^2+(z-z')^2}}}{\sqrt{a^2+(z-z')^2}}\,dz' \tag{6.41}$$

Substituting (6.39) into (6.41) yields

$$\begin{aligned} K_{mn} = &\int_{z_{n-1}}^{z_n} \frac{\sin\left[k\left(z' - z_{n-1}\right)\right]}{\sin(k\Delta z)}\left(\frac{\partial^2}{\partial z^2} + k^2\right)\frac{e^{-jk\sqrt{a^2+(z-z')^2}}}{\sqrt{a^2+(z-z')^2}}\,dz' \\ &+\int_{z_n}^{z_{n+1}} \frac{\sin\left[k\left(z_{n+1} - z'\right)\right]}{\sin(k\Delta z)}\left(\frac{\partial^2}{\partial z^2} + k^2\right)\frac{e^{-jk\sqrt{a^2+(z-z')^2}}}{\sqrt{a^2+(z-z')^2}}\,dz' \end{aligned} \tag{6.42}$$

Due to one sinusoidal pulse spanning across two divided segments, which is from z_{n-1} to z_n and another is from z_n to z_{n+1}; thus, (6.42) becomes the sum of two paths of integration. Solving the integral and the differential parts of (6.42) yields

$$\begin{aligned} K_{mn} &= \frac{1}{\sin(k\Delta z)}\left\{\begin{aligned}&\frac{e^{-jk\sqrt{a^2+(z-z_{n-1})^2}}}{\sqrt{a^2+(z-z_{n-1})^2}} - \frac{e^{-jk\sqrt{a^2+(z-z_n)^2}}}{\sqrt{a^2+(z-z_n)^2}}\cos\left[k\left(z_n - z_{n-1}\right)\right] \\ &+\frac{e^{-jk\sqrt{a^2+(z-z_{n+1})^2}}}{\sqrt{a^2+(z-z_{n+1})^2}} - \frac{e^{-jk\sqrt{a^2+(z-z_n)^2}}}{\sqrt{a^2+(z-z_n)^2}}\cos\left[k\left(z_{n+1} - z_n\right)\right]\end{aligned}\right\} \\ &= \frac{1}{\sin(k\Delta z)}\left[\begin{aligned}&\frac{e^{-jk\sqrt{a^2+(z-z_{n-1})^2}}}{\sqrt{a^2+(z-z_{n-1})^2}} + \frac{e^{-jk\sqrt{a^2+(z-z_{n+1})^2}}}{\sqrt{a^2+(z-z_{n+1})^2}} \\ &-2\frac{e^{-jk\sqrt{a^2+(z-z_n)^2}}}{\sqrt{a^2+(z-z_n)^2}}\cos(k\Delta z)\end{aligned}\right] \end{aligned} \tag{6.43}$$

where $(z_n - z_{n-1}) = (z_{n+1} - z_n) = \Delta z$. Now (6.43) is integrated by the weighting functions $F_m(z)$ over z_{m-1} to z_{m+1} as:

$$Z_{mn} = \int_{z_{m-1}}^{z_m} \frac{\sin\left[k\left(z - z_{m-1}\right)\right]}{\sin(k\Delta z)}K_{mn}dz + \int_{z_m}^{z_{m+1}} \frac{\sin\left[k\left(z_{m+1} - z\right)\right]}{\sin(k\Delta z)}K_{mn}dz \tag{6.44}$$

Equation (6.44) is similar to the induced EMF expression [16]. In fact, before the existence of the MoM, the current distribution I_z in the induced EMF formulation was assumed to be sinusoidal (as discussed in Chapter 4).

$$V_m = -j4\pi\omega\varepsilon\int_{z_m-\Delta z/2}^{z_m+\Delta z/2} F_m(z)E_z\,dz \tag{6.45}$$

Normally, E_z in (6.45) is assumed to be a delta-gap source as:

$$E_z = V_o/\Delta z \tag{6.46}$$

where V_o = 1V. The detailed description of this Galerkin's MoM procedure is available later.

6.6 Open-Ended Coaxial Probe

6.6.1 Nevels's Equation, Piecewise Uniform Function, and Point-Matching Method

For the coaxial probe, the vector voltage, V, on the aperture probe surface is assumed to be an unknown parameter that needs to be determined. In the MoM solution, the coaxial gap $a \le \rho \le b$ between inner and outer conductors, as shown in Figure 6.15, is divided into N segments (pulses) and each segment length, $\Delta\rho$, is given as:

$$\Delta\rho = \frac{(b-a)}{N} \tag{6.47}$$

From (6.47), the observation match point ρ_m is being enforced at segment m along the radial ρ-axis and is calculated as:

$$\rho_m = a + (m - 0.5)\Delta\rho \qquad m = 1,2,3,\ldots,M \tag{6.48}$$

Given that, the match point ρ_m is located in the middle of each segment. The voltage V on the surface of the aperture probe between the inner and outer conductors is discretized in terms of a finite series of known functions $F_n(\rho')$ with unknown amplitudes V_n at each segment as:

$$\begin{aligned} V &= \sum_{n=1}^{N} V_n F_n(\rho') \\ &= V_1 F_1(\rho') + V_2 F_2(\rho') + \ldots + V_N F_N(\rho') \end{aligned} \tag{6.49}$$

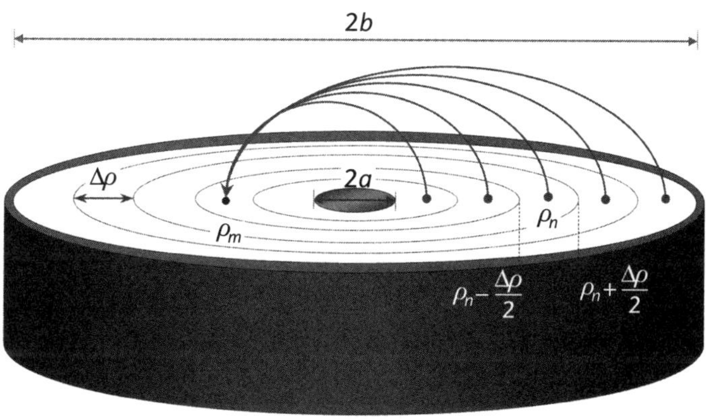

Figure 6.15 The discrete point matching ρ_m from each observation point source.

In this work, the functions $F_n(\rho')$ are chosen to be the piecewise uniform functions. Thus, (6.49) at a certain match point, ρ_m, as shown in Figure 6.15, can be reduced as:

$$V = V_1 + V_2 + \ldots + V_N \qquad (\text{at point, } \rho_m) \tag{6.50}$$

Later, the algebraic equation (6.50) for every observation point (m = 1, 2, ..., M) is found. Finally, the set of algebraic equations is written in the $M \times N$ matrix form as:

$$\left[Y_{mn}\right]\left[V_n\right] = \left[I_m\right] \tag{6.51}$$

where Y_{mn} and I_m are the discretization admittance and the discretization vector current at the surface of the aperture probe z = 0, respectively:

$$Y_{mn} = \frac{\ln\left(\dfrac{\rho_n + \Delta\rho/2}{\rho_n - \Delta\rho/2}\right)}{\eta_c \ln(b/a)} + \frac{j\omega\varepsilon_o\varepsilon_r\rho_m}{2\pi}\int_{\rho_n-\Delta\rho/2}^{\rho_n+\Delta\rho/2}\int_{-\pi}^{\pi}\cos\phi'\frac{\exp\left(-jkr_m\right)}{r_m}d\phi' d\rho' \tag{6.52a}$$

$$r_m = \sqrt{\rho_m^2 + \rho'^2 - 2\rho_m\rho'\cos\phi'} \tag{6.52b}$$

and

$$I_m = \frac{2V_o}{\eta_c \ln(b/a)} \tag{6.53}$$

The integrations with respect to ρ' over ($\rho_n - \Delta\rho/2$) to ($\rho_n + \Delta\rho/2$) in (6.52a) are solved by Simpson's rule or the Gauss-Legendre quadrature rule. The unknown vector voltages [V_n] on the aperture probe are determined by inverting the admittance matrix [Y_{mn}] and multiplying with [I_m].

$$\left[V_n\right] = \left[Y_{mn}\right]^{-1}\left[I_m\right] \tag{6.54a}$$

$$\begin{bmatrix} V_1 \\ V_2 \\ \vdots \\ V_N \end{bmatrix} = \begin{bmatrix} Y_{11} & Y_{12} & \cdots & Y_{1N} \\ Y_{21} & Y_{22} & \cdots & Y_{2N} \\ \vdots & \vdots & \ddots & \vdots \\ Y_{M1} & Y_{M2} & \cdots & Y_{MN} \end{bmatrix}^{-1} \begin{bmatrix} I_1 \\ I_2 \\ \vdots \\ I_M \end{bmatrix} \tag{6.54b}$$

Equation (6.54b) is solved using the Gaussian elimination method. For instance, the voltage V for an open-ended coaxial probe with an inner conductor radius, a = 0.65 mm, and an outer conductor radius, b = 2.05 mm, is simulated using N = 10 match points with the concentric circular ring on the aperture probe as depicted in Figure 6.16.

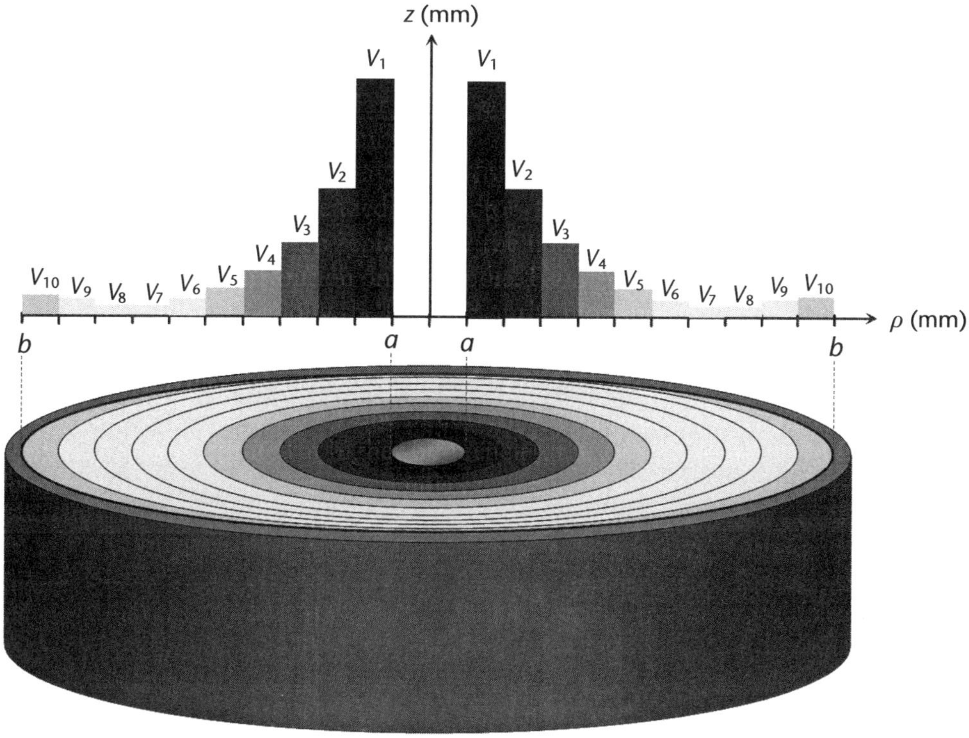

Figure 6.16 Vector voltage V_n for each grid position on the surface of the aperture probe.

In this work, the vector voltage V_n on the surface of the aperture coaxial probe is simulated using $N = 10$ match points along the radius ρ as depicted in Figure 6.17. The calculation assumes that the section of the coaxial line $a < \rho < b$ is filled with Teflon ($\varepsilon_c = 2.06$) and that the value of the ratio of b/a equals 3.298 with

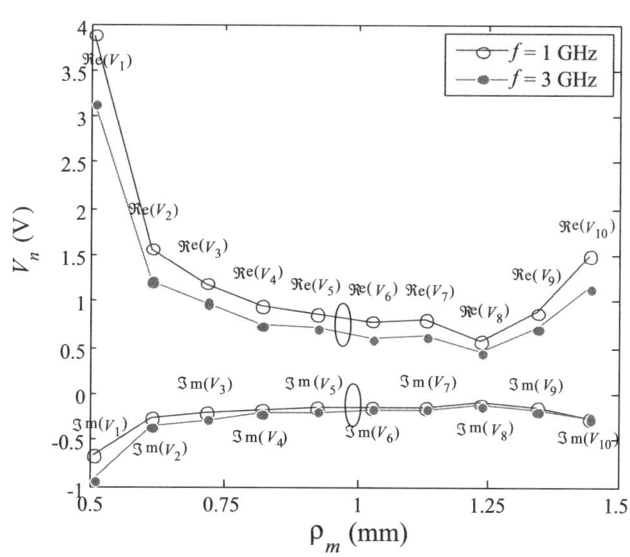

Figure 6.17 The vector voltage V_n at each matching point ρ_m axis on surface aperture ($z = z' = 0$).

a = 0.4545 mm. In addition, the external dissipative medium in the exterior region (z' > 0) is assumed to be methanol liquid, and its dispersive properties can be obtained from Cole-Cole model with $\varepsilon_s = 33.7$, $\varepsilon_\infty = 4.35$, $\tau = 49.64$ ps, and $\alpha = 0.0426$.

6.6.2 Radial Electric Field, E_ρ, Azimuth Magnetic Field, H_ϕ, and Input Admittance, Y_{in}, Interpretations

The radial electric field $E_{\rho n}$ at n match point positions on the surface of the aperture can be found by substituting the set of values of $[V_n]$ into:

$$E_{\rho n} = \begin{cases} \dfrac{V_n}{\rho} & \text{for } \left(\rho_n - \dfrac{\Delta\rho}{2}\right) < \rho < \left(\rho_n + \dfrac{\Delta\rho}{2}\right) \\ 0 & \text{other} \end{cases} \tag{6.55}$$

Once the electric field $E_{\rho n}$ at each match point has been determined, (6.23) can be reexamined by replacing (6.55) into (6.23) as:

$$H_\phi^+ = \frac{j\omega\varepsilon_o\varepsilon_r}{\pi}\sum_{n=1}^{N} V_n \int_{\rho_n-\Delta\rho/2}^{\rho_n+\Delta\rho/2}\int_0^\pi \frac{\exp(-jkr)}{r}\cos\phi' d\phi' d\rho' \tag{6.56}$$

Also, from the total value of the voltage ΣV_n, the normalized input admittance $\tilde{Y}_{in}$ of the entire surface aperture can be calculated by discretizing the integral electric field term in (6.24), yielding:

$$\tilde{Y}_{in} = \frac{2V_o}{\sum_{n=1}^{N} V_n\left(\int_{\rho_n-\Delta\rho/2}^{\rho_n+\Delta\rho/2}\frac{1}{\rho'}d\rho'\right)} - 1 = \frac{2V_o}{\sum_{n=1}^{N} V_n \ln\left(\dfrac{\rho_n+\Delta\rho/2}{\rho_n-\Delta\rho/2}\right)} - 1 \tag{6.57}$$

From Figure 6.18, the electric field decays exponentially as expected with the radial distance $E_\rho \propto 1/\rho$. The comparison between analytical and numerical results for the magnetic field H_ϕ at 3 GHz is shown in Figure 6.19. The absolute values of H_ϕ are normalized with its own maximum values $H_{\phi\max}$ to facilitate the comparison. The calculation of H_ϕ using (6.56) is performed by a 12 × 12 -order Gaussian double integral method. Figure 6.19 shows the limitation of analytical calculation using at the aperture probe $z = z' = 0$ and poor agreement with the numerical calculations [(6.56) and COMSOL simulation].

The normalized input admittance, $\tilde{Y}$, is a complex number that can be separated into normalized values of conductance $G(0)/Y_o$ and susceptance $B(0)/Y_o$ as $\tilde{Y} = G(0)/Y_o + jB(0)/Y_o$. The comparison of calculated $G(0)/Y_o$ and $B(0)/Y_o$ using models (2.11), (2.19), (2.20), and (6.57) and the COMSOL simulation and the simulation results for methanol are shown in Figure 6.20. The calculation for admittance of (2.11) is conducted by a 5 × 5 × 7-order Gaussian triple integral method, while (2.19) and (2.20) are solved by the series expansion method with nine terms in the series. Figure 6.20 clearly shows good agreement for the $G(0)/Y_o$ and $B(0)/Y_o$ results, which are obtained from (6.57) and the COMSOL simulation. Ignoring the TM_{0n} modes

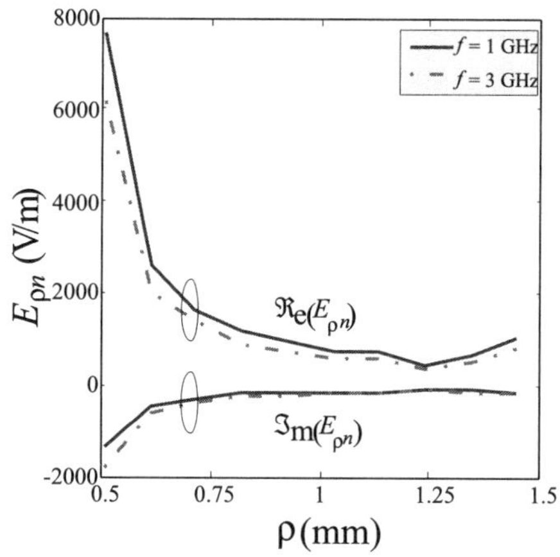

Figure 6.18 The vector electric field, $E_{\rho n}$ along the radius, ρ axis on the surface of the aperture ($z = z' = 0$).

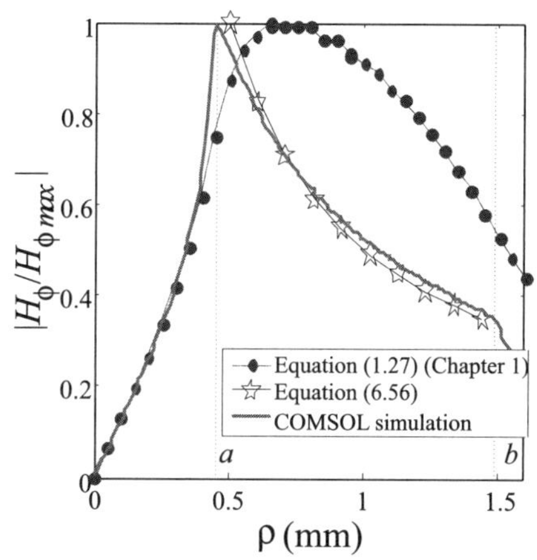

Figure 6.19 The normalized absolute magnetic field versus the radius ρ axis on the surface of the aperture ($z' = 0$).

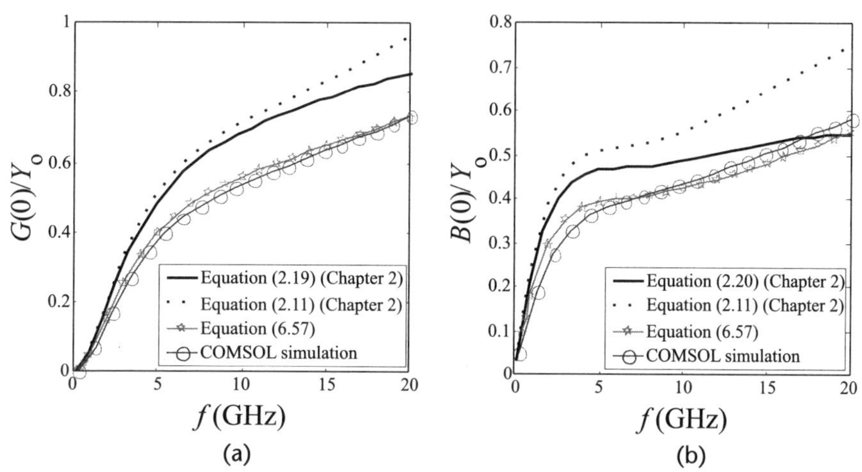

Figure 6.20 The calculated (a) normalized conductance, $G(0)/Y_o$ and (b) normalized susceptance $B(0)/Y_o$ for methanol at 25°C by considering size of probe with $b/a = 3.298$ and $a = 0.4545$ mm.

in (2.11), (2.19), and (2.20) causes low accuracy in the analytical calculation result; however, it leads to a relatively simple mathematical solution.

6.7 Center-Fed Dipole

6.7.1 Pocklington's Equation (Richmond's Expression), Piecewise Uniform Function, Magnetic Frill Sources, and Point-Matching Method

Figure 6.21 shows the cylindrical dipole extending from $z = -h$ to $z = h$ along the z-axis with the radius a and a current I_z as the filamentary line source located along the dipole center axis. For instance, matching discrete observation points at the point z_m are shown on the surface of the dipole. The current, I_z is assumed to be zero at the both ends of the dipole.

The $2h$ length of the dipole is divided into N segments (pulses) and the length of each segment Δz is given as:

$$\Delta z = \frac{2h}{N} \qquad \text{(dipole)} \tag{6.58}$$

All segments Δz have equal lengths. From (6.58), the match point z_m is being enforced at segment m along the dipole, which is calculated as:

$$z_m = (m - 0.5)\Delta z - h \qquad m = 1,2,3,\ldots,M \tag{6.59}$$

The N segments or pulses must be an odd number. The relationship between Ohm's law and the Kirchhoff's network equations is shown in Figure 6.21.

$$\sum_{n=1}^{N} Z_{mn} I_n = -E^i_{zm} \tag{6.60}$$

where the discretization of Z_{mn} in (6.60) is given as:

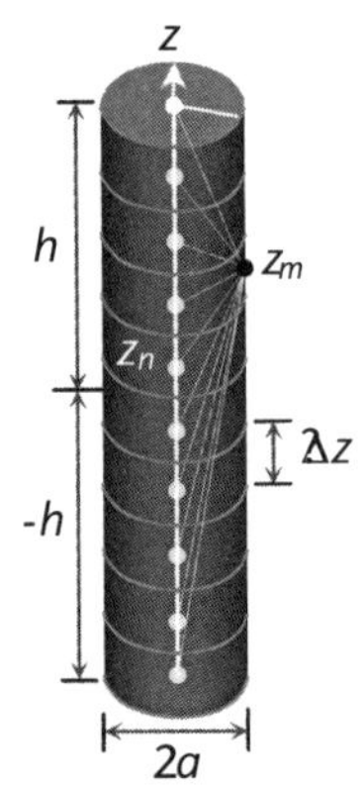

Figure 6.21 Kirchhoff's network for the matching point z_m from filamentary line current source.

$$Z_{mn} = \frac{j}{4\pi\omega\varepsilon_o} \int_{z_n-\Delta z/2}^{z_n+\Delta z/2} \frac{e^{-jkr_m}}{r_m^5} \left[\left(1 + jkr_m\right)\left(2r_m^2 - 3a^2\right) + \left(kar_m\right)^2 \right] dz' \tag{6.61}$$

where $r_m = \sqrt{a^2 + \left(z_m - z'\right)^2}$. The discretization of the electric field E_z of (4.31) along the wire can be rewritten as:

$$E_{zm}^i = \frac{V_m}{\ln(b/a)} \left[\frac{e^{-jkr_1}}{r_1} - \frac{e^{-jkr_2}}{r_2} \right] \tag{6.62}$$

where $r_1 = \sqrt{z_m^2 + a^2}$ and $r_2 = \sqrt{z_m^2 + b^2}$. The match point of (6.59) is inserted into (6.61) and (6.62). The match point z_m is located in the middle of each segment. The unknown coefficient, I_n, orthogonal over each segment, is represented as the surface current on the cylindrical dipole.

The integral part of (6.61) can be solved by numerical integration, such as Simpson's rules. Equation (6.60) produces one set of linear equations as:

$$\begin{aligned} I_1 Z_{11} + I_2 Z_{12} + I_3 Z_{13} + \ldots + I_N Z_{1N} &= -E_{z1}^i \\ I_1 Z_{21} + I_2 Z_{22} + I_3 Z_{23} + \ldots + I_N Z_{2N} &= -E_{z2}^i \\ \vdots \qquad \vdots \qquad\qquad \vdots \quad & \quad \vdots \\ I_1 Z_{N1} + I_2 Z_{N2} + I_3 Z_{N3} + \ldots + I_N Z_{NN} &= -E_{zN}^i \end{aligned} \tag{6.63}$$

Assembling (6.63) at each match point, z_m, (6.63) can be reexpressed in matrix form as:

$$\begin{bmatrix} Z_{11} & Z_{12} & Z_{13} & \cdots & Z_{1N} \\ Z_{21} & Z_{22} & Z_{23} & \cdots & Z_{2N} \\ \vdots & \vdots & \vdots & \ddots & \vdots \\ Z_{N1} & Z_{N2} & Z_{N3} & \cdots & Z_{NN} \end{bmatrix} \begin{bmatrix} I_1 \\ I_2 \\ \vdots \\ I_N \end{bmatrix} = \begin{bmatrix} -E_{z1}^i \\ -E_{z2}^i \\ \vdots \\ -E_{zN}^i \end{bmatrix} \tag{6.64a}$$

$$\left[Z_{mn}\right]\left[I_n\right] = \left[-E_{zm}^i\right] \tag{6.64b}$$

Equation (6.64) can be solved by using the Gaussian elimination routine. For instance, the current I_z for a $2h = 10$ mm cylindrical dipole conductor with a radius $a = 0.65$ mm is simulated using $N = 9$ match points along the dipole as depicted in Figure 6.22. In this linear geometry problem, the matrix $[Z_{mn}]$ in (6.64a) is symmetric Toeplitz, where only the first row of the matrix $[Z_{mn}]$ needs to be computed as:

$$\left[Z_{11}, Z_{12}, Z_{13}, Z_{14}, Z_{15}, Z_{16}, Z_{17}, Z_{18}, Z_{19}\right] \tag{6.65}$$

The values of (6.65) are arranged in a symmetric Toeplitz matrix as:

$$\begin{bmatrix} Z_{11} & Z_{12} & Z_{13} & Z_{14} & Z_{15} & Z_{16} & Z_{17} & Z_{18} & Z_{19} \\ Z_{12} & Z_{11} & Z_{12} & Z_{13} & Z_{14} & Z_{15} & Z_{16} & Z_{17} & Z_{18} \\ Z_{13} & Z_{12} & Z_{11} & Z_{12} & Z_{13} & Z_{14} & Z_{15} & Z_{16} & Z_{17} \\ Z_{14} & Z_{13} & Z_{12} & Z_{11} & Z_{12} & Z_{13} & Z_{14} & Z_{15} & Z_{16} \\ Z_{15} & Z_{14} & Z_{13} & Z_{12} & Z_{11} & Z_{12} & Z_{13} & Z_{14} & Z_{15} \\ Z_{16} & Z_{15} & Z_{14} & Z_{13} & Z_{12} & Z_{11} & Z_{12} & Z_{13} & Z_{14} \\ Z_{17} & Z_{16} & Z_{15} & Z_{14} & Z_{13} & Z_{12} & Z_{11} & Z_{12} & Z_{13} \\ Z_{18} & Z_{17} & Z_{16} & Z_{15} & Z_{14} & Z_{13} & Z_{12} & Z_{11} & Z_{12} \\ Z_{19} & Z_{18} & Z_{17} & Z_{16} & Z_{15} & Z_{14} & Z_{13} & Z_{12} & Z_{11} \end{bmatrix}$$

where $Z_{11} = Z_{22} = Z_{33} \ldots = Z_{99}$ and $Z_{21} = Z_{12}$, $Z_{31} = Z_{13}$, $Z_{41} = Z_{14}$, ..., $Z_{91} = Z_{19}$.

The current distribution calculated using the analytical sinusoidal equation (4.1) in Chapter 4 is the real value, but in an actual case, the current is a complex value, contributed to by the size of the radius of conductor a, as shown in Figure 6.24(b). This means that (4.1) is valid only for thin extended conductors. In addition, it cannot guarantee that the current distribution will always be sinusoidal in all cases of the extended conductor, especially at a high operating frequency. For a cylindrical dipole or monopole structure, the magnetic field, $\vec{H}$ around the surface conductor, which is generated by a current I_z, is in an azimuthal ϕ direction and its relationship is given as:

$$\begin{aligned} H_\phi &= \frac{I_z}{2\pi a}\hat{\phi} \\ &= \frac{1}{2\pi a}\left[\Re\text{e}\left(I_z\right) + j\Im\text{m}\left(I_z\right)\right]\hat{\phi} \\ &= \frac{1}{2\pi a}\left[\Re\text{e}\left(I_z\right)\hat{\phi} + \Im\text{m}\left(I_z\right)\hat{z}\right] \end{aligned} \tag{6.66}$$

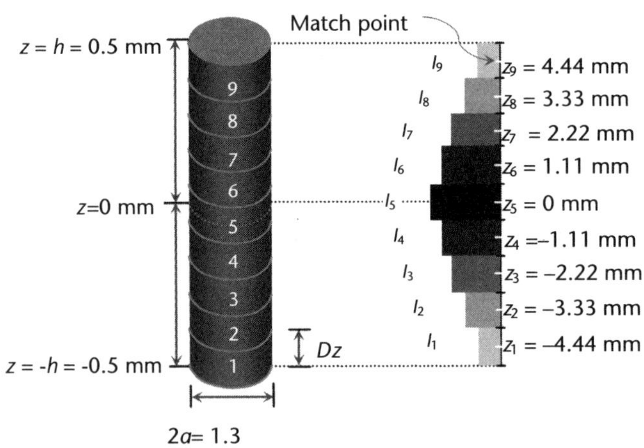

Figure 6.22 Dimensions of dipole and pulse basis function segments along a dipole with $N = 9$.

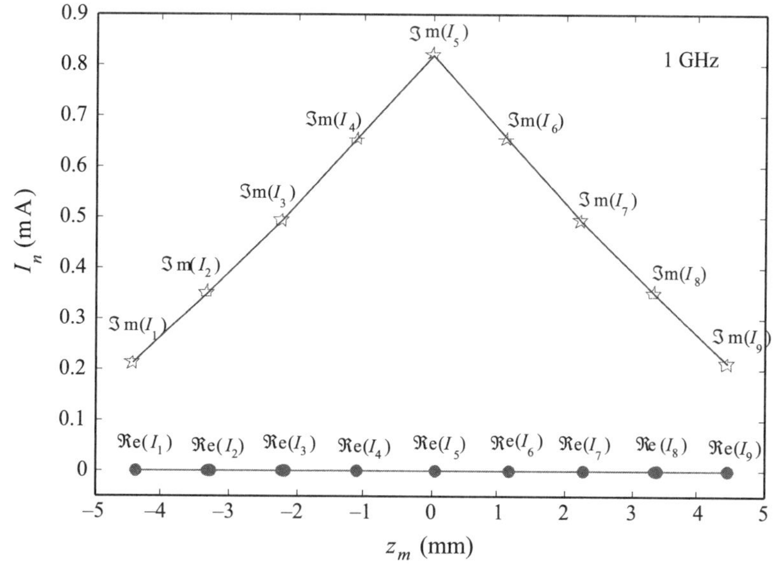

Figure 6.23 Real and imaginary parts of the current I_z for a dipole with nine segments, using Pocklington's equation, pulse basis functions, magnetic frill source, and point-matching method.

This means that the imaginary part of current $\Im m(I_z)$ refers to the current flow in the z-direction perpendicular to the ϕ direction. However, the real part of I_z represents the azimuthal current.

To date, no formulation is available for the actual distribution of the current I_z for an arbitrarily sized extended conductor. Due to the above reasons, analytical methods have recently been replaced by the numerical MoM in order to calculate the vector current distribution on the dipole. Figure 6.25 shows the variation of the current distribution I_z along the length z of the monopole. The unknown complex current I_z is determined by using point-matching MoM with Pocklington's integral equation and frill-generator source. In Figure 6.25, it is clearly shown that at low frequencies ($f < 3$ GHz), the value of current I_z approaches the real number. When the monopole is operated at higher frequencies, the current distribution I_z along the length of monopole becomes a complex number.

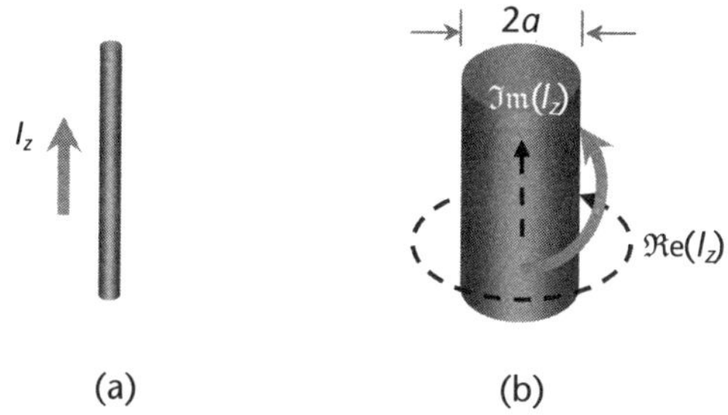

Figure 6.24 (a) Ideal analytical line current. (b) Actual situations of current distribution on a cylindrical extended conductor.

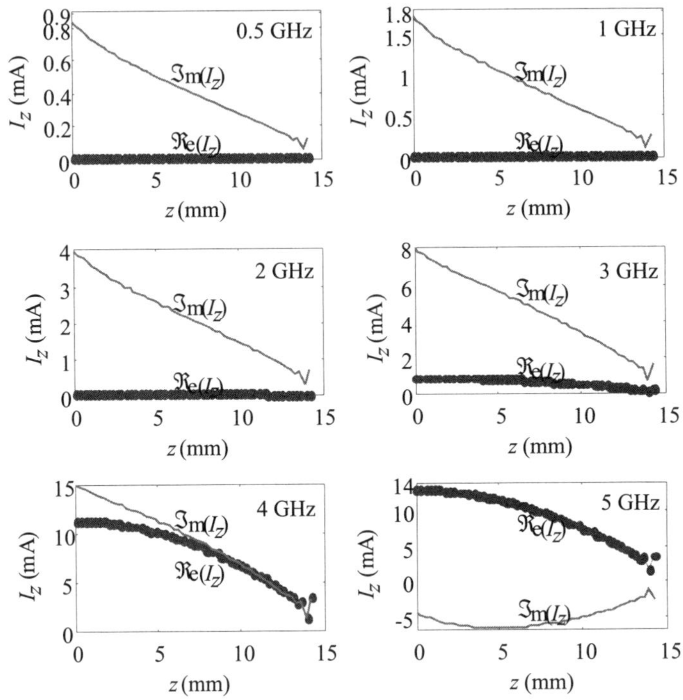

Figure 6.25 Current distribution I_z on the 1.436 cm of monopole driven from the coaxial line with $a = 0.65$ mm and $b = 2.05$ mm.

6.7.2 Input Impedance Z_{in}

The input impedance Z_{in} at $z = 0$ for the $2h$ length dipole can be calculated by using the vector current I_n at the match point ($z_m = z_{(N+1)/2}$) on the center of the dipole.

$$Z_{in} = \frac{V_o}{I_{(N+1)/2}} = \frac{1}{\Re e\left(I_{(N+1)/2}\right) + j\Im m\left(I_{(N+1)/2}\right)} = \Re e\left(Z_{in}\right) + j\Im m\left(Z_{in}\right) \quad (6.67)$$

The MoM routines are explained in detail and their MATLAB codes are listed on the accompanying CD. As mentioned in Chapter 4, the monopole has half of the input impedance value Z_{in} as compared to the dipole ($Z_{in(monopole)} = Z_{in}/2$). For example, a monopole with h =1.436 cm driven from a coaxial line, as shown in Figure 6.26, is tested using an MoM routine with $N = 39$ segments and the magnetic frills model.

Figure 6.27 shows the comparison of calculated normalized resistance, $\Re e(Z_{in})/Z_o$, and normalized reactance, $\Im m(Z_{in})/Z_o$, of a monopole with $h = 14.32$ mm in air using analytical approaches of (4.25), (4.26), and (4.27) and the MoM methods in (6.67). In the MoM computation, the monopole is uniformly divided to $N = 39$ segments, and the height of each segment is $\Delta z = 2h/N \approx 0.73$ mm. Clearly, the analytical and MoM methods are in good agreement up to 5 GHz. Pocklington's solution associated with the delta-gap voltage $V_o/\Delta z$ has a cell-size dependent error when the delta-gap size of Δz is not appropriately chosen. Normally, the delta-gap voltage model is modified as $V_o/(N_g\Delta z)$. The term N_g is the coefficient value to match

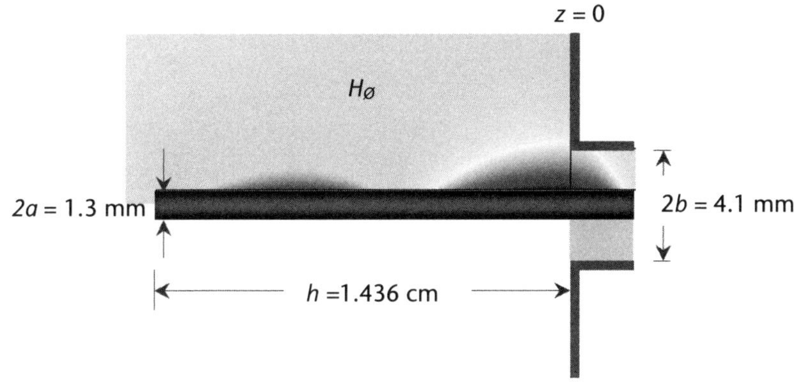

Figure 6.26 Dimensions of the monopole driven from the coaxial line.

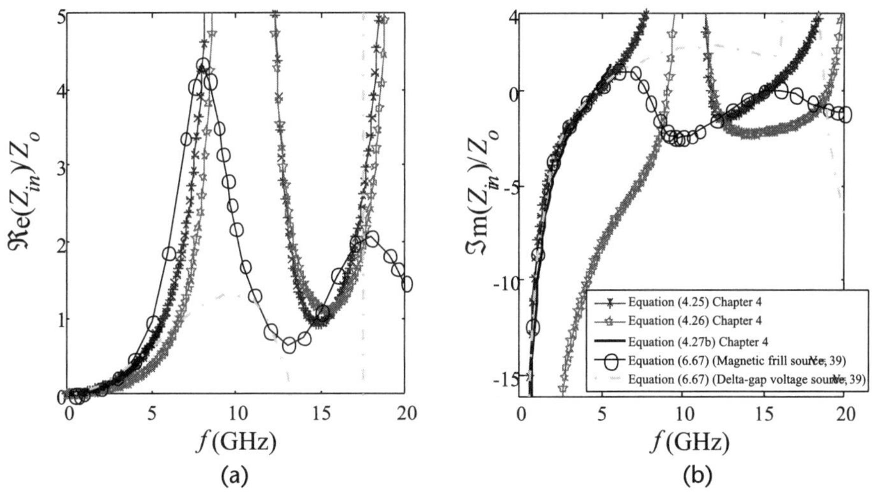

Figure 6.27 (a) Comparison of calculated normalized resistance, $\Re e(Z_{in})/Z_o$, and (b) normalized reactance, $\Im m(Z_{in})/Z_o$,of a monopole in air at 25°C using (4.25), (4.26), (4.27) and (6.67) (a = 0.65 mm, b = 2.05 mm, and h = 14.32 mm).

with the actual feed voltage. Besides, $N_g = 0.1$ is implemented in Figure 6.27. The comparison of calculated normalized resistance, $\Re e(Z_{in})/Z_o$, and reactance, $\Im m(Z_{in})/Z_o$, of the monopole driven from a coaxial line using the MoM method, including simulation results and measurements for air, are shown in Figure 6.28. The MoM computation result is in good agreement with finite element-based simulator (COMSOL Multiphysics) and finite integration-based simulator (CST Microwave Studio).

6.7.3 Far Electric Field E_θ

The study of the far-field radiation pattern is important for the dipole, which is implemented as an antenna in communication applications. The radiation pattern describes the relative strength of the radiated field through various directions, at a fixed distance r from the antenna. Thus, the transmission/reception of the RF signal by the dipole antenna from a particular direction can be identified. The radiation field can be considered far field when the observation distance r is far away from the

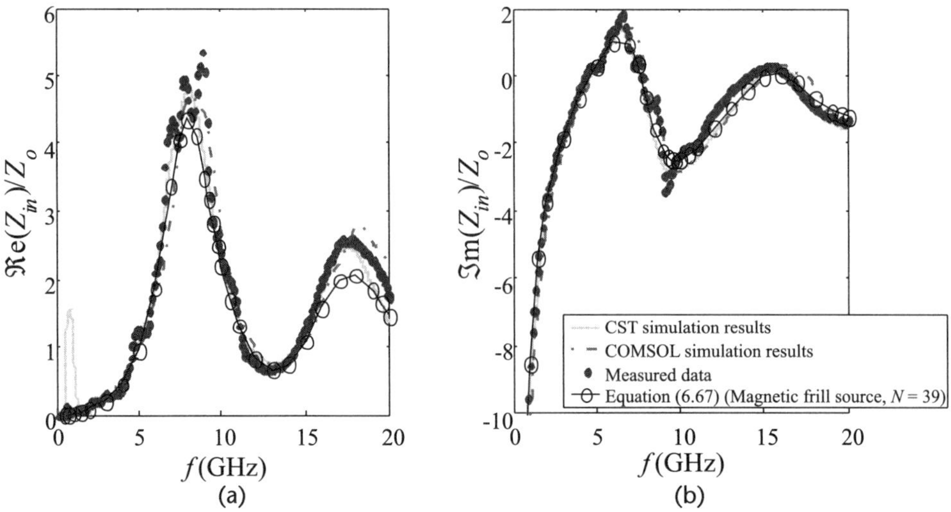

Figure 6.28 Comparison of (a) normalized resistance, $\Re e(Z_{in})/Z_o$, and (b) normalized reactance, $\Im m(Z_{in})/Z_o$, for air by considering the size of a monopole driven from the coaxial line with a = 0.65 mm, b = 2.05 mm, and h = 14.32 mm.

dipole antenna and a condition of $kr \gg 1$ is achieved. The radiation performance is often described by the E_θ and H_ϕ fields' plane pattern in spherical coordinates as shown in Figure 6.29.

When substituting (6.25) into (6.26),

$$\begin{aligned} E_\theta &\approx jk\eta_o \sin\theta \frac{e^{-jkr}}{4\pi r} \sum_{n=1}^{N} I_n \int_{z_n-\Delta z/2}^{z_n+\Delta z/2} e^{jkz'\cos\theta} dz' \\ &\approx jk\eta_o \frac{e^{-jkr}}{4\pi r} \underbrace{\sin\theta \frac{\Delta z \sin\left(\dfrac{k\Delta z\cos\theta}{2}\right)}{\left(\dfrac{k\Delta z\cos\theta}{2}\right)} \sum_{n=1}^{N} I_n e^{jkz_n\cos\theta}}_{\text{Radiation pattern}} \end{aligned} \tag{6.68}$$

Conversely, if substituting (6.38) into (6.26), the θ component of the electric field E_θ can be alternatively expressed as:

$$\begin{aligned} E_\theta &\approx \frac{jk\pi_o}{\sin\theta} \frac{e^{-jkr}}{2\pi r} \sum_{n=1}^{N} I_n \int_{z_{n-1}}^{z_{n+1}} \frac{\sin k\left(\Delta z - \left|z' - z_n\right|\right)}{\sin(k\Delta z)} e^{jkz'\cos\theta} dz' \\ &\approx j60 \frac{e^{-jkr}}{r} \underbrace{\frac{\cos(k\Delta z\cos\theta) - \cos(k\Delta z)}{\sin\theta \sin(k\Delta z)} \sum_{n=1}^{N} I_n e^{jkz_n\cos\theta}}_{\text{Radiation pattern}} \end{aligned} \tag{6.69}$$

The far field for E_θ decay is $1/r$, but the radiation pattern is dependent on the distance r ($r \approx$ the radius of the spherical boundary, ρ). The radiation patterns of E_θ with distance ρ = 5 cm from the center of the dipole in the air for 1 GHz and

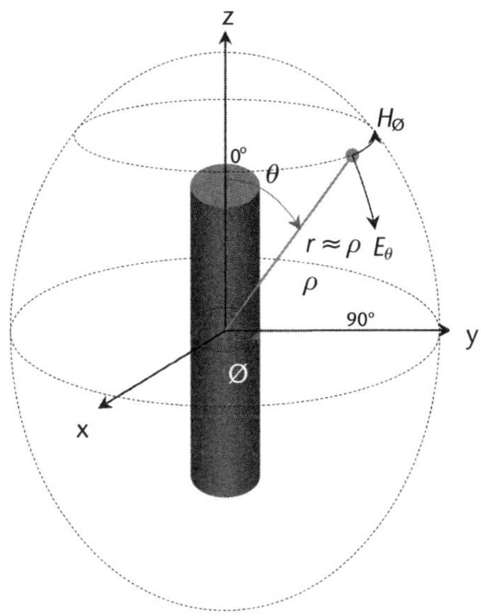

Figure 6.29 The far field of E_θ and H_ϕ components in spherical coordinates.

30 GHz are plotted in Figure 6.30. The radiation pattern with the dipole mounted horizontally is the strongest at right angles (θ = 90° and θ = 270°) to the dipole for 1 GHz and dropped to zero at the center of the dipole.

6.7.4 Hallén's Equation, Piecewise Uniform Function, and Point-Matching Method

In the MoM solution, Hallén's integral equation (4.20) for the center-fed dipole with the length $2h$ can be rewritten in the exact kernel or the reduced kernel forms:

$$\frac{1}{4\pi}\int_{-h}^{h} I_z K dz' = \xi_o \cos(kz) - \frac{jV_o}{2\eta}\sin\left(k|z|\right) \tag{6.70}$$

where the exact kernel and reduced kernel are given as:

$$K = \frac{1}{2\pi}\int_{-\pi}^{\pi} \frac{e^{-jk\sqrt{4a^2\sin^2(\phi/2)+(z-z')^2}}}{\sqrt{4a^2\sin^2(\phi/2)+(z-z')^2}} d\phi \quad \text{(Exact kernel)} \tag{6.71a}$$

and

$$K = \frac{e^{-jk\sqrt{a^2+(z-z')^2}}}{\sqrt{a^2+(z-z')^2}} \quad \text{(Reduced kernel)} \tag{6.71b}$$

The $2h$ length dipole is divided into N segments (pulses) and the length of each segment Δz is given as:

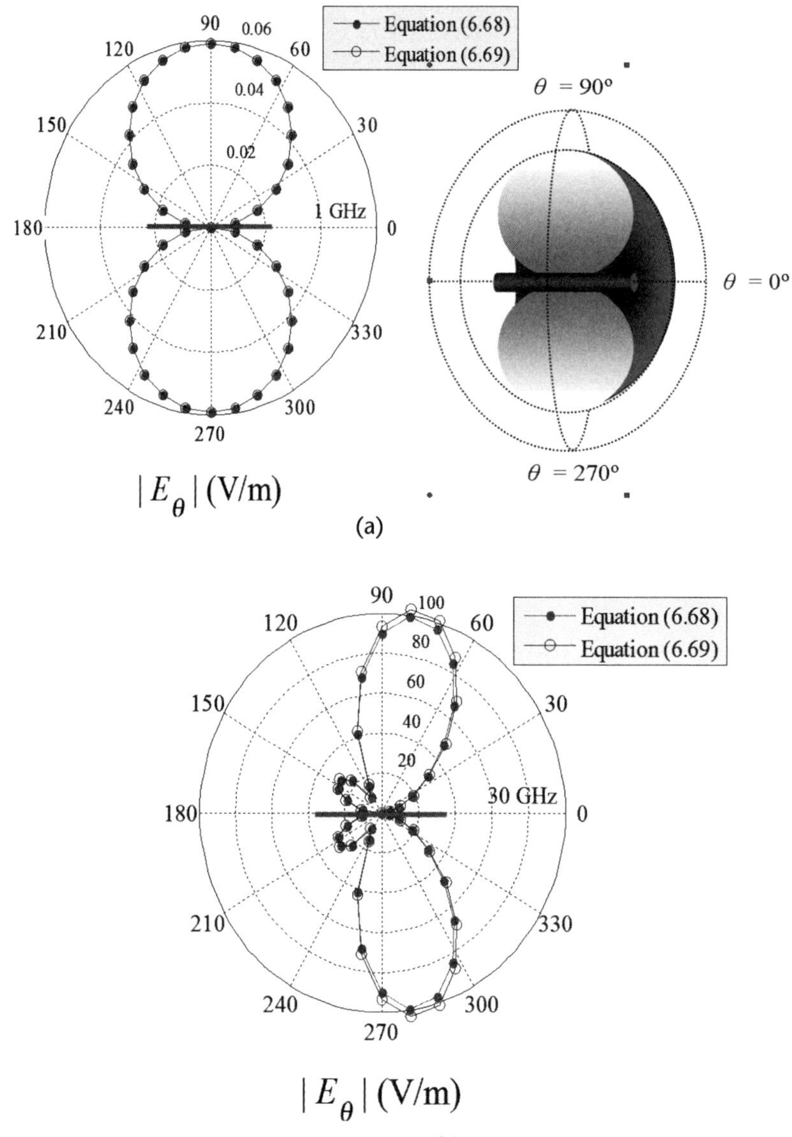

Figure 6.30 Far-field pattern of E_θ with ρ = 5 cm from $2h$ = 2.864 cm length of dipole for (a) 1 GHz and (b) 30 GHz.

$$\Delta z = \frac{2h}{N} \tag{6.72}$$

From (6.72), the match point z_m is being enforced at the segment m along the dipole, which is calculated as:

$$z_m = (m - 0.5)\Delta z - h \qquad m = 1,2,3,\ldots,N \tag{6.73}$$

The integral part of (6.70) can be solved by numerical integration, such as Simpson's rules. Equation (6.70) produces one set of linear equations as:

$$\begin{aligned} I_1A_{11} + I_2A_{12} + \ldots + I_{N-1}A_{1,N-1} + I_NA_{1N} &= \xi_o B_1 + C_1 \\ I_1A_{21} + I_2A_{22} + \ldots + I_{N-1}A_{2,N-1} + I_NA_{2N} &= \xi_o B_2 + C_2 \\ \vdots \quad \vdots \quad \vdots \quad \vdots \quad &\vdots \\ I_1A_{N1} + I_2A_{N2} + \ldots + I_{N-1}A_{N,N-1} + I_NA_{NN} &= \xi_o B_N + C_N \end{aligned} \tag{6.74}$$

where the A_{mn} integral kernel is given as:

$$A_{mn} = \frac{1}{4\pi}\int_{z_n-\Delta z/2}^{z_n+\Delta z/2}\int_{-\pi}^{\pi} \frac{e^{-jk\sqrt{4a^2\sin^2(\phi/2)+\left(z_m-z'\right)^2}}}{2\pi\sqrt{4a^2\sin^2(\phi/2)+\left(z_m-z'\right)^2}}\,d\phi dz' \tag{6.75a}$$

or

$$A_{mn} = \frac{1}{4\pi}\int_{z_n-\Delta z/2}^{z_n+\Delta z/2} \frac{e^{-jk\sqrt{a^2+\left(z_m-z'\right)^2}}}{\sqrt{a^2+\left(z_m-z'\right)^2}}\,dz' \qquad \text{(Reduced kernel)} \tag{6.75b}$$

The B_m and C_m elements on the left side of (6.74) are expressed as:

$$B_m = \cos\left(kz_m\right) \tag{6.76}$$

$$C_m = -\frac{j}{2\eta}\sin\left(k\left|z_m\right|\right) \tag{6.77}$$

Subsequently, (6.74) can be expressed in a matrix form as:

$$\begin{bmatrix} A_{11} & A_{12} & \cdots & A_{1,N-1} & A_{1N} \\ A_{21} & A_{22} & \cdots & A_{2,N-1} & A_{2N} \\ \vdots & \vdots & \ddots & \vdots & \vdots \\ A_{N1} & A_{N2} & \cdots & A_{N,N-1} & A_{NN} \end{bmatrix} \begin{bmatrix} I_1 \\ I_2 \\ \vdots \\ I_N \end{bmatrix} = \xi_o \begin{bmatrix} B_1 \\ B_2 \\ \vdots \\ B_N \end{bmatrix} + \begin{bmatrix} C_1 \\ C_2 \\ \vdots \\ C_N \end{bmatrix} \tag{6.78}$$

As mentioned in Section 4.1.4, for analytical analysis, the ξ_o term is the constant coefficient that requires the current to vanish ($I_z = 0$) at both ends of the dipole. In the MoM solution, the ξ_o term in (6.70) is treated as an unknown coefficient in the matrix of (6.78), where the value of ξ_o is used to correct the condition at the last match point z_M when I_N is assumed to be zero at the point. The MATLAB calculation program of (6.70) is shown on the accompanying CD.

Wilton and Butler [11] had reported that Hallén's equation offers a good convergence of the three common basis functions (piecewise uniform, piecewise linear, and piecewise sinusoidal functions) compared to Pocklington's equation. Pocklington's equation associated with a reduced kernel is also known as Richmond's equation. Figure 6.31 shows the $\mathfrak{Re}(I_z)$ and $\mathfrak{Im}(I_z)$ parts of calculated current along the $2h$ = 1 cm length of dipole with N = 51 using Richmond's equation, and Hallén's equation with the exact and reduced kernels, respectively. From Figures 6.31(a, c), the

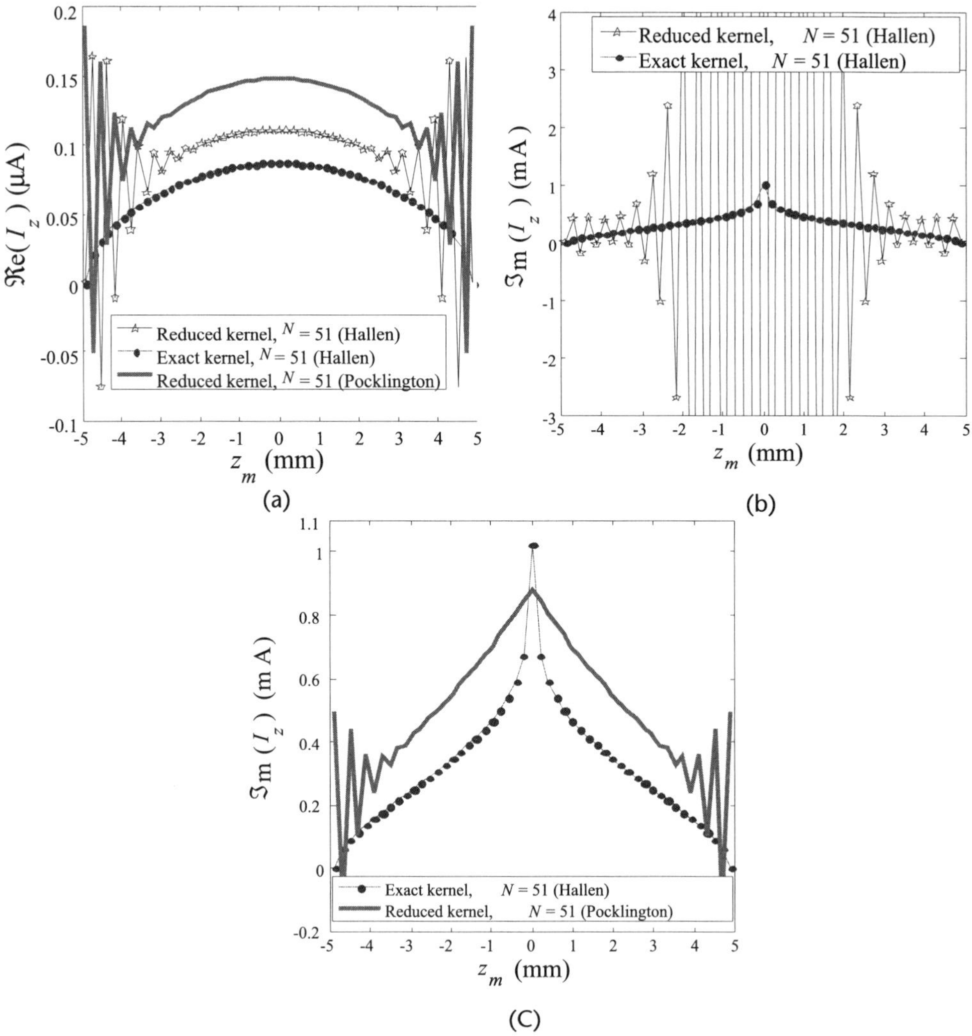

Figure 6.31 Comparison of (a) calculated $\Re e(I_z)$, (b) $\Im m(I_z)$ using Hallén's equation, and (c) $\Im m(I_z)$ using Pocklington's equation.

calculated current, [$\Re e(I_z)$ and $\Im m(I_z)$], near both ends of the dipole at $z = h$ using Richmond's equation did not converge. However, Hallén's equation with a reduced kernel requires a step size Δz that is greater than the radius of the dipole, a ($\Delta z \gg a$), in order to obtain convergence of computation. For instance, in the case $\Delta z < a$ (with $a = 0.00065$m and $\Delta z \approx 0.0002$m), instability in the current computation occurred, and the calculated imaginary current $\Im m(I_z)$ oscillated, as shown in Figure 6.31(b).

6.7.5 Richmond's Method of Moments, Sinusoidal Pulse Functions, Delta-Gap Sources, and Galerkin's Method

In this section, we try to analyze another MoM routine that is widely studied; it is slightly different than the previous method. The $2h$ length of the dipole is divided into N sinusoidal segments (pulses) and the length of each segment Δz is given as:

$$\Delta z = \frac{2h}{N+1} \qquad \text{(dipole)} \tag{6.79}$$

The N sinusoidal segments must be an odd number.

$$z_m = (m-1)\Delta z - h \quad m = 1,2,3,\ldots,N+2 \tag{6.80}$$

The match point z_m is located in the middle of each sinusoidal segment. As in the previous routine, we need to solve the matrix of:

$$\left[Z_{mn}\right]\left[I_n\right] = \left[V_m\right] \tag{6.81}$$

where

$$\begin{aligned} Z_{mn} &= \frac{j\omega\mu_o}{4\pi k \sin^2(k\Delta z)} \times \\ &\left\{\int_{z_{m-1}}^{z_m} K_{mn} \sin\left[k\left(z - z_{m-1}\right)\right] dz + \int_{z_m}^{z_{m+1}} K_{mn} \sin\left[k\left(z_{m+1} - z\right)\right] dz\right\} \\ &= \frac{j\omega\mu_o}{4\pi k \sin^2(k\Delta z)} \int_{z_{m-1}}^{z_{m+1}} K_{mn} \sin\left[k\left(\Delta z - \left|z' - z_m\right|\right)\right] dz \end{aligned} \tag{6.82}$$

K_{mn} is expressed as:

$$K_{mn} = \frac{e^{-jkr_{n+1}}}{r_{n+1}} + \frac{e^{-jkr_{n-1}}}{r_{n-1}} - 2\cos(k\Delta z)\frac{e^{-jkr_n}}{r_n} \tag{6.83}$$

and

$$r_{n-1} = \sqrt{a^2 + \left(z - z_{n-1}\right)^2}; \quad r_n = \sqrt{a^2 + \left(z - z_n\right)^2}; \quad r_{n+1} = \sqrt{a^2 + \left(z - z_{n+1}\right)^2}$$

r_{n-1} and r_{n+1} are the distances from the observation point to the end points of the sinusoidal interval, and r_n is the distance to the midpoint of the sinusoidal interval as shown in Figure 6.32. The discretization electric field E_z along the wire can be rewritten as:

$$\begin{aligned} V_m &= \frac{-jk}{4\pi\omega\varepsilon_o\varepsilon_r \sin(k\Delta z)} \left\{ \begin{aligned} &\int_{z_{m-1}}^{z_m} E_{zm} \sin\left[k\left(z - z_{m-1}\right)\right] dz \\ &+ \int_{z_m}^{z_{m+1}} E_{zm}^i \sin\left[k\left(z_{m+1} - z\right)\right] dz \end{aligned} \right\} \\ &= \frac{-jk}{4\pi\omega\varepsilon_o\varepsilon_r \sin(k\Delta z)} \int_{z_{m-1}}^{z_{m+1}} E_{zm} \sin\left[k\left(\Delta z - \left|z - z_m\right|\right)\right] dz \end{aligned} \tag{6.84}$$

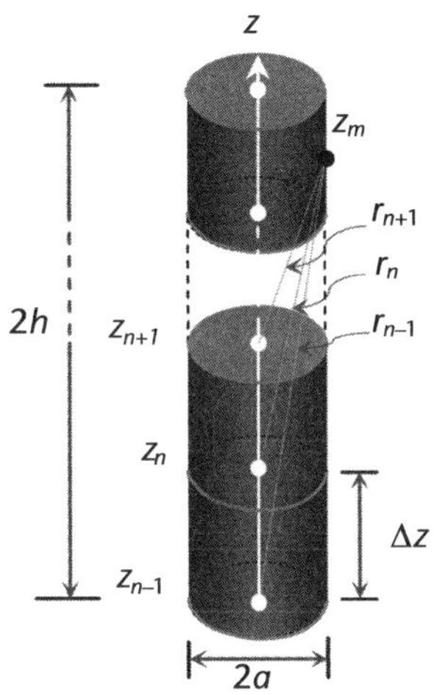

Figure 6.32 Richmond's MoM for matching the discrete point from the observation point source along the filamentary line.

The z component of the electric field E_{zm} in (6.84) is reexpressed as:

$$E_{zm}^{i} = \frac{-jk}{4\pi\omega\varepsilon_o\varepsilon_r \sin(k\Delta z)}\left[\frac{e^{-jkr_{m+1}}}{r_{m+1}} + \frac{e^{-jkr_{m-1}}}{r_{m-1}} - 2\cos(k\Delta z)\frac{e^{-jkr_m}}{r_m}\right] \quad (6.85)$$

and

$$r_{m-1} = \sqrt{a^2 + \left(z - z_{m-1}\right)^2};\ r_m = \sqrt{a^2 + \left(z - z_m\right)^2};\ r_{m+1} = \sqrt{a^2 + \left(z - z_{m+1}\right)^2}$$

The match point of (6.71) is inserted into both (6.73) and (6.75). The unknown coefficient, I_n, orthogonal over each sinusoidal segment, is represented as the surface current on the cylindrical dipole as shown in Figure 6.33.

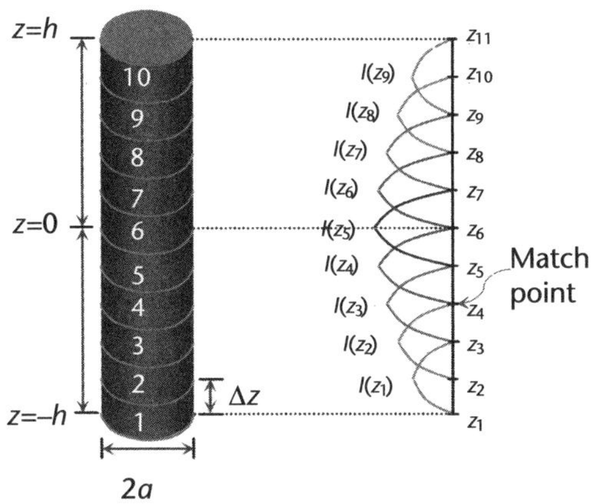

Figure 6.33 Dimensions of dipole and pulse basis function segments along a dipole using $N = 9$ sinusoidal current pulses with $(N + 2) = 11$ match points.

6.8 Dielectric-Coated Dipole

In 1976, Richmond and Newman [17] extended the application of the MoM using the piecewise sinusoidal approximation of the current on the dipole into a dielectric-coated dipole. In the MoM solution, for a dielectric-coated dipole, the algebraic equation is given as:

$$\sum_{n=1}^{N} I_n\left(Z_{mn} + Z'_{mn}\right) = V_m \tag{6.86}$$

where Z_{mn} and V_m in (6.86) are the same as (6.61) and (6.62), respectively. Subsequently, Z'_{mn} is formulated as [17]:

$$Z'_{mn} = \begin{cases} \dfrac{jk}{2\pi\omega\varepsilon_o}\dfrac{\left(\varepsilon_r - \varepsilon_{r2}\right)}{\varepsilon_r\varepsilon_{r2}}\ln\left(\dfrac{b}{a}\right)\left[\dfrac{k\Delta z}{\sin^2(k\Delta z)} + \dfrac{1}{\tan(k\Delta z)}\right] & m = n \\ \dfrac{-jk}{4\pi\omega\sin(k\Delta z)}\dfrac{\left(\varepsilon_r - \varepsilon_{r2}\right)}{\varepsilon_r\varepsilon_{r2}}\ln\left(\dfrac{b}{a}\right)\left[1 + \dfrac{k\Delta z}{\tan(k\Delta z)}\right] & m \neq n \end{cases} \tag{6.87}$$

where ε_{r2} and ε_r are the relative permittivities of the dielectric coating and the exterior medium, respectively. In fact, Z'_{mn} is the additional term that is contributed by the dielectric coating insulation along the dipole.

6.9 Dipole Array

Analytical solutions for a dipole array were discussed in Section 4.5. However, the computation of current flow and induced current on each dipole element cannot be handled by analytical methods. Thus, matrix methods such as the MoM are used to solve these kinds of problems. In general, for the N dipole element array, the mutual coupling relationship can be written in a set of linear equations as:

$$\begin{aligned} I_1Z_{11} + I_2Z_{12} + \ldots + I_NA_{1,N} &= V_1 \\ I_1Z_{21} + I_2Z_{22} + \ldots + I_NZ_{2,N} &= V_2 \\ \vdots \qquad \vdots \qquad\qquad \vdots \\ I_1Z_{N1} + I_2Z_{N2} + \ldots + I_NZ_{N,N} &= V_N \end{aligned} \tag{6.88a}$$

or

$$\begin{bmatrix} Z_{11} & Z_{12} & \cdots & Z_{1N} \\ Z_{21} & Z_{22} & \cdots & Z_{2N} \\ \vdots & \vdots & \ddots & \vdots \\ Z_{N1} & Z_{N2} & \cdots & Z_{NN} \end{bmatrix}\begin{bmatrix} I_1 \\ I_2 \\ \vdots \\ I_N \end{bmatrix} = \begin{bmatrix} V_1 \\ V_2 \\ \vdots \\ V_N \end{bmatrix} \tag{6.88b}$$

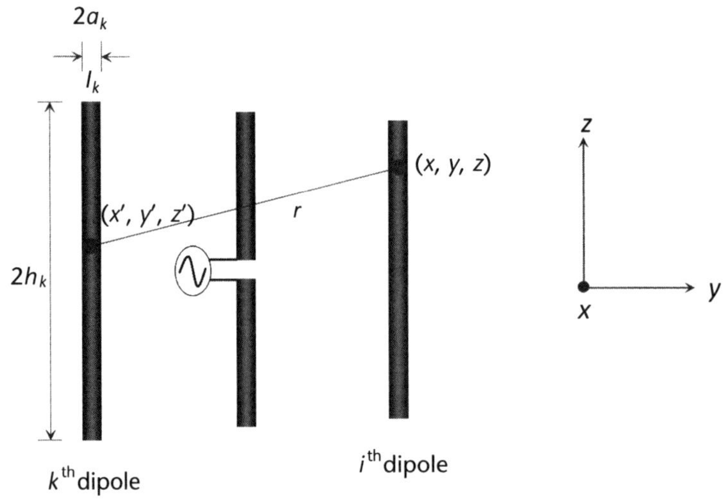

Figure 6.34 Coordinate distance from a matching point on *k*th dipole to *i*th dipole [18].

where Z_{kk} (Z_{11}, Z_{22}, ..., Z_{NN}) is the self-impedance for the kth dipole element. However, Z_{ki} is the mutual impedance between the kth dipole and the ith dipole in the array (k = 1, 2, ..., N and i = 1, 2, ..., N) as shown in Figure 6.34. Z_{kk} or Z_{ki} in (6.88) can be calculated using (4.69). If the kth dipole element does not connect to a voltage source, V_k = 0V (except the driven dipole element).

References

[1] Harrington, R. F., "Matrix Methods for Field Problems," *Proceedings of the IEEE*, Vol. 55, 1967, pp. 136–149.

[2] Nevels, R. D., C. M. Butler, and W. Yablon, "The Annular Slot Antenna in a Lossy Biological Medium," *IEEE Trans. on Microwave Theory Tech.*, Vol. 33, 1985, pp. 314–319.

[3] Pocklington, H. C., "Electrical Oscillations in Wire," *Cambridge Philos. Soc. Proc.*, Vol. 4, 1897, pp. 324–332.

[4] Hallén, E., "Theoretical Investigations into Transmitting and Receiving Antennae," *Nova Acta Regiae Sco. Sci. Upsaliensis*, Ser. IV, Vol. 4, No. 2, 1938, pp. 1–44.

[5] Gajda, G. B., and S. S. Stuchly, "Numerical Analysis of Open-Ended Coaxial Lines," *IEEE Trans. on Microwave Theory Tech.*, Vol. 31, 1983, pp. 380–384.

[6] Butler, C. M., and D. R. Wilton, "Analysis of Various Numerical Techniques Applied to Thin-Wire Scatterers," *IEEE Trans. on Antennas Propagat.*, Vol. 23, 1975, pp. 534–540.

[7] Imbriale, W. A., and P. G. Ingerson. "On Numerical Convergence of Moment Solutions of Moderately Thick Wire Antennas Using Sinusoidal Basis Functions," *IEEE Trans. on Antennas Propagat.*, Vol. 21, 1973, pp. 363–366.

[8] Klein, C. A., and R. Mittra, "The Effect of Different Testing Functions in the Moment Method Solution of Thin-Wire Antenna Problems," *IEEE Trans. on Antennas Propagat.*, Vol. 23, 1975, pp. 258–261.

[9] Pearson, L. W., and C. M. Butler, "Inadequacies of Collocation Solutions to Pocklington-Type Models of Thin-Wire Structures," *IEEE Trans. on Antennas Propagat.*, Vol. 23, 1975, pp. 295–298.

[10] Richmond, J. H., and N. H. Geary, "Mutual Impedance Between Coplanar-Skew Dipoles," *IEEE Trans. on Antennas Propagat.*, Vol. 18, 1970, pp. 414–416.

[11] Wilton, D. R., and C. M. Butler, "Efficient Numerical Techniques for Solving Pocklington's Equation and Their Relationships to Other Methods," *IEEE Trans. on Antennas Propagat.*, Vol. 24, 1976, pp. 83–86.

[12] Richmond, J. H., "Digital Computer Solutions of the Rigorous Equations for Scattering Problems," *Proc. IEEE*, Vol. 53, 1965, pp. 796–804.

[13] Mittra, R., *Numerical and Asymptotic Techniques in Electromagnetics, Topics in Applied Physics*, Vol. 3, Berlin, Germany: Springer, 1975.

[14] Moore, J., and R. Pizer, *Moment Methods in Electromagnetics: Techniques and Applications*, New York: John Wiley and Sons, 1984.

[15] Burke, G. J., and A. J. Poggio, *Numerical Electromagnetics Code (NEC)—Method of Moments, Part II: Program Description-Code*, Report UCID-18834, Lawrence Livermore Laboratory, 1981.

[16] Brillouin, L., "Origin of Radiation Resistance," *Radioélectricité*, Vol. 3, 1922, pp. 147–152.

[17] Richmond, J. H., and E. H. Newman, "Dielectric Coated Wire Antennas," *Radio Sci.*, Vol. 11, 1976, pp. 13–20.

[18] Thiele, G. A., "Analysis of Yagi-Uda Type Antennas," *IEEE Trans. on Antennas Propagat.*, Vol. 17, 1969, pp. 24–31.

Selected Bibliography

Aburwein, A. A., A. A. B. Othman, and A. S. A. Kassim, "Input Admittance and Aperture Field of Flanged Coaxial Line Radiating into Layered Half-Space," *Int. J. Electronics*, Vol. 87, 2000, pp. 961–971.

Adopley, J. A. K., D. G. Dudley, and R. Taherian, "Analytical Models for Determination of Complex Permittivity," *Progress in Electromagnetics Research*, Vol. 15, 1997, pp. 109–139.

Anderson, L. S., G. B. Gajda, and S. S. Stuchly, "Analysis of an Open-Ended Coaxial Line Sensor in Layered Dielectric," *IEEE Trans. on Instrum. Meas.*, Vol. 35, 1986, pp. 13–18.

Balanis, C. A., *Antenna Theory Analysis and Design*, 2nd ed., New York: John Wiley & Sons, 1997.

Baker, J. J., et al., "Analysis of an Open-Ended Coaxial Probe with Lift-Off for Nondestructive Testing," *IEEE Trans. on Instrum. Meas.*, Vol. 43, 1994, pp. 1093–1099.

Bancroft, R., *Understanding Electromagnetic Scattering Using the Moment Method: A Practical Approach*, Norwood, MA: Artech House, 1996.

Esselle, K. P. A. P., and S. S. Stuchly, "Capacitive Sensors for In-Vivo Measurements of the Dielectric Properties of Biological Materials," *IEEE Trans. on Instrum. Meas.*, Vol. 37, 1988, pp. 101–105.

Garg, R., *Analytical and Computational Methods in Electromagnetic*, Norwood, MA: Artech House, 2008.

Gibson, W. C., *The Method of Moments in Electromagnetics*, London, U.K.: Chapman & Hall/CRC, 2008.

Lee, J. H., H. J. Eom, and K. H. Jun, "Reflection of a Coaxial Line Radiating into a Parallel Plate," *IEEE Microwave Guided Wave Letters*, Vol. 6, 1996, pp. 135–137.

Levine, H., and C. H. Papas, "Theory of the Circular Diffraction Antenna," *J. Appl. Phys.*, Vol. 22, 1951, pp. 29–43.

Li, C. L., and K. M. Chen, "Determination of Electromagnetic Properties of Materials Using Flanged Open-Ended Coaxial Probe-Full-Wave Analysis," *IEEE Trans. on Instrum. Meas.*, Vol. 44, 1995, pp. 19–27.

Li, L. L., et al., "Flanged Coaxial Microwave Probes for Measuring Thin Moisture Layers," *IEEE Trans. on Biomedical Engineering*, Vol. 39, 1992, pp. 49–57.

Lim, C. P., L. W. Li, and E. P. Li, "Fast Full-Wave Analysis of a Cylindrical Antenna Using a Single Integral with an Exact Kernel," *IEEE Antennas and Wireless Propagation Letters*, Vol. 1, 2002, pp. 43–45.

Misra, D., "A Quasi-Static Analysis of Open-Ended Coaxial Lines," *IEEE Trans. on Microwave Theory Tech.*, Vol. 35, 1987, pp. 925–928.

Misra, D. K., "A Study on Coaxial Line Excited Monopole Probes for In Situ Permittivity Measurements," *IEEE Trans. on Instrum. Meas.*, Vol. 36, 1987, pp. 1015–1019.

Mohammad, A. S., S. M. Riad, and W. A. Davis, "Wideband Dielectric Characterization Using a Dielectric Filled Cavity Adapted to the End of a Transmission Line," *IEEE Trans. on Instrum. Meas.*, Vol. 39, 1990, pp. 485–491.

Noh, Y. C., and H. J. Eom, "Radiation from a Flanged Coaxial Line into a Dielectric Slab," *IEEE Trans. on Microwave Theory Tech.*, Vol. 47, 1999, pp. 2158–2161.

Pournaropoulos, C. L., and D. K. Misra, "A Study on the Coaxial Aperture Electromagnetic Sensor and Its Application in Material Characterization," *IEEE Trans. on Instrum. Meas.*, Vol. 43, 1994, pp. 111–115.

Scott, L. D., "Polynomial-Exponential-Product Theory for Antennas in Homogeneous Isotropic Media," *Div. Eng. and Appl. Phys., Harvard Univ., Cambridge, Mass., NASA Sci. Rep. 7*, 1969.

Tsai, L. L., "A Numerical Solution for the Near and Far Fields of an Annular Ring of Magnetic Current," *IEEE Trans. on Antennas Propagat.*, Vol. 20, 1972, pp. 569–576.

Werner, D. H., P. L. Werner, and J. K. Breakall, "Some Computational Aspects of Pocklington's Electric Field Integral Equation for Thin Wires," *IEEE Trans. on Antennas Propagat.*, Vol. 42, 1994, pp. 561–563.

Weiner, M. M., et al., *Monopole Elements on Circular Ground Plane*, Norwood, MA: Artech House, 1987.

CHAPTER 7

Numerical Modeling: The Finite-Difference Time-Domain Method

In this chapter, the analysis of the finite-difference method (FDM) and finite-difference time-domain (FDTD) method for open-ended coaxial probes, dipoles, and monopoles will be reviewed. The FDTD method was first reported by Yee [1], in order to solve Maxwell's curl equation on grids staggered in space and time. A more detailed description of Yee's algorithm is available in Section 7.1.1. In fact, the FDTD method is a general class of FDMs used to solve electrodynamics problems in the time domain. The complete review of this method can be found in several textbooks [2–6]. In this chapter, the FDTD treatment is focused on the subcell model for the dipole (monopole) and open-ended coaxial probe. For a thin dipole (or monopole), a subcell model is used in the FDTD computation without reducing the dimension of the local grid, in which the radius a of a thin dipole must be less than the grid cell dimension Δz. The application of three-dimensional (3-D) subcell FDTD method on a dipole was first published by Umashankar et al. [7]. However, the analysis of the monopole driven from the coaxial line using the two-dimensional (2-D) subcell FDTD method was first studied by Maloney et al. [8]. The derivation of the governing equation for the FDTD method is straightforward compared to the MoM method. However, the boundary condition in the FDTD solution is required to be defined. The comparison between the MoM and FDTD features is tabulated in Table 7.1.

7.1 Mathematical Background of the FDTD

7.1.1 Yee's Algorithm

The governing partial differential equations in the FDTD for electromagnetic solutions are the Maxwell's curl equations in the time domain:

$$\vec{\nabla} \times \vec{H} = \vec{J} + \varepsilon_o \varepsilon_r \frac{\partial \vec{E}}{\partial t} \quad \text{(Ampere's law)} \tag{7.1}$$

$$\vec{\nabla} \times \vec{E} = -\mu_o \mu_r \frac{\partial \vec{H}}{\partial t} \quad \text{(Faraday's law)} \tag{7.2}$$

Table 7.1 Comparison of the MoM and FDTD Techniques

MoM Method	*FDTD Method*
1. Normally, the MoM solution for a linear dipole (or monopole) is in one or two dimensions.	1. The FDTD solution for a linear dipole (or monopole) is in two or three dimensions.
2. Normally, the governing equation is in a frequency-domain integral form.	2. The governing equation is in time-domain partial difference equation (PDE).
3. The governing equation is derived from the integral form of Maxwell's equation by integrating along the length of the dipole (or over the surface of the probe aperture).	3. The governing equation is derived using the time-domain partial difference form of Maxwell's equations in cylindrical coordinates (or Cartesian coordinates in three dimensions).
4. Does not require any treatment for an unbounded scattering field region. Green's function in the governing integral equation is naturally satisfied with an unbounded radiated field region.	4. Need to define the boundary conditions at the edges of the boundary grid points. For an unbounded field radiation region [perfectly matched layers (PML)], a modified difference equation for boundary grid point is required to ensure the boundary can absorb the radiation waves without any reflection. For conducting boundary [perfect electric conducting (PEC)], the field values at the boundary grid point are set to be zero.
5. The unknown voltage or current is the parameter required to be solved. All unknown currents on the conductor along the length of the dipole are coupled together and matched with a known voltage at a certain observation point (matching point). Express the relationship between the unknown current and known voltage in a linear equation form. The process is repeated for other matching points. All linear equations for each field matching point are arranged in a matrix form and are solved by the Gaussian elimination method.	5. The unknown electric field or magnetic field is the parameter required to be solved. The interactions between the electric field and the magnetic field in the space of the radiation region are matched with the time of the wave propagation. Field values at one grid point are updated in terms of neighboring values. The electric fields at the grid point in a volume of space are solved at a given instant in time. The corresponding magnetic fields at the next grid point in the same spatial volume are solved at the next instant in time. The process is repeated until the desired transient or steady-state electromagnetic field behavior is fully evolved. No linear algebra or matrix inversion is required.

Equations (7.1) and (7.2) can be rewritten in scalar forms with the aid of Cartesian coordinate as:

$$\begin{aligned}
\frac{\partial H_z}{\partial y} - \frac{\partial H_y}{\partial z} &= \varepsilon_o \varepsilon_r \frac{\partial E_x}{\partial t} \\
\frac{\partial H_x}{\partial z} - \frac{\partial H_z}{\partial x} &= \varepsilon_o \varepsilon_r \frac{\partial E_y}{\partial t} \\
\frac{\partial H_y}{\partial x} - \frac{\partial H_x}{\partial y} &= \varepsilon_o \varepsilon_r \frac{\partial E_z}{\partial t}
\end{aligned} \tag{7.3}$$

and

$$\begin{aligned}\frac{\partial E_z}{\partial y} - \frac{\partial E_y}{\partial z} &= -\mu_o \mu_r \frac{\partial H_x}{\partial t} \\ \frac{\partial E_x}{\partial z} - \frac{\partial E_z}{\partial x} &= -\mu_o \mu_r \frac{\partial H_y}{\partial t} \\ \frac{\partial E_y}{\partial x} - \frac{\partial E_x}{\partial y} &= -\mu_o \mu_r \frac{\partial H_z}{\partial t}\end{aligned} \tag{7.4}$$

Each electric field component (E_x, E_y, and E_z) is located at the midway point between a pair of magnetic field components (H_x, H_y, and H_z) and conversely is shown in Figure 7.1.

To have a better understanding of the mathematics of the FDTD, a simple one-dimensional (1-D) propagation wave is used as an example to link the electromagnetic theory with the FDTD. Assume that the polarized electromagnetic wave propagates in a z-direction, and the both electric and magnetic energy fields (E_x and H_y components) are mutually perpendicular and vibrate with a phase difference of 90°, as shown in Figure 7.2. As discussed in Chapter 1, the propagated electromagnetic wave is caused by the mutual changing of electric energy and magnetic energy. When the E_x component reaches a maximum value, the value of H_y is equal to zero and vice versa. For the 1-D case, (7.3) and (7.4) can be reduced as (only E_x and H_y):

$$\frac{\partial E_x}{\partial z} = -\mu_o \mu_r \frac{\partial H_y}{\partial t} \tag{7.5}$$

$$-\frac{\partial H_y}{\partial z} = \varepsilon_o \varepsilon_r \frac{\partial E_x}{\partial t} \tag{7.6}$$

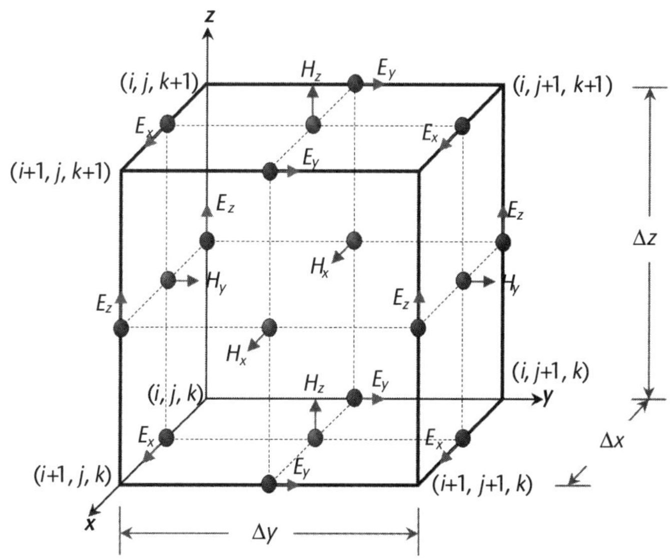

Figure 7.1 Position and direction of the field components within a rectangular unit cell of the Yee lattice in three dimensions.

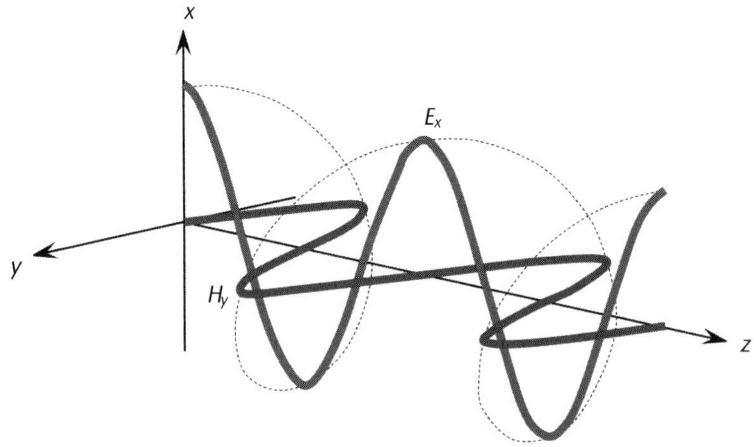

Figure 7.2 Polarization of E_x and H_y components that propagate in the z-direction.

In the FDTD solution, (7.5) and (7.6) are discretized using central-difference approximations to the space and time partial first derivatives as:

$$-\frac{E_x^n(k+1) - E_x^n(k)}{\Delta z} = \mu_o \mu_r \frac{H_y^{n+0.5}(k+0.5) - H_y^{n-0.5}(k+0.5)}{\Delta t} \tag{7.7}$$

$$-\frac{H_y^{n+0.5}(k+0.5) - H_y^{n+0.5}(k-0.5)}{\Delta z} = \varepsilon_o \varepsilon_r \frac{E_x^{n+1}(k) - E_x^n(k)}{\Delta t} \tag{7.8}$$

where superscript n represents the number of the time step and indicates that n + 1 is the next step forward in time. k in parentheses is the number of the spatial step. Δt and Δz are the time grid size and spatial grid size, respectively. Sometimes Δt and Δz are also known as the spatial and time increments, respectively. First, a rectangular grid in space z and time t needs to be identified. The points (z_k, t_n) on the grid can be defined as:

$$z_k = (k-1)\Delta z, \qquad k = 1,2,\ldots,N_z \tag{7.9a}$$

$$t_n = (n-1)\Delta t, \qquad n = 1,2,\ldots,N_t \tag{7.9b}$$

However, the grid points $(z_{k+0.5}, t_{n+0.5})$ at half-space and half-time can be defined as:

$$z_{k+0.5} = (k-0.5)\Delta z, \qquad k = 1,2,\ldots,N_z - 1 \tag{7.10a}$$

$$t_{n+0.5} = (n-0.5)\Delta t, \qquad n = 1,2,\ldots,N_t - 1 \tag{7.10b}$$

where the rectangular grid sizes (Δz and Δt) are determined as:

$$\Delta z = \frac{h}{N_z - 1} \tag{7.11}$$

where h is the total length of the simulation space. From (7.11), the time grid size can be determined as:

$$\Delta t = \frac{\Delta z}{2c} \tag{7.12}$$

where c is the velocity of light in free space. In fact, the electromagnetic field propagation is excited by a source. Various types of excitation source models have been implemented in the FDTD algorithms. One common and simple source model is the Gaussian pulse model:

$$E_{exc} = \exp\left[-\frac{\left(t - t_o\right)^2}{2\tau^2}\right] \tag{7.13}$$

where t_o ($\approx \pi \times \tau$) is the amount of time shift or time delay. τ is a parameter that determines the width of the Gaussian pulse in both the time domain and the frequency domain. The relationship between the τ and the maximum operation frequency $f_{\max}$ is given as [9]:

$$\tau = \frac{\sqrt{2.3}}{\pi f_{\max}} \tag{7.14a}$$

or

$$\tau = \frac{2\sqrt{2.3}}{\pi\left(f_{\max} - f_{\min}\right)} \tag{7.14b}$$

From (7.7) and (7.8), the update equations are given as:

$$H_y^{n+0.5}(k + 0.5) = H_y^{n-0.5}(k + 0.5) - \frac{\Delta t}{\mu_o \mu_r \Delta z}\left[E_x^n(k + 1) - E_x^n(k)\right] \tag{7.15}$$

$$E_x^{n+1}(k) = E_x^n(k) - \frac{\Delta t}{\varepsilon_o \varepsilon_r \Delta z}\left[H_y^{n+0.5}(k + 0.5) - H_y^{n+0.5}(k - 0.5)\right] \tag{7.16}$$

7.1.2 MATLAB Codes for 1-D FDTD

The different amplitudes of E_x and H_y may occur in (7.14) and (7.15). In fact, the relationship between the amplitudes of E_x and H_y are given as:

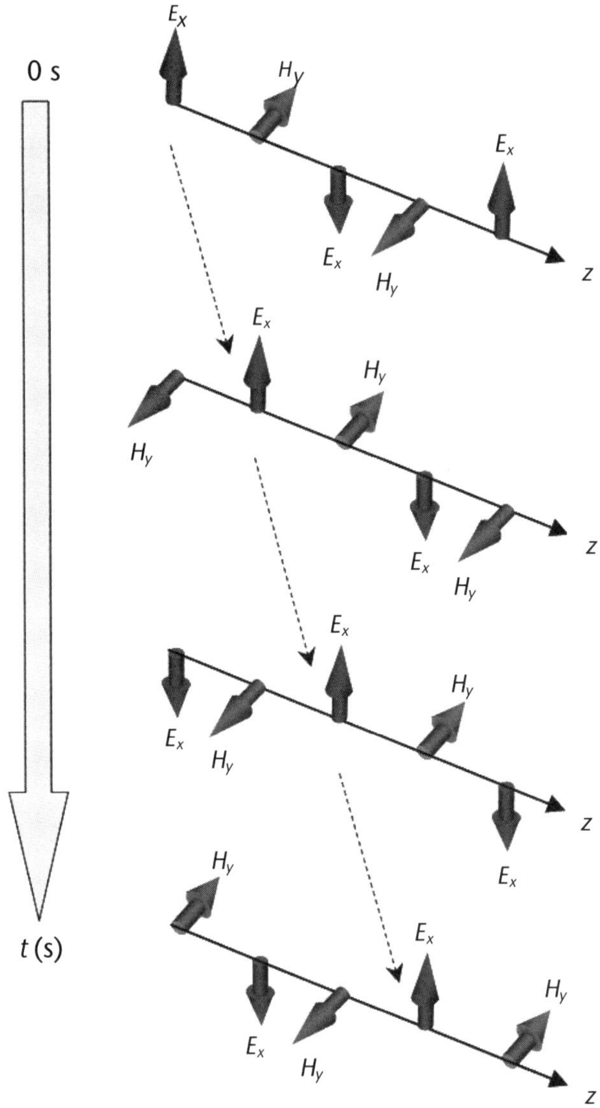

Figure 7.3 The shift of maximum E_x components with time t in the z-direction.

$$H_y = \frac{E_x}{\eta_o} = \sqrt{\frac{\varepsilon_o}{\mu_o}} E_x \tag{7.17}$$

Consequently, Sullivan [5] and Taflove and Hagness [6] introduced a normalization of the E_x field in FDTD (E_x is divided by intrinsic impedance of free space, $\eta_o = \sqrt{\mu_o/\varepsilon_o}$):

$$E_x = \sqrt{\frac{\mu_o}{\varepsilon_o}} \tilde{E}_x \tag{7.18}$$

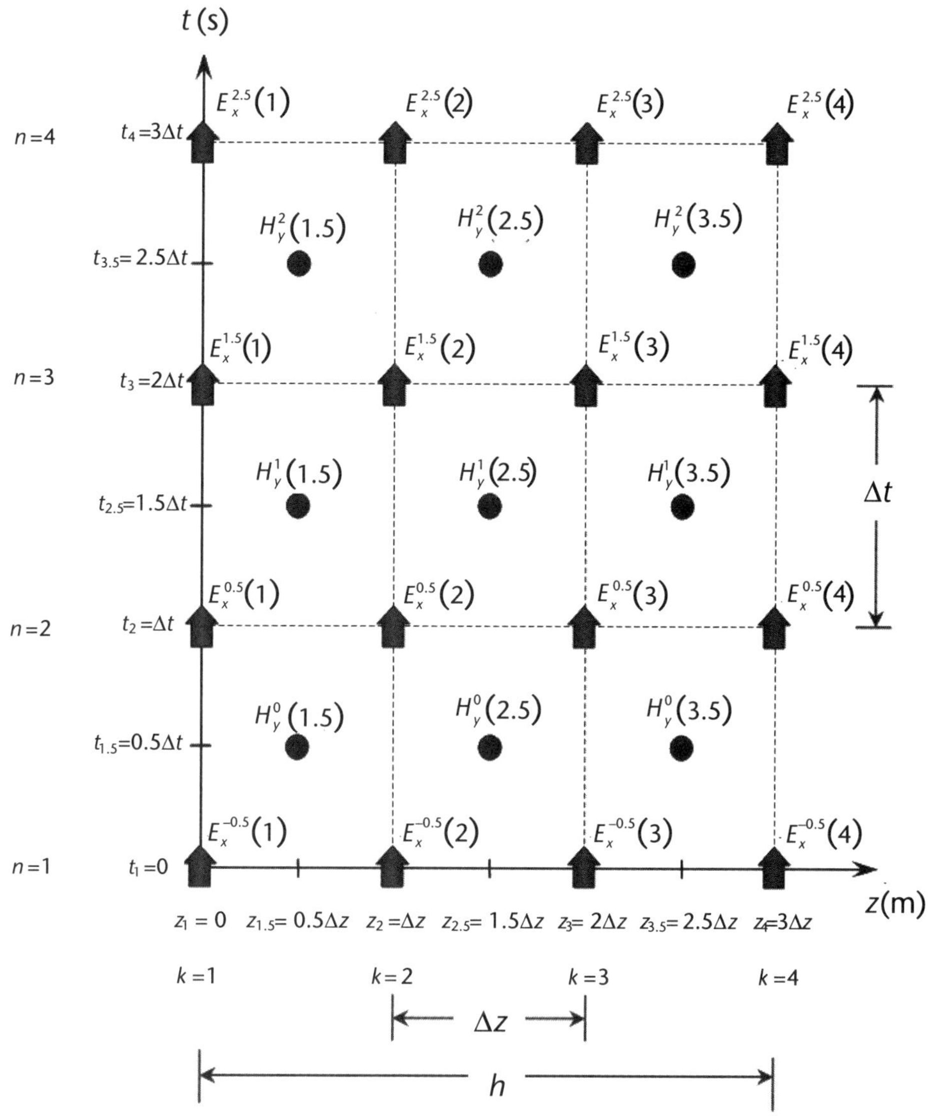

Figure 7.4 The corresponding FDTD grids for Figures 7.2 and 7.3.

Substitute (7.18) into (7.15) and (7.16), yielding

$$
\begin{aligned}
H_y^{n+0.5}(k+0.5) &= H_y^{n-0.5}(k+0.5) - \frac{\Delta t}{\mu_o \mu_r \Delta z}\sqrt{\frac{\mu_o}{\varepsilon_o}}\left[\tilde{E}_x^n(k+1) - \tilde{E}_x^n(k)\right] \\
&= H_y^{n-0.5}(k+0.5) - \frac{\Delta t}{\sqrt{\mu_o \varepsilon_o}\,\mu_r \Delta z}\left[\tilde{E}_x^n(k+1) - \tilde{E}_x^n(k)\right] \qquad (7.19) \\
&= H_y^{n-0.5}(k+0.5) - \frac{\Delta t}{c\mu_r \Delta z}\left[\tilde{E}_x^n(k+1) - \tilde{E}_x^n(k)\right]
\end{aligned}
$$

and

$$
\begin{aligned}
\sqrt{\frac{\mu_o}{\varepsilon_o}}\tilde{E}_x^{n+1}(k) &= \sqrt{\frac{\mu_o}{\varepsilon_o}}\tilde{E}_x^{n}(k) - \frac{\Delta t}{\varepsilon_o\varepsilon_r\Delta z}\left[H_y^{n+0.5}(k+0.5) - H_y^{n+0.5}(k-0.5)\right] \\
\tilde{E}_x^{n+1}(k) &= \tilde{E}_x^{n}(k) - \frac{\Delta t}{\varepsilon_o\varepsilon_r\Delta z}\sqrt{\frac{\varepsilon_o}{\mu_o}}\left[H_y^{n+0.5}(k+0.5) - H_y^{n+0.5}(k-0.5)\right] \\
&= \tilde{E}_x^{n}(k) - \frac{\Delta t}{\sqrt{\mu_o\varepsilon_o}\varepsilon_r\Delta z}\left[H_y^{n+0.5}(k+0.5) - H_y^{n+0.5}(k-0.5)\right] \\
&= \tilde{E}_x^{n}(k) - \frac{\Delta t}{c\varepsilon_r\Delta z}\left[H_y^{n+0.5}(k+0.5) - H_y^{n+0.5}(k-0.5)\right]
\end{aligned}
\tag{7.20}
$$

where $c = 1/\sqrt{\mu_o\varepsilon_o}$ is the velocity of light in free space. There are some issues that need to be considered when turning the FDTD or Yee's algorithm into MATLAB code, since MATLAB is only able to handle integer array indexes. Thus, the half of a cell shift $(k + 0.5)$ in (7.19) and (7.20) are ignored in the MATLAB codes, which are rewritten as:

$$
\underbrace{H_y(k)}_{\text{Current time step}} = \underbrace{H_y(k) - \frac{\Delta t}{c\mu_r\Delta z}\left[E_x(k+1) - E_x(k)\right]}_{\text{Previous half time step}}, \quad k = 1 \text{ to } N_z \tag{7.21}
$$

$$
\underbrace{E_x(k)}_{\text{Current time step}} = \underbrace{E_x(k) - \frac{\Delta t}{c\varepsilon_r\Delta z}\left[H_y(k) - H_y(k-1)\right]}_{\text{Previous half time step}}, \quad k = 2 \text{ to } N_z \tag{7.22}
$$

There are no time subscripts in (7.21) and (7.22). A cycle loop ("*for*" loop in MATLAB code) is implemented to compute the $H_y(k)$ and $E_x(k)$ fields for a certain number of time steps. The various $H_y(k)$ based on (7.21) are computed using the loops from $k = 1$ to $k = N_z$. However, $E_x(k)$ in (7.22) must run from $k = 2$ to $k = N_z$, rather than from 1 to N_z, because the $E_x(k)$ in (7.22) is required to be updated after initialization; the initial $H_y(1)$ value in (7.22) is computed from (7.21).

Figure 7.5 shows the simulation of propagation of E_x and H_y field components for 2 ns of time using the 1-D FDTD routine on the accompanying CD. The propagations of E_x and H_y are excited by a Gaussian pulse in the center of the simulation domain at the point $k_c = 25$. Later, the positive E_x field is propagated outward from the center in both direction. Simultaneously, the negative H_y field is propagated in the opposite direction of the z-axis, and the positive H_y field is in the z-axis.

7.2 Center-Fed Dipole

7.2.1 3-D FDTD Subcell Model

The subcell model normally is used in the FDTD computation for a thin dipole with a radius a, which is much smaller than its length, $2h$, and also much smaller as compared to the FDTD cell size as shown Figure 7.6.

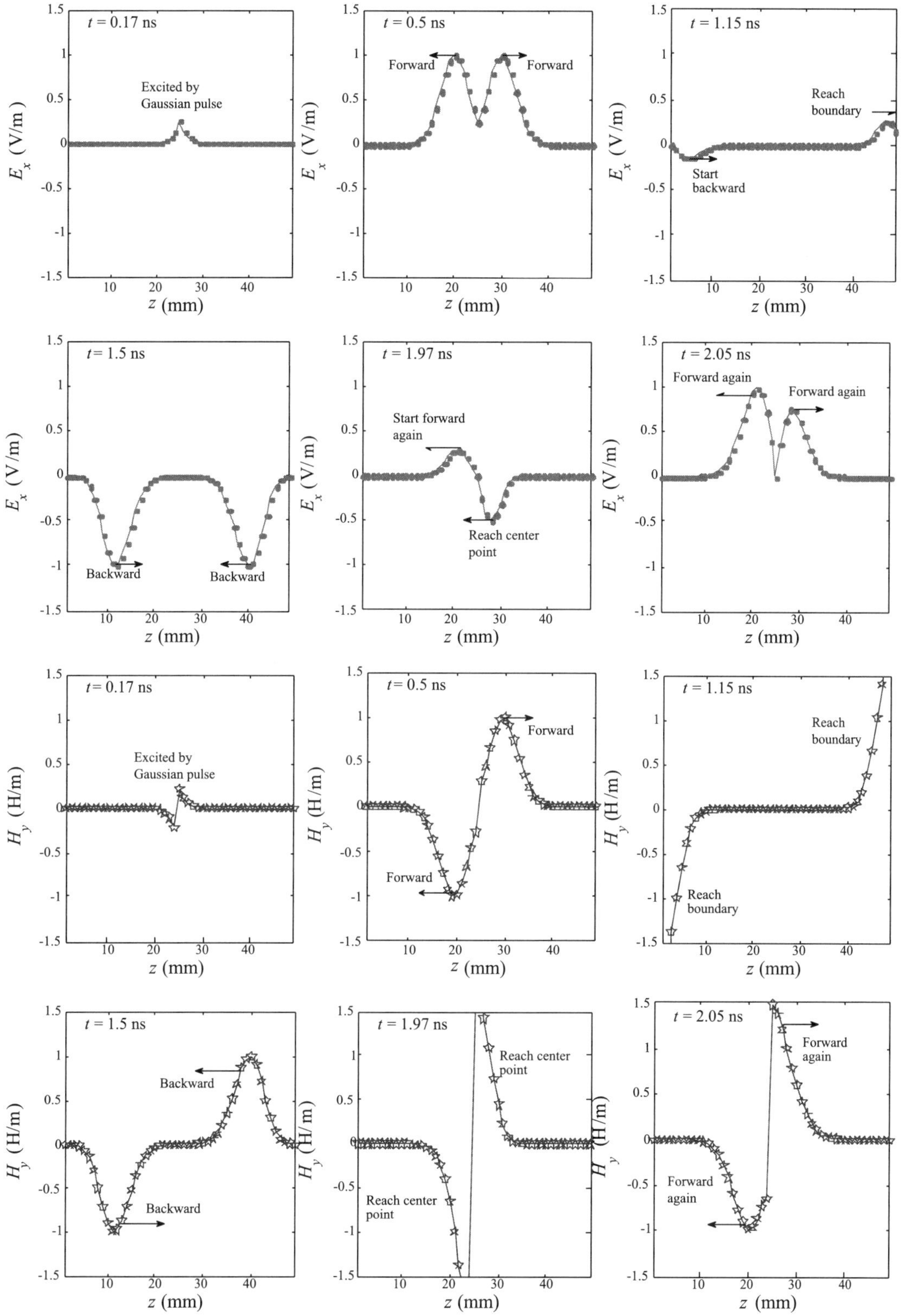

Figure 7.5 The 1-D FDTD simulation of propagation wave in free space for 2 ns of time.

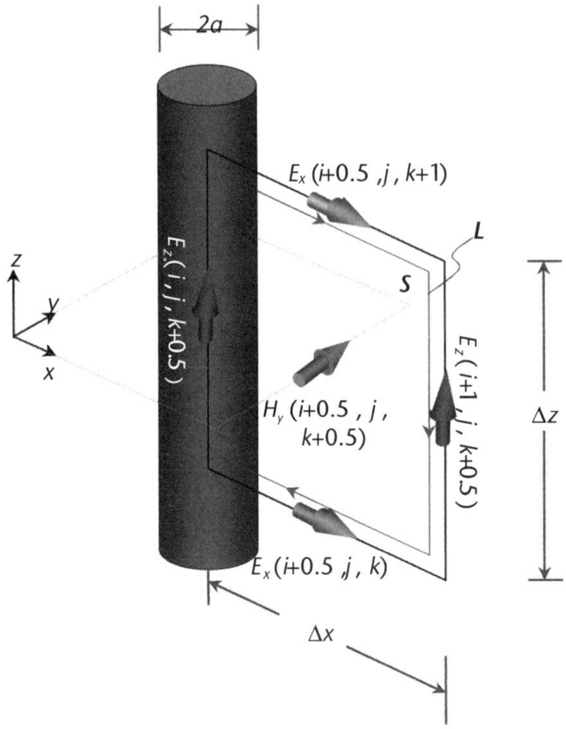

Figure 7.6 Faraday's law contour path for a thin-wire model.

As discussed in Chapter 1, the decay of a radial electric field E_x and an azimuth magnetic field H_y around the thin cylindrical conductor is proportional to $1/\rho$, where ρ is the radial distance from the center of the cylindrical conductor. Thus, H_y can be expressed as:

$$H_y = \frac{H_o}{\rho} \tag{7.23}$$

Similarly, E_x is given as:

$$E_x = \frac{E_o}{\rho} \tag{7.24}$$

where H_o and E_o are the amplitude constants. However, the axial electric field, E_z does not change in x and y directions. By referring to Figure 7.8, the unknown $H_y(i + 0.5, j, k + 0.5)$ component, which is circulated by E_z and E_x, is located at $\rho = \Delta x/2$; thus, (7.23) becomes:

$$H_y(i + 0.5, j, k + 0.5) = \frac{2H_o}{\Delta x} \tag{7.25}$$

From (7.25), the amplitude H_o for the local azimuthal magnetic field H_y can be written as:

$$H_o = \frac{\Delta x}{2} H_y(i + 0.5, j, k + 0.5) \tag{7.26}$$

By incorporating (7.26) into (7.23), H_y at any point ρ on the surface of the x-y plane within the local grid space can be predicted based on the value of $H_y(i + 0.5, j, k + 0.5)$ as:

$$H_y = \frac{H_o}{\rho} = \frac{\Delta x}{2\rho} H_y(i + 0.5, j, k + 0.5) \tag{7.27}$$

Similarly, the two radial electric fields E_x at the upper and lower H_y can be, respectively, expressed as:

$$E_x = \frac{\Delta x}{2\rho} E_x(i + 0.5, j, k + 1) \qquad \text{(upper)} \tag{7.28a}$$

$$E_x = \frac{\Delta x}{2\rho} E_x(i + 0.5, j, k) \qquad \text{(lower)} \tag{7.28b}$$

The two axial electric fields E_z at the both sides of vertical edge are written as:

$$E_z = E_z(i + 1, j, k + 0.5) \qquad \text{(right)} \tag{7.29a}$$

$$E_z = E_z(i, j, k + 0.5) \qquad \text{(left)} \tag{7.29b}$$

The update equations for the subcell FDTD solution can be derived by applying an integral form of Faraday's law to the rectangular surface S (x-z plane surface) with the bounding contour L [5]:

$$-\mu \iint_S \frac{\partial \vec{H}}{\partial t} \cdot d\vec{S} = \oint_L \vec{E} \cdot d\vec{l} \tag{7.30}$$

Incorporating (7.27) and (7.28) into (7.30) yields

$$\begin{aligned} &-\mu \int_0^{\Delta z} \int_a^{\Delta x} \frac{\partial}{\partial t} \frac{\Delta x}{2\rho} H_y(i + 0.5, j, k + 0.5) \cdot d\rho dz \\ &= \int_0^{\Delta z} E_z(i, j, k + 0.5)\, dz + \int_{\Delta z}^{0} E_z(i + 1, j, k + 0.5)\, dz \\ &+ \int_a^{\Delta x} \frac{\Delta x}{2\rho} E_x(i + 0.5, j, k + 1)\, d\rho + \int_{\Delta x}^{a} \frac{\Delta x}{2\rho} E_x(i + 0.5, j, k)\, d\rho \end{aligned} \tag{7.31}$$

$E_z(i, j, k + 0.5)$ in (7.31) is set to zero, since the electric field vanishes inside the conductor. The integral patch of E_x is limited from a to Δx in order to exclude the region inside the thin conductor. Thus,

$$-\mu\frac{\Delta z\Delta x}{2}\ln\left(\frac{\Delta x}{a}\right)\frac{\partial H_y(i+0.5,j,k+0.5)}{\partial t}$$

$$= 0 - \Delta z E_z(i+1,j,k+0.5) + \frac{\Delta x}{2}\ln\left(\frac{\Delta x}{a}\right)\left\{E_x(i+0.5,j,k+1) - E_x(i+0.5,j,k)\right\} \tag{7.32}$$

Discretizing the time derivative of H_y in (7.32) yields

$$-\mu\frac{\Delta z\Delta x}{2}\ln\left(\frac{\Delta x}{a}\right)\frac{H_y^{n+0.5}(i+0.5,j,k+0.5) - H_y^{n-0.5}(i+0.5,j,k+0.5)}{\Delta t}$$

$$= -\Delta z E_z^n(i+1,j,k+0.5) + \frac{\Delta x}{2}\ln\left(\frac{\Delta x}{a}\right)\left[E_x^n(i+0.5,j,k+1) - E_x^n(i+0.5,j,k)\right] \tag{7.33}$$

Rearranging (7.33) yields

$$H_y^{n+0.5}(i+0.5,j,k+0.5)$$

$$= H_y^{n-0.5}(i+0.5,j,k+0.5) - \frac{\Delta t}{\mu\Delta z}\left[E_x^n(i+0.5,j,k+1) - E_x^n(i+0.5,j,k)\right] \tag{7.34}$$

$$+ \frac{2\Delta t}{\mu\Delta x\ln(\Delta x/a)}E_z^n(i+1,j,k+0.5)$$

The same procedure can be used to obtain the update equation of H_x:

$$H_x^{n+0.5}(i,j+0.5,k+0.5)$$

$$= H_x^{n-0.5}(i,j+0.5,k+0.5) + \frac{\Delta t}{\mu\Delta z}\left[E_y^n(i,j+0.5,k+1) - E_y^n(i,j+0.5,k)\right] \tag{7.35}$$

$$- \frac{2\Delta t}{\mu\Delta y\ln(\Delta x/a)}E_z^n(i,j+1,k+0.5)$$

From (7.34) and (7.35), the current I through the center (i_c, j_c, k_c) of the thin conductor (dipole), which is normalized with free space impedance, $\sqrt{\mu_o/\varepsilon_o} \approx 376.73\Omega$, can be determined as:

$$
\begin{aligned}
\tilde{I} &= \sqrt{\frac{\varepsilon_o}{\mu_o}} \oint \vec{H} \cdot d\vec{l} \\
&= \sqrt{\frac{\varepsilon_o}{\mu_o}} \left\{ \begin{aligned} &\int_0^{\Delta x} H_x^{n+0.5}(i_c, j_c - 0.5, k_c + 0.5) + \int_0^{\Delta y} H_y^{n+0.5}(i_c + 0.5, j_c, k_c + 0.5) \\ &+ \int_{\Delta x}^{0} H_x^{n+0.5}(i_c, j_c + 0.5, k_c + 0.5) + \int_{\Delta y}^{0} H_y^{n+0.5}(i_c - 0.5, j_c, k_c + 0.5) \end{aligned} \right\} \\
&= \sqrt{\frac{\varepsilon_o}{\mu_o}} \left\{ \begin{aligned} &\left[H_x^{n+0.5}(i_c, j_c - 0.5, k_c + 0.5) - H_x^{n+0.5}(i_c, j_c + 0.5, k_c + 0.5)\right]\Delta x \\ &+ \left[H_y^{n+0.5}(i_c + 0.5, j_c, k_c + 0.5) - H_y^{n+0.5}(i_c - 0.5, j_c, k_c + 0.5)\right]\Delta y \end{aligned} \right\}
\end{aligned}
\tag{7.36}
$$

To ensure the stability of the 3-D FDTD computation, the time step Δt (time increment) in (7.34) and (7.35) must satisfy the following condition:

$$
\Delta t \leq \frac{1}{c\sqrt{(1/\Delta x)^2 + (1/\Delta y)^2 + (1/\Delta z)^2}} \tag{7.37}
$$

For a cubical spatial grid ($\Delta x = \Delta y = \Delta z$), (7.37) can be reduced as a Courant condition:

$$
\Delta t \leq \frac{\Delta x}{c\sqrt{\alpha}} \tag{7.38}
$$

where α is the dimension of the FDTD simulation. Generally, the value of Δt is determined as [4, 6, 10]:

$$
\Delta t = \frac{\Delta x}{2c} \tag{7.39}
$$

7.2.2 Improvement of the 3-D FDTD Subcell Model

It is known that when a current flows through the dipole, it will induce a tangential azimuthal magnetic field H_ϕ along a circular patch around the dipole as shown in Figure 7.7.

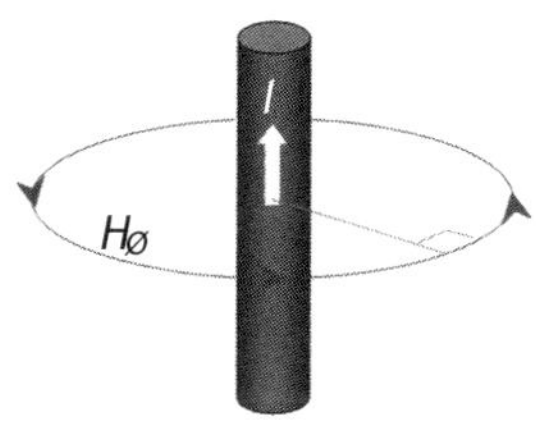

Figure 7.7 Circular path of H_ϕ around the dipole.

The corresponding magnetic field components H_x and H_y in Cartesian coordinates of the azimuth magnetic field, H_ϕ, are shown in Figure 7.8(a). However, the edge of the Cartesian cell in the FDTD solution is in a rectangular shape as shown in Figure 7.8(b). Thus, the actual H_x and H_y components in Figure 7.8(c) are required in order to project them onto the Cartesian cell edges by multiplying their fields with scaling factor K_H, as shown in Figure 7.8(d).

The curvature improvement in the FDTD is reported by Mäkinen et al. [11]. The expression of the scaling factor K_H is derived using the relative length of the integration circular path C_1 and the rectangle path C_2. K_H is given as:

$$K_H = \frac{\oint_{C_1} dl_1}{\oint_{C_2} dl_2} = \left(\frac{\Delta x}{\Delta y}\right)\int_0^{\Delta y/2} \frac{\cos\phi}{\sqrt{(\Delta x/2)^2 + y^2}} \cdot dy = \left(\frac{\Delta x}{\Delta y}\right)\int_0^{\Delta y/2} \frac{\Delta x/2}{(\Delta x/2)^2 + y^2} \cdot dy$$
$$= \left(\frac{\Delta x}{\Delta y}\right)\tan^{-1}\left(\frac{\Delta y}{\Delta x}\right) \quad (7.40)$$

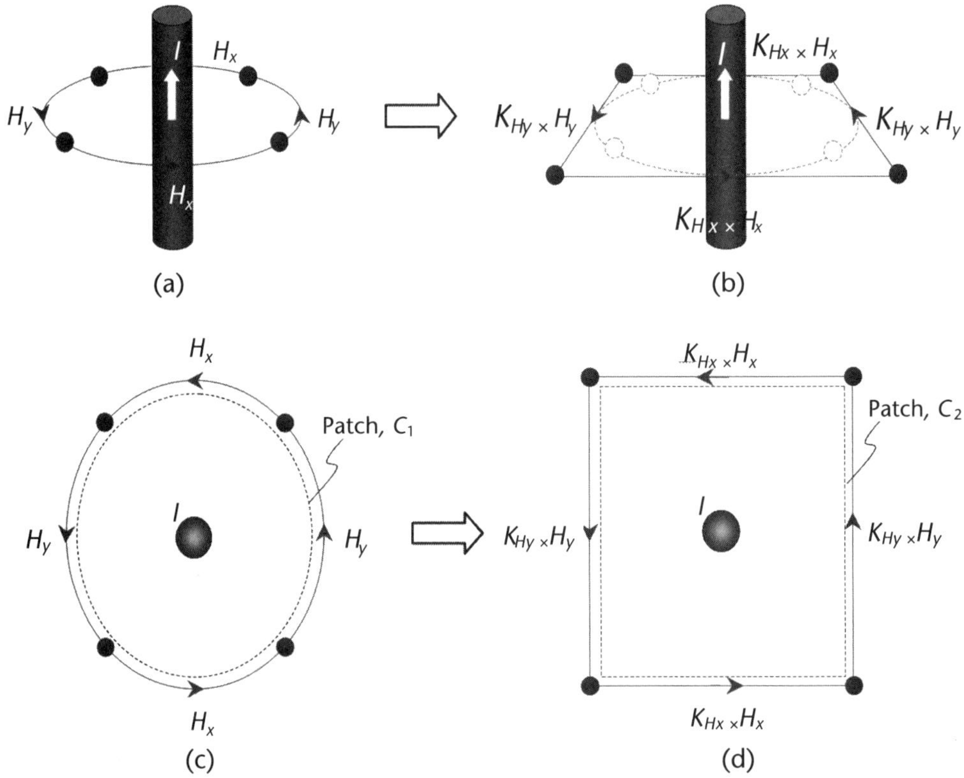

Figure 7.8 Projection of (a) looping magnetic field component to (b) Cartesian cell edges. A cross-sectional view of (c) the circular path and (d) Cartesian cell edge integrations.

The derivation of (7.40) refers to a magnified quarter of the FDTD cell edge as shown in Figure 7.9. The cross-sectional view of the quarter cell displays a clear projection of the circular H_y to the Cartesian cell edge.

Hence, the H_y component in the update equation of E_x is projected to a Cartesian cell edge by multiplying H_y with K_{Hy}. Finally, the corrected update equation of E_x is reexpressed as (7.41d) in which the average value for the radial electric field flowing through a Cartesian cell surface is computed.

However, the E_x component in the updated equation of H_y in (7.34) is projected from the Cartesian cell surface to a cylindrical surface as shown in Figure 7.10. Now,

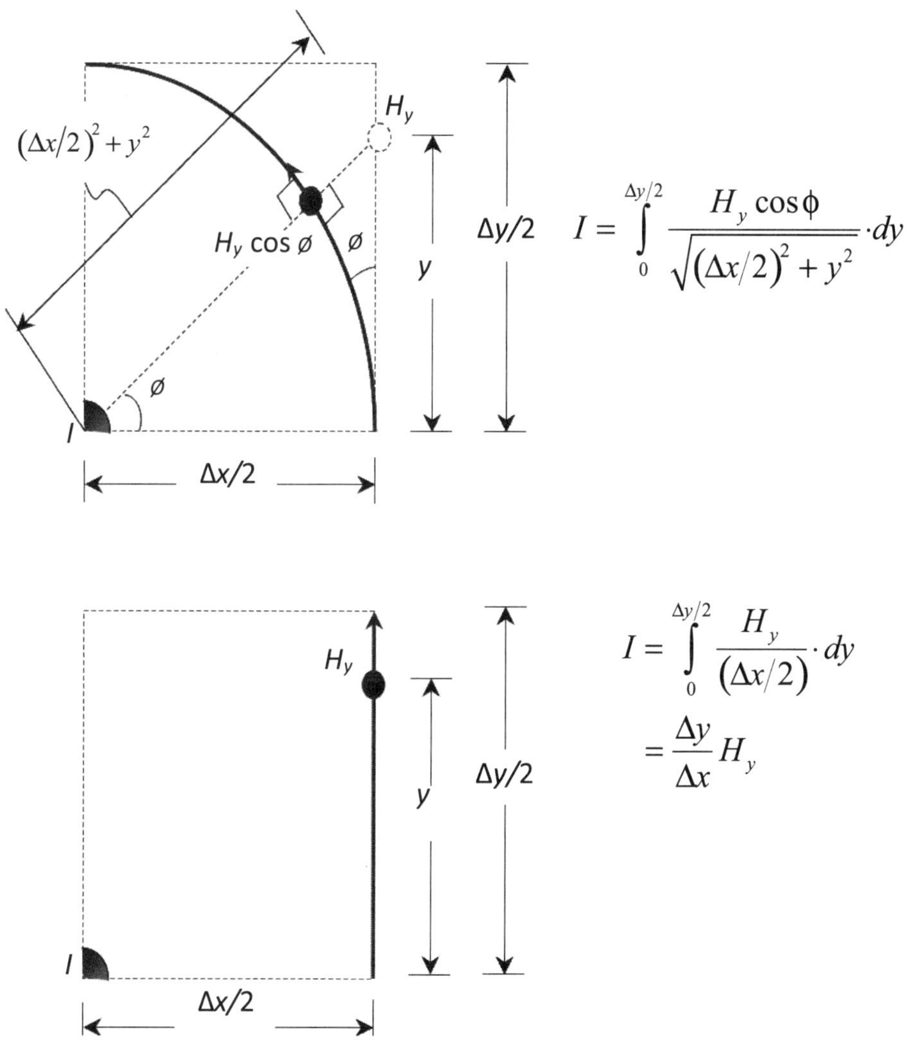

Figure 7.9 The projection of a magnetic field component H_y that is on a circular edge to a magnetic field component now along the Cartesian cell edge aided by trigonometry.

the modified update equation of H_y component is rewritten as (7.41b), which represents the value of circular magnetic field around the dipole as shown in Figure 7.8(c).

The six field components in the 3-D FDTD solution (E_x, E_y, E_z, H_x, H_y, and H_z) are shown in Figure 7.11.

For magnetic field components, these are:

$$\begin{aligned} H_x^{n+0.5}(i,j+0.5,k+0.5) &= H_x^{n-0.5}(i,j+0.5,k+0.5) \\ &+\frac{K_{Ey}\Delta t}{\mu \Delta z}\left[E_y^n(i,j+0.5,k+1)-E_y^n(i,j+0.5,k)\right] \\ &-\frac{2\Delta t}{\mu \Delta y \ln(\Delta x/a)}\left[E_z^n(i,j+1,k+0.5)-E_z^n(i,j,k+0.5)\right] \end{aligned} \tag{7.41a}$$

$$\begin{aligned} H_y^{n+0.5}(i+0.5,j,k+0.5) &= H_y^{n-0.5}(i+0.5,j,k+0.5) \\ &-\frac{K_{Ex}\Delta t}{\mu \Delta z}\left[E_x^n(i+0.5,j,k+1)-E_x^n(i+0.5,j,k)\right] \\ &+\frac{2\Delta t}{\mu \Delta x \ln(\Delta x/a)}\left[E_z^n(i+1,j,k+0.5)-E_z^n(i,j,k+0.5)\right] \end{aligned} \tag{7.41b}$$

$$\begin{aligned} H_z^{n+0.5}(i+0.5,j+0.5,k) &= H_z^{n-0.5}(i+0.5,j+0.5,k) \\ &+\frac{\Delta t}{\mu \Delta y}\left[E_x^n(i+0.5,j+1,k)-E_x^n(i+0.5,j,k)\right] \\ &-\frac{\Delta t}{\mu \Delta x}\left[E_y^n(i+1,j+0.5,k)-E_y^n(i,j+0.5,k)\right] \end{aligned} \tag{7.41c}$$

For electric field components, these are:

$$\begin{aligned} E_x^{n+1}(i+0.5,j,k) &= E_x^n(i+0.5,j,k) \\ &-\frac{K_{Hy}\Delta t}{\varepsilon \Delta z}\left[H_y^{n+0.5}(i+0.5,j,k+0.5)-H_y^{n+0.5}(i+0.5,j,k-0.5)\right] \\ &+\frac{\Delta t}{\varepsilon \Delta y}\left[H_z^{n+0.5}(i+0.5,j+0.5,k)-H_z^{n+0.5}(i+0.5,j-0.5,k)\right] \end{aligned} \tag{7.41d}$$

$$\begin{aligned} E_y^{n+1}(i,j+0.5,k) &= E_y^n(i,j+0.5,k) \\ &+\frac{K_{Hx}\Delta t}{\varepsilon \Delta z}\left[H_x^{n+0.5}(i,j+0.5,k+0.5)-H_x^{n+0.5}(i,j+0.5,k-0.5)\right] \\ &-\frac{\Delta t}{\varepsilon \Delta x}\left[H_z^{n+0.5}(i+0.5,j+0.5,k)-H_z^{n+0.5}(i-0.5,j+0.5,k)\right] \end{aligned} \tag{7.41e}$$

$$E_z^{n+1}(i,j,k+0.5) = E_z^{n}(i,j,k+0.5) + \frac{\Delta t}{\varepsilon \Delta x}\left[H_y^{n+0.5}(i+0.5,j,k+0.5) - H_x^{n+0.5}(i-0.5,j,k+0.5)\right] - \frac{\Delta t}{\varepsilon \Delta y}\left[H_x^{n+0.5}(i,j+0.5,k+0.5) - H_x^{n+0.5}(i,j-0.5,k+0.5)\right] \quad (7.41f)$$

where the scaling factors (K_{Hy}, K_{Hx}, K_{Ex}, and K_{Ey}) in (7.41a) through (7.41f) are expressed as:

$$K_{Hy} = \left(\frac{\Delta x}{\Delta y}\right)\tan^{-1}\left(\frac{\Delta y}{\Delta x}\right) \quad (7.42a)$$

$$K_{Hx} = \left(\frac{\Delta y}{\Delta x}\right)\tan^{-1}\left(\frac{\Delta x}{\Delta y}\right) \quad (7.42b)$$

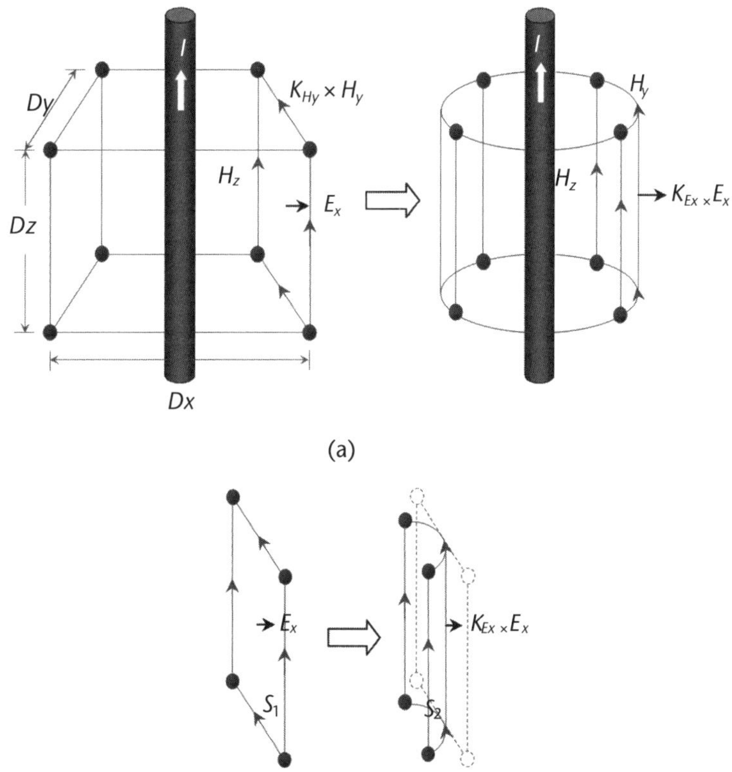

Figure 7.10 (a) Relationship between Cartesian surface and cylindrical surface. (b) Projection of the radial electric field E_x component to a cylindrical surface.

$$K_{Ex} = \left[\left(\frac{\Delta x}{\Delta y}\right)\tan^{-1}\left(\frac{\Delta y}{\Delta x}\right)\right]^{-1} \tag{7.42c}$$

$$K_{Ey} = \left[\left(\frac{\Delta y}{\Delta x}\right)\tan^{-1}\left(\frac{\Delta x}{\Delta y}\right)\right]^{-1} \tag{7.42d}$$

Finally, the normalized total current around the excitation grid cell (i_c, j_c, k_c) can be rewritten as:

$$\begin{aligned}\tilde{I}^{n+0.5}\left(i_c, j_c, k_c + 0.5\right) = &\sqrt{\frac{\varepsilon_o}{\mu_o}}\begin{bmatrix} H_x^{n+0.5}\left(i_c, j_c - 0.5, k_c + 0.5\right) \\ -H_x^{n+0.5}\left(i_c, j_c + 0.5, k_c + 0.5\right)\end{bmatrix} K_{Hx}\Delta x \\ &+\sqrt{\frac{\varepsilon_o}{\mu_o}}\begin{bmatrix} H_y^{n+0.5}\left(i_c + 0.5, j_c, k_c + 0.5\right) \\ -H_y^{n+0.5}\left(i_c - 0.5, j_c, k_c + 0.5\right)\end{bmatrix} K_{Hy}\Delta y\end{aligned} \tag{7.43}$$

Normally, the current flow at the dipole or monopole end is assumed to be zero in order to simplify the numerical computational. In fact, the currents at the both ends of the dipole are not equal to zero due to a charge distributed on the end cap. The accumulated charges on both dipole ends will produce a small current flow to the end cap, which will affect the local electromagnetic fields (reflection of the electromagnetic fields), especially for the dipole with a thicker radius. Mäkinen et al. [11] corrected the end cap current in the FDTD computation using an electrostatic line

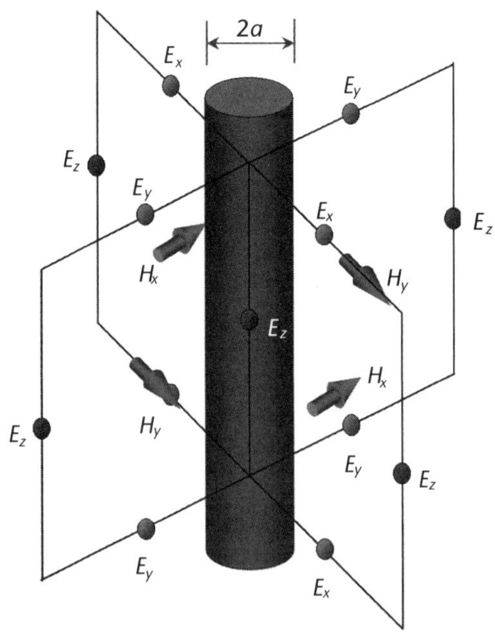

Figure 7.11 Electric field and magnetic field components surrounding the thin conductor.

charge approach (uniformly charged ring model), in which the charge is uniformly distributed over the ring as shown in Figure 7.12(a). The modified update equations with auxiliary correction terms near the end cap of the dipole proposed by Mäkinen et al. [11] are summarized next. Hyun et al. [12] later treated this problem again using an electrostatic surface charge approach [a uniformly charged disk model as shown in Figure 7.12(b)] in order to reduce the complexity of the formulations. The modified updated equations without auxiliary correction terms are described below.

The updated equations at the end of the dipole (i, j, k_{end}) proposed by Mäkinen et al. [11] are:

$$\begin{aligned} E_x^{n+1}\left(i+0.5, j, k_{\text{end}}\right) &= E_x^n\left(i+0.5, j, k_{\text{end}}\right) + \frac{k_M}{k_E} E_{x,ave\_q}^{n+0.5}\left(i+0.5, j, k_{\text{end}}\right) \\ &- \frac{\Delta t}{\varepsilon \Delta z}\left[H_y^{n+0.5}\left(i+0.5, j, k_{\text{end}}+0.5\right) - H_y^{n+0.5}\left(i+0.5, j, k_{\text{end}}-0.5\right)\right] \\ &+ \frac{\Delta t}{\varepsilon \Delta y}\left[H_z^{n+0.5}\left(i+0.5, j+0.5, k_{\text{end}}\right) - H_z^{n+0.5}\left(i+0.5, j-0.5, k_{\text{end}}\right)\right] \end{aligned} \tag{7.44a}$$

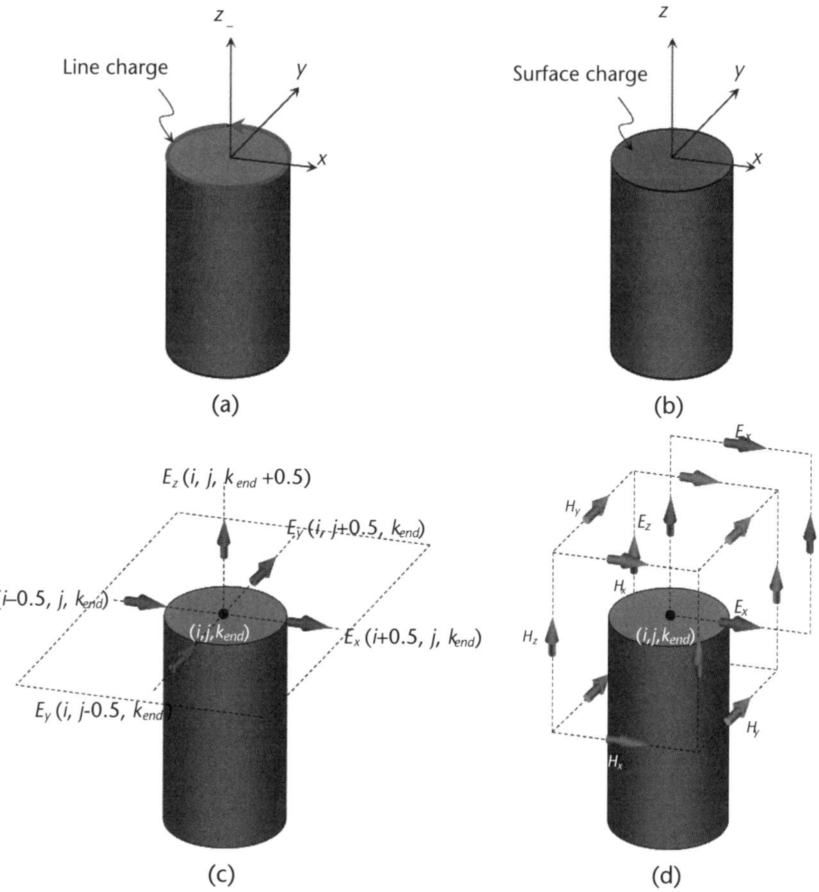

Figure 7.12 (a) Ring line charge [11]. (b) Disk surface charge [12]. (c) Electric and (d) magnetic field components around the finite end of the dipole.

$$E_z^{n+1}\left(i,j,k_{\text{end}}+0.5\right)=E_z^{n}\left(i,j,k_{\text{end}}+0.5\right)+E_{z,ave\_q}^{n+0.5}\left(i,j,k_{\text{end}}+0.5\right)$$
$$+\frac{\Delta t}{\varepsilon\Delta x}\left[H_y^{n+0.5}\left(i+0.5,j,k_{\text{end}}+0.5\right)-H_x^{n+0.5}\left(i-0.5,j,k_{\text{end}}+0.5\right)\right] \quad (7.44b)$$
$$-\frac{\Delta t}{\varepsilon\Delta y}\left[H_x^{n+0.5}\left(i,j+0.5,k_{\text{end}}+0.5\right)-H_x^{n+0.5}\left(i,j-0.5,k_{\text{end}}+0.5\right)\right]$$

K_M, K_E, $E_{x,ave\_q}^{n+0.5}(i+0.5, j, k_{\text{end}})$, and $E_{z,ave\_q}^{n+0.5}(i, j, k_{\text{end}}+0.5)$ terms in (7.44a) and (7.44b) are given as:

$$K_M=\left(1-\frac{a}{\Delta y}\right) \quad (7.45a)$$

$$K_E=\left[\left(\frac{\Delta x}{\Delta y}\right)\tan^{-1}\left(\frac{\Delta y}{\Delta x}\right)\right]^{-1} \quad (7.45b)$$

$$E_{x,ave\_q}^{n+0.5}\left(i+0.5,j,k_{\text{end}}\right)=\begin{bmatrix}\left(1/\sqrt{2}\right)+0.166(a/\Delta x)\\ -3.65(a/\Delta x)^2+8.91(a/\Delta x)^3\\ -16.1(a/\Delta x)^4\end{bmatrix}E_{\rho,q}^{n+0.5} \quad (7.45c)$$

$$E_{z,ave\_q}^{n+0.5}\left(i,j,k_{\text{end}}+0.5\right)=\begin{bmatrix}(\pi/6)-0.111(a/\Delta z)\\ +3.31(a/\Delta z)^2-2.51(a/\Delta z)^3\end{bmatrix}E_{z,q}^{n+0.5} \quad (7.45d)$$

$$E_{z,q}^{n+0.5}=\frac{\Delta q^{n+0.5}}{4\pi\varepsilon_o}\frac{(\Delta z/2)}{\left[a^2+(\Delta z/2)^2\right]^{3/2}} \quad (7.45e)$$

$$E_{\rho,q}^{n+0.5}=\frac{\Delta q^{n+0.5}}{4\pi\varepsilon_o}\frac{(\Delta x/2-a)\mathrm{K}\left(k_1\right)+(\Delta x/2+a)\mathrm{E}\left(k_1\right)}{\pi(\Delta x/2)\left[(\Delta x/2)^2-a^2\right]} \quad (7.45f)$$

where $K(k_1)$ and $E(k_1)$ are the complete elliptic integrals of the first and second kind, respectively. The parameter k_1 is expressed as:

$$k_1=\frac{2\sqrt{a(\Delta x/2)}}{a+(\Delta x/2)} \quad (7.46)$$

The change Δq in the total charge q at the dipole end is written as:

$$\Delta q^{n+0.5} = -\left(I_{k_{\text{end}}-0.5}^{n+0.5}\right)\Delta t$$
$$= -\Delta t \left\{ \begin{bmatrix} H_y^{n+0.5}\left(i+0.5, j, k_{\text{end}}-0.5\right) \\ -H_y^{n+0.5}\left(i-0.5, j, k_{\text{end}}-0.5\right) \end{bmatrix} K_{Hy}\Delta y - \begin{bmatrix} H_x^{n+0.5}\left(i, j+0.5, k_{\text{end}}-0.5\right) \\ -H_x^{n+0.5}\left(i, j-0.5, k_{\text{end}}-0.5\right) \end{bmatrix} K_{Hx}\Delta x \right\} \quad (7.47)$$

and $I_{k_{\text{end}}-0.5}^{n+0.5}$ is the current at the one-half space cell below the dipole end.

On the other hand, the updated equations at the end of the dipole (i, j, k_{end}) proposed by Hyun et al. [12]:

$$\begin{aligned} H_y^{n+0.5}\left(i+0.5, j, k_{\text{end}}+0.5\right) &= H_y^{n-0.5}\left(i+0.5, j, k_{\text{end}}+0.5\right) \\ &- \frac{\Delta t}{\mu\Delta z}\left[E_x^n\left(i+0.5, j, k_{\text{end}}+1\right) - E_x^n\left(i+0.5, j, k_{\text{end}}\right)L_{Ex}\right] \\ &+ \frac{\Delta t}{\mu\Delta x}\left[E_z^n\left(i+1, j, k_{\text{end}}+0.5\right) - E_z^n\left(i, j, k_{\text{end}}+0.5\right)L_{Ez}\right] \end{aligned} \quad (7.48a)$$

$$\begin{aligned} E_x^{n+1}\left(i+0.5, j, k_{\text{end}}\right) &= E_x^n\left(i+0.5, j, k_{\text{end}}\right) \\ &- \frac{\Delta t}{\varepsilon\Delta z}\frac{1}{S_{Ex}}\left[H_y^{n+0.5}\left(i+0.5, j, k_{\text{end}}+0.5\right) - H_y^{n+0.5}\left(i+0.5, j, k_{\text{end}}-0.5\right)L_{Hy}\right] \\ &+ \frac{\Delta t}{\varepsilon\Delta y}\frac{1}{S_{Ex}}\left[H_z^{n+0.5}\left(i+0.5, j+0.5, k_{\text{end}}\right) - H_z^{n+0.5}\left(i+0.5, j-0.5, k_{\text{end}}\right)\right] \end{aligned}$$

(7.48b)

where the parameters L_{Ex}, L_{Ez}, L_{Hy}, and S_{Ex} in (7.48) are given as:

$$L_{Ex} = 0.25\left(\frac{\Delta x}{a} - 1\right) \quad (7.49a)$$

$$L_{Ez} = \frac{\sqrt{a^2 + 0.5\Delta z)^2}}{\sqrt{a^2 + (0.5\Delta z)^2} - 0.5\Delta z}\left(a^2 + \Delta z^2 - \sqrt{a^2 + \Delta z^2}\right) \quad (7.49b)$$

$$L_{Hy} = \left(\frac{\Delta x}{\Delta y}\right)\tan^{-1}\left(\frac{\Delta y}{\Delta x}\right) \quad (7.49c)$$

$$S_{Ex} = \frac{0.5}{\sqrt{(0.5\Delta x)^2 + (0.5\Delta y)^2}} \quad (7.49d)$$

Apart from the six update equations, there are two more issues that need to be identified, namely the field excitation source and the edge boundary of the solution domain. The overall configuration of the FDTD solutions for a dipole is shown in Figure 7.13. The excitation source needs to be focused on before discussing the issues on the domain boundary.

7.2.3 Excitation Source

The source field is conventionally assumed to be fed at the center gap of the dipole with one spatial interval Δz in length as shown in Figure 7.13. Although this model

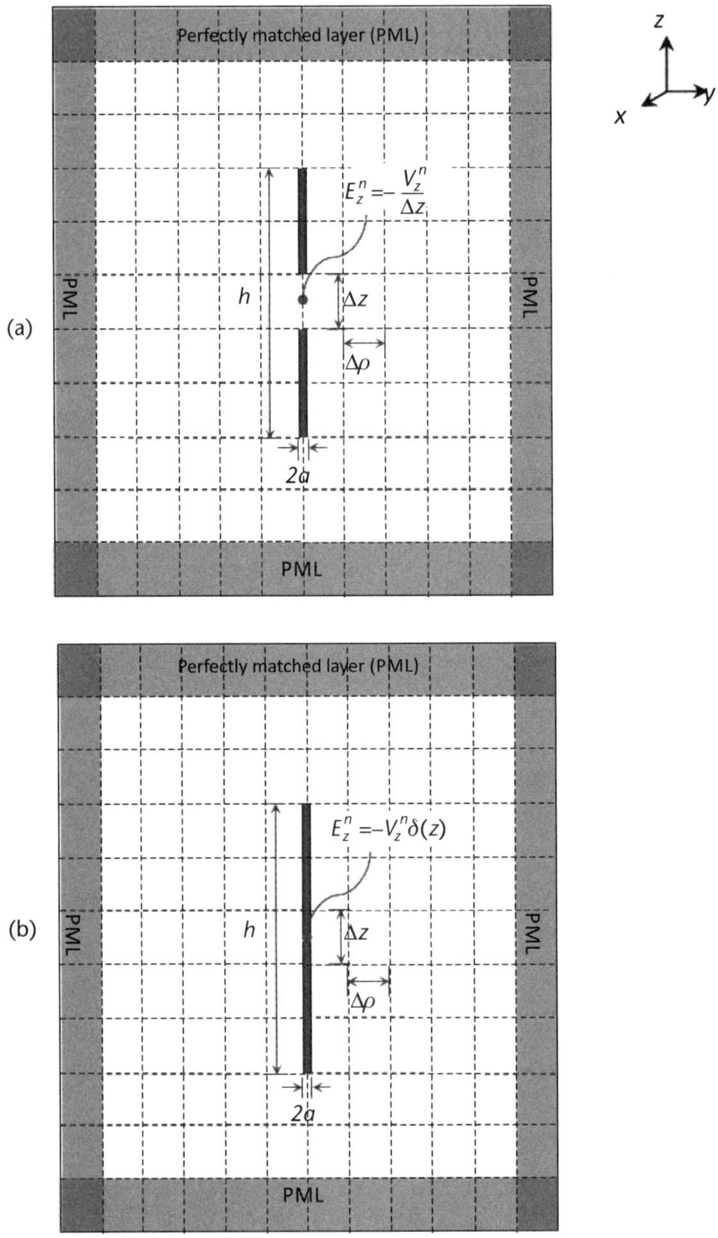

Figure 7.13 (a) Gap-feed and (b) infinitesimal-gap feed located at the center of dipole.

does not correspond exactly to the geometry of the feed region for the actual dipole, it is simple and saves computational resources. The excitation electric field E_{exc} in this gap is related to the terminal voltage V_{inc} between the drive-point terminals by:

$$V_{inc} = -\int_0^{\Delta z} E_{exc}\, dz \tag{7.50}$$

and

$$E_{exc}^{n}\left(i_c, j_c, k_c\right) = -\frac{V_{inc}^{n}\left(i_c, j_c, k_c\right)}{\Delta z} \tag{7.51}$$

Equation (7.51) used for the FDTD corresponds to the MoM delta-gap source model. However, this model will cause cell-size dependent errors when the chosen size of Δz is not sufficiently small. Normally, the actual feeding gap must be smaller than the cell size Δz and the fed region is assumed to be gapless. Watanabe et al. [13] proposed the infinitesimal-gap source as:

$$E_{exc} = -V_{inc}\delta(z) \tag{7.52}$$

where $\delta(z)$ is the Dirac delta function. Equation (7.52) has significantly reduced the errors that were caused by the size of the FDTD cell.

7.2.4 Boundary Conditions for 3-D FDTD

It is known that the scattering field of the dipole and the open-ended coaxial probe radiate into an unbounded space. However, in an FDTD solution, the solution domain, in which the field is computed, must be limited in dimension. Thus, a large simulation space is required to cover the entire scattering field in order to avoid scattering field reflections back into the simulation domain due to the mismatch at the spatial boundary. However, a large simulation domain will cause high computational cost. Thus, Mur [14] was the first who proposed an artificial condition at the simulation boundary. This absorbing boundary condition (ABC) reduces the effect of the scattering field reflection and at the same time truncates the simulation space. Later, the most efficient ABC was developed by Berenger [15], who assumed that the simulation space is surrounded by a thin artificial absorbing layer, the perfectly matched layer (PML), as shown in Figure 7.13. Subsequently, many researchers modified Mur and Berenger's work to improve the simplicity and the efficiency of absorption of the ABC or PML [2, 6, 10, 16].

In this section, the uniaxial perfectly matched layer (UPML) is the only method that will be discussed in detail [5, 16]. The UPML is modeled as a lossy uniaxial medium to absorb incident waves at any angle to it without reflection. The relative permittivity ε_r of the high lossy anisotropic medium can be written as:

$$\tilde{\varepsilon}_r = \tilde{\varepsilon}_r' - j\tilde{\varepsilon}_r'' = \left(\tilde{\varepsilon}_r' - j\frac{\sigma}{\omega\varepsilon_o}\right) \tag{7.53}$$

The real part of relative permittivity ε_r' is the dielectric constant, and the imaginary part, ε_r'' is the dielectric loss factor. In FDTD computation, the value of ε_r' is set to be unity or adjusted to improve the numerical performance of the UPML. However, the loss factor, ε_r'' influences the energy absorption or attenuation of the material, which can be expressed in terms of conductivity, σ. The updated equations in the UPML region are derived using modified Maxwell's curl equations in the frequency domain as:

$$\vec{\nabla} \times \vec{H} = j\omega\varepsilon_o\varepsilon_r\tilde{\varepsilon}\vec{E} \tag{7.54a}$$

$$\vec{\nabla} \times \vec{E} = -j\omega\mu_o\mu_r\tilde{\mu}\vec{H} \tag{7.54b}$$

where the relative permittivity ε and the relative permeability μ in (7.54) are written in tensors form as:

$$\begin{aligned}\tilde{\varepsilon} = \tilde{\mu} &= \begin{bmatrix} s_x^{-1} & 0 & 0 \\ 0 & s_x & 0 \\ 0 & 0 & s_x \end{bmatrix}\begin{bmatrix} s_y & 0 & 0 \\ 0 & s_y^{-1} & 0 \\ 0 & 0 & s_y \end{bmatrix}\begin{bmatrix} s_z & 0 & 0 \\ 0 & s_z & 0 \\ 0 & 0 & s_z^{-1} \end{bmatrix} \\ &= \begin{bmatrix} s_y s_z / s_x & 0 & 0 \\ 0 & s_x s_z / s_y & 0 \\ 0 & 0 & s_x s_y / s_z \end{bmatrix}\end{aligned} \tag{7.55}$$

where

$$s_x = 1 + \frac{\sigma_x}{j\omega\varepsilon_o},\ s_y = 1 + \frac{\sigma_y}{j\omega\varepsilon_o},\ \text{and } s_z = 1 + \frac{\sigma_z}{j\omega\varepsilon_o} \tag{7.56}$$

The UPML terminated with an outer perfect electric conductor (PEC) boundary is shown in Figure 7.14.

Theoretically, the propagating E_x field along the normal x-axis of the UPML will be absorbed without any reflection if the σ_x variation is only along the x-direction while σ_y and σ_z remain constant along the y- and the z-directions. This behavior is similar for σ_y and σ_z. In fact, there are four conditions of the 3-D UPML that are required to be considered:

1. A normal x-direction towards the UPML, the $\sigma_x \neq 0$ and $\sigma_y = \sigma_z = 0$.

$$s_x = 1 + \frac{\sigma_x}{j\omega\varepsilon_o}$$
$$s_y = 1$$
$$s_z = 1$$

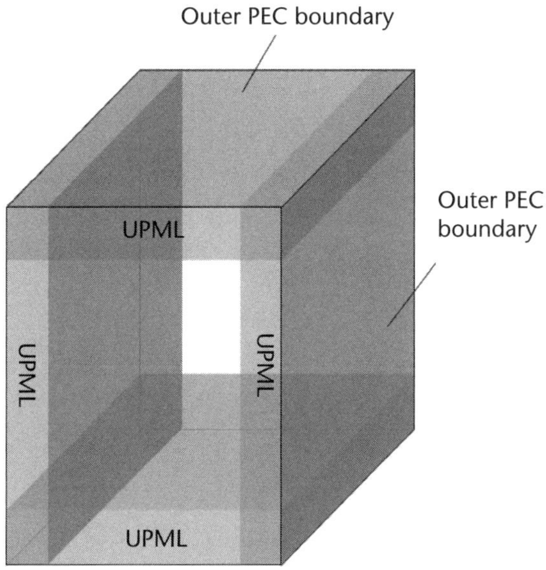

Figure 7.14 Lossy uniaxial perfectly matched layer (UPML) with outer PEC boundary.

2. A normal z-direction towards the UPML, the $\sigma_z \neq 0$ and $\sigma_x = \sigma_y = 0$.

$$s_x = 1$$
$$s_y = 1 + \frac{\sigma_y}{j\omega\varepsilon_o}$$
$$s_z = 1$$

3. A normal y-direction towards the UPML, the $\sigma_y \neq 0$ and $\sigma_x = \sigma_z = 0$.

$$s_x = 1$$
$$s_y = 1$$
$$s_z = 1 + \frac{\sigma_z}{j\omega\varepsilon_o}$$

4. Within the overlapping UPML at corner regions, all parameters, $\sigma_x \neq \sigma_z \neq \sigma_y \neq 0$

$$s_x = 1 + \frac{\sigma_x}{j\omega\varepsilon_o}$$
$$s_y = 1 + \frac{\sigma_y}{j\omega\varepsilon_o}$$
$$s_z = 1 + \frac{\sigma_z}{j\omega\varepsilon_o}$$

In the 3-D case, (7.54a) and (7.54b) are rewritten as:

$$\begin{bmatrix} \frac{\partial H_z}{\partial y} - \frac{\partial H_y}{\partial z} \\ \frac{\partial H_x}{\partial z} - \frac{\partial H_z}{\partial x} \\ \frac{\partial H_y}{\partial x} - \frac{\partial H_x}{\partial y} \end{bmatrix} = j\omega\varepsilon_o\varepsilon_r \begin{bmatrix} s_y s_z/s_x & 0 & 0 \\ 0 & s_x s_z/s_y & 0 \\ 0 & 0 & s_x s_y/s_z \end{bmatrix} \begin{bmatrix} E_x \\ E_y \\ E_z \end{bmatrix} \tag{7.57a}$$

$$\begin{bmatrix} \frac{\partial E_z}{\partial y} - \frac{\partial E_y}{\partial z} \\ \frac{\partial E_x}{\partial z} - \frac{\partial E_z}{\partial x} \\ \frac{\partial E_y}{\partial x} - \frac{\partial E_x}{\partial y} \end{bmatrix} = -j\omega\mu_o\mu_r \begin{bmatrix} s_y s_z/s_x & 0 & 0 \\ 0 & s_x s_z/s_y & 0 \\ 0 & 0 & s_x s_y/s_z \end{bmatrix} \begin{bmatrix} H_x \\ H_y \\ H_z \end{bmatrix} \tag{7.57b}$$

From (7.57), the update equations in the UPML can be derived. Let us consider an electromagnetic wave that is only in the UPML with a normal directed along the x-direction ($\sigma_x \neq 0$, $\sigma_y = 0$, $\sigma_z = 0$); subsequently, (7.57) can be reduced as:

$$j\omega\left(1 + \frac{\sigma_x}{j\omega\varepsilon_o}\right)\varepsilon_o\varepsilon_r E_z = \frac{\partial H_y}{\partial x} - \frac{\partial H_x}{\partial y} \tag{7.58a}$$

$$j\omega\left(1 + \frac{\sigma_x}{j\omega\varepsilon_o}\right)\varepsilon_o\varepsilon_r E_y = \frac{\partial H_x}{\partial z} - \frac{\partial H_z}{\partial x} \tag{7.58b}$$

$$j\omega\left(1 + \frac{\sigma_x}{j\omega\varepsilon_o}\right)^{-1}\varepsilon_o\varepsilon_r E_x = \frac{\partial H_z}{\partial y} - \frac{\partial H_y}{\partial z} \tag{7.58c}$$

$$j\omega\left(1 + \frac{\sigma_x}{j\omega\varepsilon_o}\right)\mu_o\mu_r H_z = \frac{\partial E_x}{\partial y} - \frac{\partial E_y}{\partial x} \tag{7.58d}$$

$$j\omega\left(1 + \frac{\sigma_x}{j\omega\varepsilon_o}\right)\mu_o\mu_r H_y = \frac{\partial E_z}{\partial x} - \frac{\partial E_x}{\partial z} \tag{7.58e}$$

$$j\omega\left(1 + \frac{\sigma_x}{j\omega\varepsilon_o}\right)^{-1}\mu_o\mu_r H_x = \frac{\partial E_y}{\partial z} - \frac{\partial E_z}{\partial y} \tag{7.58f}$$

Substitute $\varepsilon_o\varepsilon_r E = D$ into the left side of (7.58a) and discretize the term in the time domain as [5]:

$$\begin{aligned} j\omega\left(1+\frac{\sigma_x}{j\omega\varepsilon_o}\right)D_z &= \frac{\partial D_z}{\partial t}+\frac{\sigma_x}{\varepsilon_o}D_z \\ &\approx \frac{D_z^{n+0.5}(i,j,k)-D_z^{n-0.5}(i,j,k)}{\Delta t}+\frac{\sigma_x(i)}{\varepsilon_o}\frac{D_z^{n+0.5}(i,j,k)-D_z^{n-0.5}(i,j,k)}{2} \\ &= \frac{D_z^{n+0.5}(i,j,k)}{\Delta t}\left(1+\frac{\sigma_x(i)\Delta t}{2\varepsilon_o}\right)-\frac{D_z^{n-0.5}(i,j,k)}{\Delta t}\left(1-\frac{\sigma_x(i)\Delta t}{2\varepsilon_o}\right) \end{aligned} \tag{7.59}$$

The discretized right side of (7.58a) can be rewritten as:

$$\begin{aligned} &\frac{\partial H_y}{\partial x}-\frac{\partial H_x}{\partial y} \\ &\approx \frac{H_y^n(i+0.5,j,k)-H_y^n(i-0.5,j,k)}{\Delta x}-\frac{H_x^n(i,j+0.5,k)-H_x^n(i,j-0.5,k)}{\Delta y} \end{aligned} \tag{7.60}$$

Thus, the discretization of (7.58a) can be rewritten as:

$$\begin{aligned} &\frac{D_z^{n+0.5}(i,j,k)}{\Delta t}\left(1+\frac{\sigma_x(i)\Delta t}{2\varepsilon_o}\right)-\frac{D_z^{n-0.5}(i,j,k)}{\Delta t}\left(1-\frac{\sigma_x(i)\Delta t}{2\varepsilon_o}\right) \\ &= \frac{H_y^n(i+0.5,j,k)-H_y^n(i-0.5,j,k)}{\Delta x}-\frac{H_x^n(i,j+0.5,k)-H_x^n(i,j-0.5,k)}{\Delta y} \end{aligned} \tag{7.61}$$

Rearranging (7.61) yields

$$D_z^{n+0.5}(i,j,k) = g_3(i)D_z^{n-0.5}(i,j,k)+g_2(i)\Delta t\left\{\begin{aligned} &\frac{H_y^n(i+0.5,j,k)-H_y^n(i-0.5,j,k)}{\Delta x} \\ &-\frac{H_x^n(i,j+0.5,k)-H_x^n(i,j-0.5,k)}{\Delta y} \end{aligned}\right\} \tag{7.62}$$

where $g_3(i)$ and $g_2(i)$ in (7.62), respectively, represent the function as:

$$g_3(i) = \frac{\left(1-\dfrac{\sigma_x(i)\Delta t}{2\varepsilon_o}\right)}{\left(1+\dfrac{\sigma_x(i)\Delta t}{2\varepsilon_o}\right)}, \text{ and } g_2(i) = \frac{1}{\left(1+\dfrac{\sigma_x(i)\Delta t}{2\varepsilon_o}\right)} \tag{7.63}$$

Similar derivation routines can be applied to (7.58a), (7.58b), (7.58d), and (7.58e). For instance, the updated equation of (7.58e) is given as:

$$\begin{aligned} H_y^{n+0.5}(i+0.5,j,k+0.5) &= f_3(i+0.5)H_y^{n-0.5}(i+0.5,j,k+0.5) \\ &-f_2(i+0.5)\frac{\Delta t}{\mu_o\mu_r\Delta z}\left[E_x^n(i+0.5,j,k+1)-E_x^n(i+0.5,j,k)\right] \\ &+f_2(i+0.5)\frac{\Delta t}{\mu_o\mu_r\Delta x}\left[E_z^n(i+1,j,k+0.5)-E_z^n(i,j,k+0.5)\right] \end{aligned} \tag{7.64}$$

where $f_3(i + 0.5)$ and $f_2(i + 0.5)$ in (7.64) are given as:

$$f_3(i) = \frac{\left(1-\dfrac{\sigma_x(i+0.5)\Delta t}{2\varepsilon_o}\right)}{\left(1+\dfrac{\sigma_x(i+0.5)\Delta t}{2\varepsilon_o}\right)} \text{ and } f_2(i) = \frac{1}{\left(1+\dfrac{\sigma_x(i+0.5)\Delta t}{2\varepsilon_o}\right)} \tag{7.65}$$

However, (7.58c) and (7.58f) require more complex treatment compared to (7.58a), (7.58b), (7.58d), and (7.58e). Here, (7.58c) is used as an example to show the derivation steps of the update equation. First, rewrite (7.58c) as:

$$j\omega D_x = \left(1+\frac{\sigma_x}{j\omega\varepsilon_o}\right)\left(\frac{\partial H_z}{\partial y}-\frac{\partial H_y}{\partial z}\right) = \left(\frac{\partial H_z}{\partial y}-\frac{\partial H_y}{\partial z}\right)+\frac{\sigma_x}{j\omega\varepsilon_o}\left(\frac{\partial H_z}{\partial y}-\frac{\partial H_y}{\partial z}\right) \tag{7.66}$$

Transform (7.66) into the time domain, the $(1/j\omega)$ term in (7.66) may be treated as an integral operator over time and $j\omega$ as a derivative over time:

$$\frac{\partial D_x}{\partial t} = \left(\frac{\partial H_z}{\partial y}-\frac{\partial H_y}{\partial z}\right)+\frac{\sigma_x}{\varepsilon_o}\int_0^T\left(\frac{\partial H_z}{\partial y}-\frac{\partial H_y}{\partial z}\right)dt \tag{7.67}$$

Discretizing (7.67) yields

$$\frac{D_x^{n+1}(i+0.5,j,k)-D_x^n(i+0.5,j,k)}{\Delta t} = \frac{\vec{\nabla}\times\vec{H}}{\Delta y}+\frac{\sigma_x}{\varepsilon_o}\left(\Delta t\sum_{n=0}^{T}\frac{\vec{\nabla}\times\vec{H}}{\Delta y}\right) \tag{7.68}$$

where $\Delta y = \Delta z$ and

$$\vec{\nabla}\times\vec{H} = \left\{\begin{aligned} &H_z^{n+0.5}(i+0.5,j+0.5,k)-H_z^{n+0.5}(i+0.5,j-0.5,k) \\ &-K_{Hy}H_y^{n+0.5}(i+0.5,j,k+0.5)+K_{Hy}H_y^{n+0.5}(i+0.5,j,k-0.5) \end{aligned}\right\} \tag{7.69}$$

Rearrange (7.68) and implement this into an FDTD algorithm giving:

$$\begin{aligned} D_x^{n+1}(i+0.5,j,k) &= D_x^n(i+0.5,j,k)+\frac{\Delta t}{\Delta y}\left(\vec{\nabla}\times\vec{H}\right)+\frac{\Delta t\sigma_x}{\varepsilon_o}\left(\frac{\Delta t}{\Delta y}\sum_{n=0}^{T}\vec{\nabla}\times\vec{H}\right) \\ &= D_x^n(i+0.5,j,k)+\frac{\Delta t}{\Delta y}\left(\vec{\nabla}\times\vec{H}\right)+\frac{\Delta t}{\Delta y}I_{D_x}^{n+1}(i+0.5,j,k) \end{aligned} \tag{7.70}$$

where

$$I_{D_x}^{n+1}(i+0.5,j,k) = I_{D_x}^{n}(i+0.5,j,k) + \frac{\Delta t}{\Delta y} g_1(i+0.5)\left(\vec{\nabla}\times\vec{H}\right) \quad (7.71)$$

and $g_1(i + 0.5)$ in (7.71) is given as:

$$g_1(i+0.5) = \frac{\sigma_x(i+0.5)\Delta t}{2\varepsilon_o} \quad (7.72)$$

The variation of σ_x along the x-direction in the UPML is assumed to be:

$$\sigma_x(i) = \sigma_{\max}\left(\frac{L-i}{L}\right)^m \qquad 0 \le x \le L \quad (7.73)$$

where i is the x-coordinate at the boundary between the UPML and the scattering field region. The parameter L is the total thickness of the UPML. Typical values for the parameter m are between 2 and 4. In fact, σ_y and σ_z along the y- and z-axes can be calculated using the similar expression in (7.73). The function σ_x has a maximum value at the front of the UPML interface and decays to zero within the UPML as shown in Figure 7.15. In Sullivan [5], m in (7.73) is set to be 3 and $\sigma_{\max}$ is found empirically to be equal to 0.33.

In a 3-D FDTD computation, besides x-direction, the y-direction and z-direction dependent of UPML are also required to incorporate into (7.58) [5]:

$$j\omega\left(1+\frac{\sigma_x}{j\omega\varepsilon_o}\right)\left(1+\frac{\sigma_y}{j\omega\varepsilon_o}\right)\left(1+\frac{\sigma_z}{j\omega\varepsilon_o}\right)^{-1}\varepsilon_o\varepsilon_r E_z = \frac{\partial H_y}{\partial x} - \frac{\partial H_x}{\partial y} \quad (7.74a)$$

$$j\omega\left(1+\frac{\sigma_x}{j\omega\varepsilon_o}\right)\left(1+\frac{\sigma_y}{j\omega\varepsilon_o}\right)^{-1}\left(1+\frac{\sigma_z}{j\omega\varepsilon_o}\right)\varepsilon_o\varepsilon_r E_y = \frac{\partial H_x}{\partial z} - \frac{\partial H_z}{\partial x} \quad (7.74b)$$

$$j\omega\left(1+\frac{\sigma_x}{j\omega\varepsilon_o}\right)^{-1}\left(1+\frac{\sigma_y}{j\omega\varepsilon_o}\right)\left(1+\frac{\sigma_z}{j\omega\varepsilon_o}\right)\varepsilon_o\varepsilon_r E_x = \frac{\partial H_z}{\partial y} - \frac{\partial H_y}{\partial z} \quad (7.74c)$$

$$j\omega\left(1+\frac{\sigma_x}{j\omega\varepsilon_o}\right)\left(1+\frac{\sigma_y}{j\omega\varepsilon_o}\right)\left(1+\frac{\sigma_z}{j\omega\varepsilon_o}\right)^{-1}\mu_o\mu_r H_z = \frac{\partial E_x}{\partial y} - \frac{\partial E_y}{\partial x} \quad (7.74d)$$

$$j\omega\left(1+\frac{\sigma_x}{j\omega\varepsilon_o}\right)\left(1+\frac{\sigma_y}{j\omega\varepsilon_o}\right)^{-1}\left(1+\frac{\sigma_z}{j\omega\varepsilon_o}\right)\mu_o\mu_r H_y = \frac{\partial E_z}{\partial x} - \frac{\partial E_x}{\partial z} \quad (7.74e)$$

$$j\omega\left(1+\frac{\sigma_x}{j\omega\varepsilon_o}\right)^{-1}\left(1+\frac{\sigma_y}{j\omega\varepsilon_o}\right)\left(1+\frac{\sigma_z}{j\omega\varepsilon_o}\right)\mu_o\mu_r H_x = \frac{\partial E_y}{\partial z} - \frac{\partial E_z}{\partial y} \quad (7.74f)$$

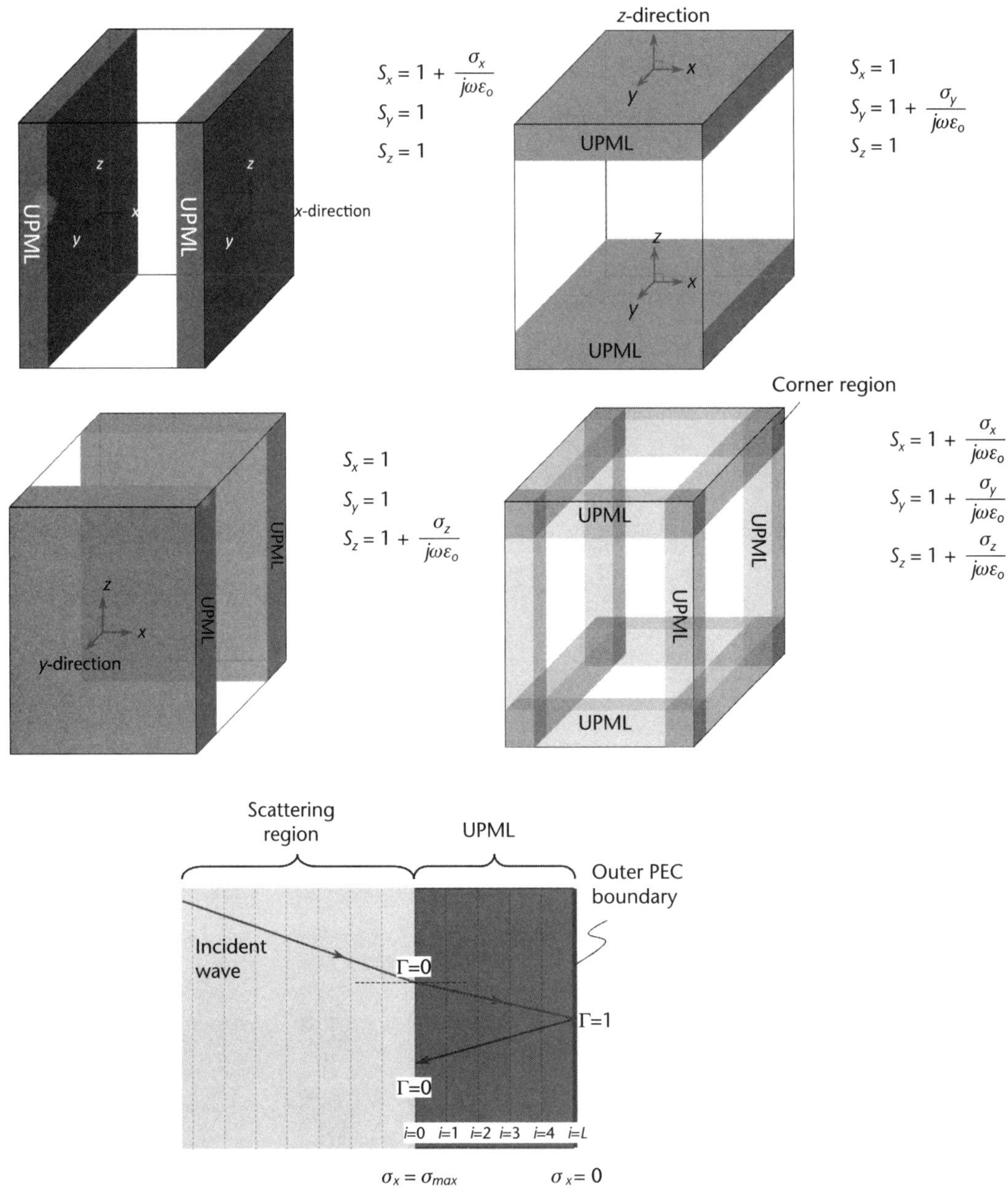

Figure 7.15 Reflection coefficient Γ of the incident wave at the surface between scattering and UPML is equal to zero. The incident wave is attenuated in lossy UPML before the wave reaches the outer PEC boundary. The remained incident wave is reflected from PEC boundary and decay further before back to the scattering region. Finally, the reflected wave reaches the surface between scattering region and UPML is almost zero.

The detailed derivation of the updated equations using (7.74a) through (7.74f) can be found in Sullivan [5, 10].

7.2.5 Postprocessing

FDTD results can be used to identify the frequency-domain output using fast Fourier transform (FFT). The algorithm for the FFT can be found in the "*fft*" MATLAB command. For instance, the variation of the input impedance Z_{in} of a dipole at the driving point (the center of the dipole) with an operating frequency can be calculated using the time-domain current I^n and voltage V^n values resulted from the FDTD computations:

$$Z_{\text{in}} = \frac{\text{FFT}\left(V^n\right)}{\text{FFT}\left(\tilde{I}^n\right)} \tag{7.75}$$

where I^n in (7.43) is the current flowing through the center of the dipole. However, the V^n in (7.88) is the voltage pulse excited at the center of the dipole. The corresponding operating frequency f for each calculated input impedance Z_{in} can be determined using the number of the discrete time step N_t or the maximum time t_{max} value used in the FDTD simulation. The frequency step Δf can be determined as:

$$\Delta f = \frac{1}{t_{\text{max}}} = \frac{1}{N_t \Delta t} \tag{7.76}$$

where Δt is the time step size. For an even number of N_t, the value of frequencies f_n at the corresponding nth step of time are defined as:

$$f_n = (n-1)\Delta f, \qquad n = 1,2,\ldots,N_t/2 \tag{7.77}$$

and

$$f_n = \left(n - N_t - 1\right)\Delta f, \qquad n = \left(N_t/2\right)+1,\left(N_t/2\right)+2,\ldots,N_t \tag{7.78}$$

Equation (7.77) will give the positive values of the frequencies, while (7.78) will give the dual Z_{in} of (7.75) in negative value due to the symmetric feature of Fourier transform. Thus, the frequency range is normally extracted only from $n = 1$ to $n = N_t/2$ using (7.77).

7.2.6 MATLAB Codes for the 3-D Subcell FDTD

The FDTD routines in Sections 7.2.1 through 7.2.5 are implemented in MATLAB. As mentioned in Section 7.1.2, the MATLAB can only handle integer array indexes; thus, the original FDTD algorithm implementation of the subcell model in Figure 7.16(a) must be modified to the MATLAB code implementation as shown in Figure

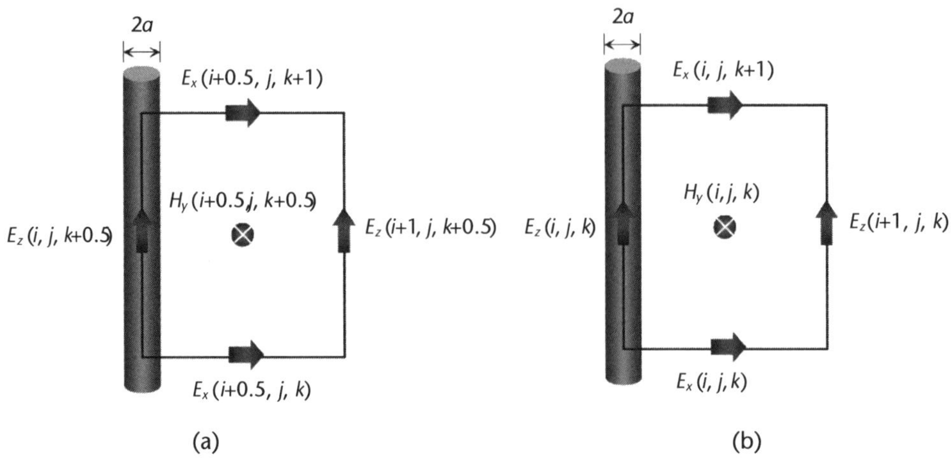

Figure 7.16 (a) FDTD algorithm implementation. (b) MATLAB code implementation.

7.16(b). Figure 7.17 shows the simulated propagation outward of E_z fields after 1.5 ns, which is excited by the Gaussian pulse source at the center (i_c, j_c, z_c) of the dipole with 28.64 mm of length. The comparison of calculated normalized resistance, $\mathfrak{Re}(Z_{in})/Z_o$, and reactance, $\mathfrak{Im}(Z_{in})/Z_o$, of the monopole using two different MoM methods (in Chapter 6) and the FDTD method is shown in Figure 7.18. The difference among those simulated impedances mainly resulted from an inaccurate source model.

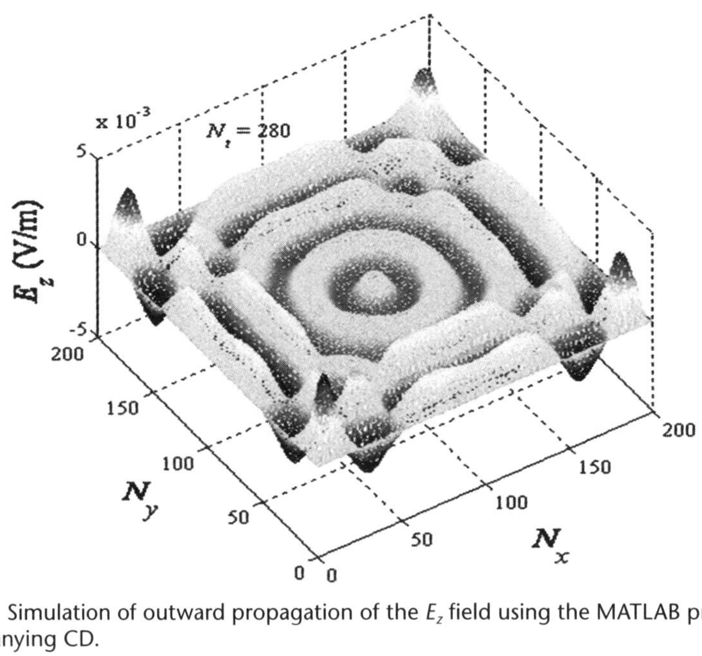

Figure 7.17 Simulation of outward propagation of the E_z field using the MATLAB program on the accompanying CD.

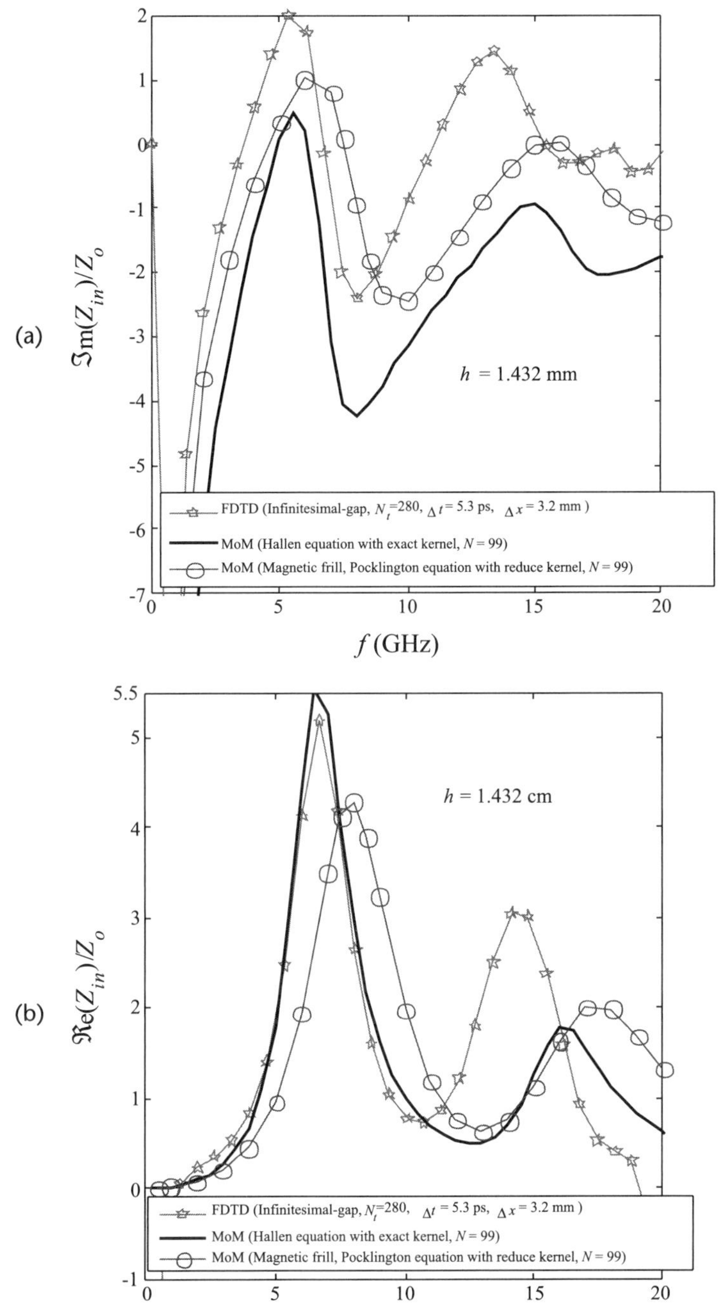

Figure 7.18 Comparison of (a) normalized resistance, $\Re e(Z_{in})/Z_o$, and (b) normalized reactance, $\Im m(Z_{in})/Z_o$, in air by considering the size of monopole with a = 0.65 mm, b = 2.05 mm, and h = 14.32 mm.

7.3 Monopole Driven from a Coaxial Line and an Open-Ended Coaxial Probe

7.3.1 1-D Coaxial Transmission-Line Feed Model

Normally, a monopole is fed by the coaxial transmission line as discussed in Chapter 4. Maloney et al. [17] modeled the propagation wave along the coaxial line by a 1-D equivalent circuit instead of a 3-D FDTD algorithm as shown in Figure 7.19. The equivalent circuit is matched with the 3-D FDTD updated equation at the driven point, (i_c, j_c, k_c).

The normalized telegraphist's equations of (1.37) and (1.38) for lossless transmission line ($G = 0$ and $R = 0$) can be expressed in the time domain:

$$\frac{\partial V}{\partial t} = -vZ_o \frac{dI}{dz} \tag{7.79}$$

$$\frac{dV}{dz} = -\frac{Z_o}{v}\frac{\partial I}{\partial t} \tag{7.80}$$

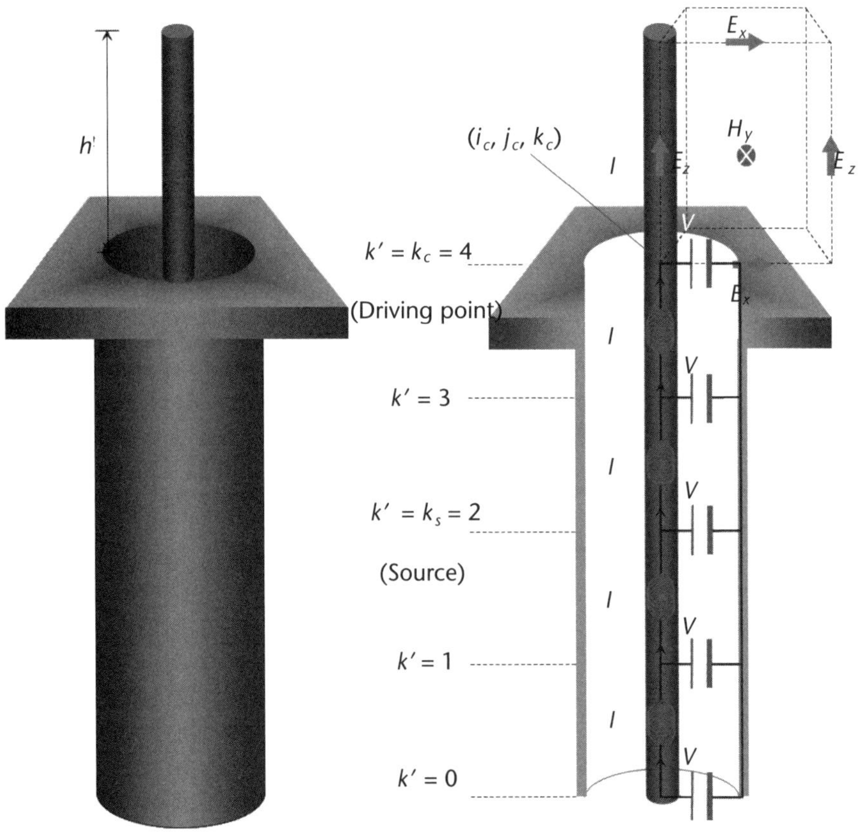

Figure 7.19 A 1-D transmission line model for feeding the monopole from a coaxial line and 3-D FDTD lattice at the driving point of the monopole.

where v ($\approx 3\times10^8/\sqrt{\varepsilon_c}$ ms^{-1}) and Z_o ($\approx 50\Omega$) are the velocity of the wave propagation and the characteristic impedance in the coaxial transmission line, respectively. ε_c is the relative dielectric constant of a material filled in the coaxial line. The discretzation of (7.79) and (7.80) can be written as:

$$\frac{V^{n+1}(k)-V^{n}(k)}{\Delta t} = -vZ_o\frac{I^{n+0.5}(k+0.5)-I^{n+0.5}(k-0.5)}{\Delta z} \tag{7.81}$$

$$\frac{I^{n+0.5}(k+0.5)-I^{n-0.5}(k+0.5)}{\Delta t} = -\left(\frac{v}{Z_o}\right)\frac{V^{n}(k+1)-V^{n}(k)}{\Delta z} \tag{7.82}$$

Rearrange (7.81), and the updated equation for V^{n+1} is given as:

$$V^{n+1}(k) = V^{n}(k) - Z_o\left(\frac{v\Delta t}{\Delta z}\right)\left[I^{n+0.5}(k+0.5)-I^{n+0.5}(k-0.5)\right] \tag{7.83}$$

The expression (7.83) can be used for $k = 2, 3, \ldots, N_z - 1$ and $N_t \geq 2$. Similarly, from (7.82), the updated current for $I^{n+0.5}$ can be obtained as:

$$I^{n+0.5}(k+0.5) = I^{n-0.5}(k+0.5) - \frac{1}{Z_o}\left(\frac{v\Delta t}{\Delta z}\right)\left[V^{n}(k+1)-V^{n}(k)\right] \tag{7.84}$$

Equation (7.84) can be used for $k = 1, 3, \ldots, N_z - 1$ and $N_t \geq 2$. The k values specify definitions at positions $k' = 0$, $k' = 1$, $k' = 2$, and $k' = 4$ along the coaxial line are respectively summarized next [17].

At $k' = 0$, the simple ABC is considered. V^{n+1} at $k' = 0$ is set to be equal to V^{n-1} at $k' = 1$.

$$V^{n+1}(0) = V^{n-1}(1) \tag{7.85}$$

At $k = k_S = 2$, the updated equations including incident voltage source V_{inc} are written as:

$$\begin{aligned} V^{n+1}(k_S) = {} & V^{n}(k_S) - Z_o\left(\frac{v\Delta t}{\Delta z}\right)\left[I^{n+0.5}(k_S+0.5)-I^{n+0.5}(k_S-0.5)\right] \\ & +\left(\frac{v\Delta t}{\Delta z}\right)V_{inc}^{n+0.5}(k_S+0.5) \end{aligned} \tag{7.86}$$

$$\begin{aligned} I^{n+0.5}(k_S+0.5) = {} & I^{n-0.5}(k_S+0.5) - \frac{1}{Z_o}\left(\frac{v\Delta t}{\Delta z}\right)\left[V^{n}(k_S+1)-V^{n}(k_S)\right] \\ & +\frac{1}{Z_o}\left(\frac{v\Delta t}{\Delta z}\right)V_{inc}^{n+0.5}(k_S) \end{aligned} \tag{7.87}$$

where excitation voltage V_{inc} is represented by a Gaussian pulse as:

$$V_{inc} = V_o \exp\left(-t^2/2\tau_p^2\right) \tag{7.88}$$

where V_o is the constant voltage amplitude on the excitation boundary, which is assumed to be equal to 1V. The term τ_p is the width of the pulse, which can be estimated by [17]:

$$\tau_p = 1.61 \times 10^{-1}\tau \tag{7.89}$$

where τ (= h/c) is the time required for the light to propagate through the length h of the monopole.

At $k' = k_c = 4$, the 1-D transmission line grid is matched with point (i_c, j_c, k_c) of the 3-D Yee's lattice. Thus, the update equation for H_y is obtained by applying Faraday's law, which is given as:

$$\begin{aligned} H_y^{n+0.5}\left(i_c + 0.5, j_c, k_c + 0.5\right) &= H_y^{n-0.5}\left(i_c + 0.5, j_c, k_c + 0.5\right) \\ &- \frac{K_{Ex}\Delta t}{\mu \Delta z}\left[E_x^n\left(i_c + 0.5, j_c, k_c + 1\right) - \frac{2}{\ln(\Delta x/a)}\frac{V^n\left(k_c\right)}{\Delta x}\right] \\ &+ \frac{2\Delta t}{\mu \Delta x \ln(\Delta x/a)} E_z^n\left(i_c + 1, j_c, k_c + 0.5\right) \end{aligned} \tag{7.90}$$

However, the updated equation for a normalized current I at $k_c + 0.5$ is obtained by applying Ampere's law, which is given as:

$$\begin{aligned} \tilde{I}^{n+0.5}\left(k_c + 0.5\right) &= \sqrt{\frac{\varepsilon_o}{\mu_o}}\begin{bmatrix} H_x^{n+0.5}\left(i_c, j_c - 0.5, k_c + 0.5\right) \\ -H_x^{n+0.5}\left(i_c, j_c + 0.5, k_c + 0.5\right) \end{bmatrix} K_{Hx}\Delta x \\ &+ \sqrt{\frac{\varepsilon_o}{\mu_o}}\begin{bmatrix} H_y^{n+0.5}\left(i_c + 0.5, j_c, k_c + 0.5\right) \\ -H_y^{n+0.5}\left(i_c - 0.5, j_c, k_c + 0.5\right) \end{bmatrix} K_{Hy}\Delta y \end{aligned} \tag{7.91}$$

7.3.2 2-D FDTD Subcell Model

Maloney et al. [8] were the first to implement a 2-D FDTD dipole (monopole) model instead of the 3-D FDTD. The dipole structure is rotationally symmetric, and the field distribution around the dipole is independent on an azimuth ϕ coordinate as shown in Figure 7.20. The relevant time-domain Maxwell's equations used in the 2-D FDTD are expressed as:

$$\frac{\partial E_\rho}{\partial z} - \frac{\partial E_z}{\partial \rho} = -\mu_o \frac{\partial H_\phi}{\partial t} \tag{7.92a}$$

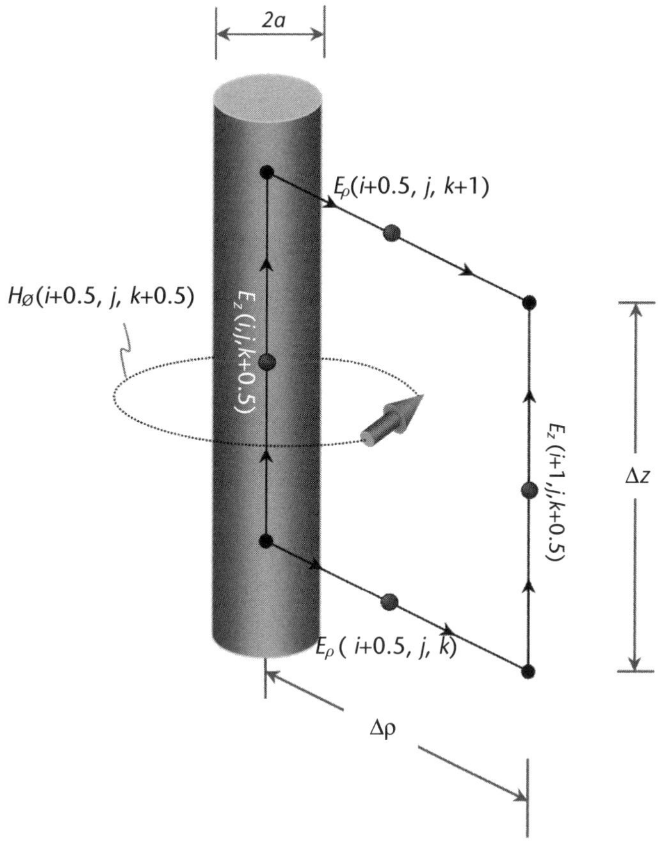

Figure 7.20 Symmetrical field of H_ϕ surrounding the FDTD discrete cell at $(i + 0.5, j, k + 0.5)$.

$$-\frac{\partial H_\phi}{\partial z} = \varepsilon_o \frac{\partial E_\rho}{\partial t} \tag{7.92b}$$

$$\frac{1}{\rho}\frac{\partial\left(\rho H_\phi\right)}{\partial \rho} = \varepsilon_o \frac{\partial E_z}{\partial t} \tag{7.92c}$$

The updated equations from the discretization of (7.92a) through (7.92c) are given as:

$$\begin{aligned} H_\phi^{n+0.5}(i,j) = H_\phi^{n-0.5}(i,j) + \frac{\Delta t}{\mu_o \Delta\rho}\left[E_z^n(i+0.5,j) - E_z^n(i-0.5,j)\right] \\ -\frac{\Delta t}{\mu_o \Delta z}\left[E_\rho^n(i,j+0.5) - E_\rho^n(i,j-0.5)\right] \end{aligned} \tag{7.93a}$$

$$E_\rho^{n+1}(i,j-0.5) = E_\rho^n(i,j-0.5) - \frac{\Delta t}{\varepsilon_o \Delta z}\left[H_\phi^{n+0.5}(i,j) - H_\phi^{n+0.5}(i,j-1)\right] \tag{7.93b}$$

$$E_z^{n+1}(i+0.5,j) = E_z^n(i+0.5,j) - \frac{\Delta t}{\varepsilon_o \Delta\rho}\frac{1}{\rho_{i+0.5}}\left[\rho_{i+1}H_\phi^{n+0.5}(i+1,j) - \rho_i H_\phi^{n+0.5}(i,j)\right] \tag{7.93c}$$

The time step size Δt is chosen to satisfy the condition:

$$\Delta t < \frac{1}{c}\frac{\Delta\rho\Delta z}{\sqrt{(\Delta\rho)^2 + (\Delta z)^2}} \tag{7.94}$$

where c (= 299,792,458 ms^{-1}) is the velocity of light in free space. In Maloney et al. [8], the FDTD method is used to analyze the cylindrical monopole that is driven from a coaxial line through a ground plane. The geometry and boundary surface of the monopole are shown in Figure 7.21. The excitation or incident electric field E_{exc} at the cross-section of coaxial line AA' is expressed as [8]:

$$E_{exc} = \frac{V_{inc}}{\rho\ln(b/a)} \tag{7.95}$$

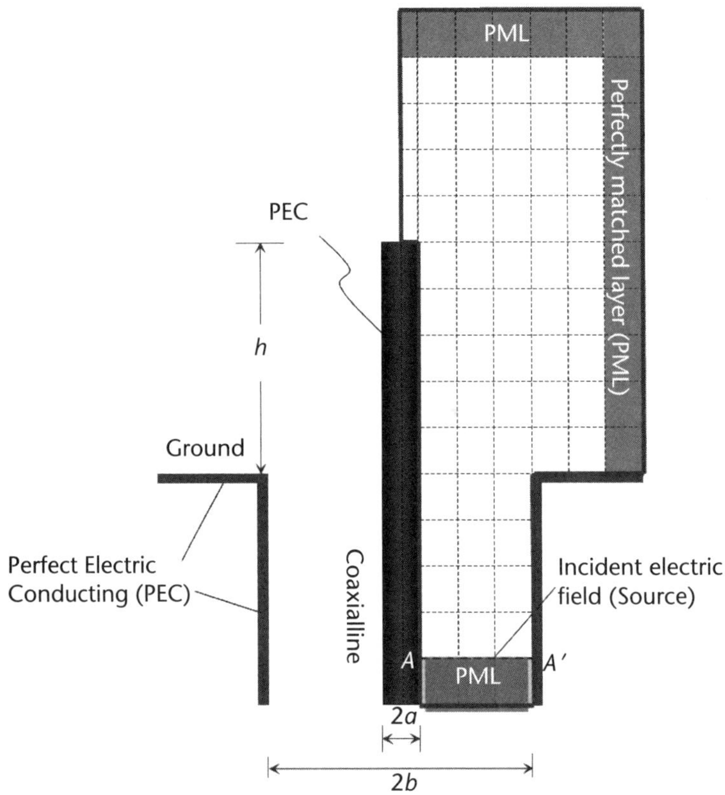

Figure 7.21 The rotationally symmetric monopole fed from a coaxial line.

The excitation voltage V_{inc} in (7.95) is assumed to be Gaussian pulse (7.88). Figure 7.22 shows the magnified part of the FDTD cell edge with labeled field components.

In 2009, Hyun et al. [18] combined both the infinitesimal-gap voltage source model and 2-D transmission line theory to represent the feed source E^n_{exc} at the coaxial aperture as:

$$E^n_{exc} = -\left(V^n_{inc} + V^n_{refl}\right)\delta(z) \tag{7.96}$$

Equation (7.96) takes into account the reflected waves between the monopole and the coaxial feed line at the coaxial aperture. Thus, the total feed voltage V^n_z is the linear superposition of the incident voltage V^n_{inc} and the reflected voltage V^n_{refl}, which is expressed as:

$$V^n_{refl} = V^n_{inc} - Z_o I^n \approx V^n_{inc} - Z_o H^{n-0.5}_{\phi}(i + 0.5, j + 0.5)\pi\Delta\rho \tag{7.97}$$

where Z_o ($\approx 50\Omega$) is the characteristic impedance of the coaxial feed line [see (4.29).

The total current I^n at the coaxial aperture is given as:

$$I^n = H^{n-0.5}_{\phi}\left(i_c + 0.5, j_c + 0.5\right)\pi\Delta\rho \tag{7.98}$$

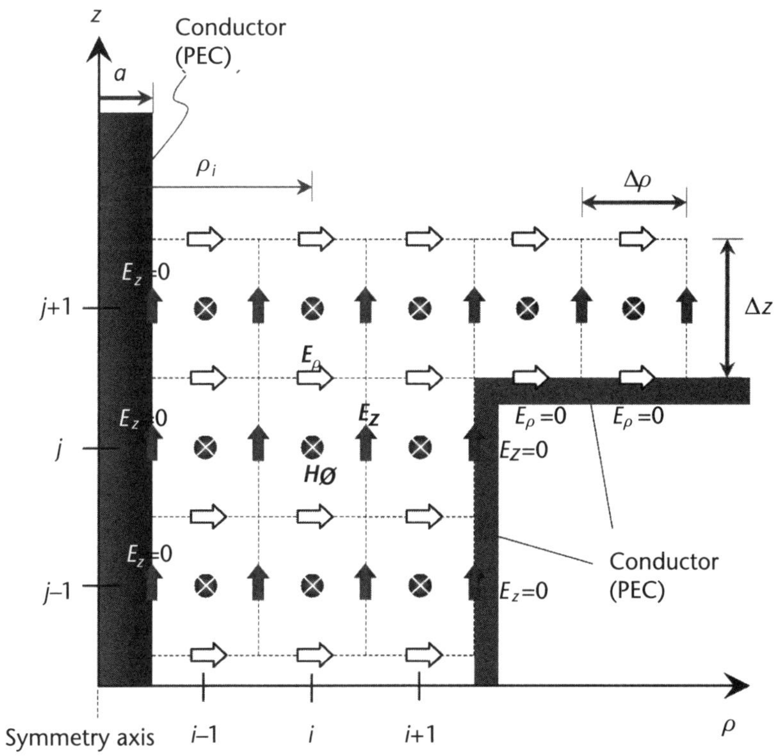

Figure 7.22 Spatial grid ($\Delta\rho$ and Δz) and field components (H_ϕ, E_ρ, and E_z) for a 2-D coaxial problem [8].

and the magnetic field update equation around the feed cell is given as:

$$
\begin{aligned}
H_\phi^{n+0.5}\left(i_c+0.5, j_c+0.5\right) &= H_\phi^{n-0.5}\left(i_c+0.5, j_c+0.5\right) \\
&+\frac{\Delta t}{\mu \Delta \rho} \frac{2}{\ln(\Delta \rho / a)} E_z^n\left(i_c+1, j_c+0.5\right)-\frac{\Delta t}{\mu \Delta z} E_\rho^n\left(i_c+0.5, j_c+1\right) \\
&+\frac{\Delta t}{\mu \Delta \rho} \frac{2}{\ln(\Delta \rho / a)} \frac{V_{inc}^n+V_{ref}^n}{\Delta z}
\end{aligned} \tag{7.99}
$$

Substituting (7.97) into (7.99) yields

$$
\begin{aligned}
H_\phi^{n+0.5}\left(i_c+0.5, j_c+0.5\right) &= H_\phi^{n-0.5}\left(i_c+0.5, j_c+0.5\right) \\
&+\frac{\Delta t}{\mu \Delta \rho} \frac{2}{\ln(\Delta \rho / a)} E_z^n\left(i_c+1, j_c+0.5\right)-\frac{\Delta t}{\mu \Delta z} E_\rho^n\left(i_c+0.5, j_c+1\right) \\
&+\frac{\Delta t}{\mu \Delta \rho} \frac{2}{\ln(\Delta \rho / a)} \frac{2V_{inc}^n-Z_o H_\phi^{n-0.5}\left(i_c+0.5, j_c+0.5\right)\pi\Delta\rho}{\Delta z}
\end{aligned} \tag{7.100}
$$

The configuration of the model proposed by Hyun et al. [18] is shown in Figure 7.23. Clearly, the number of the FDTD cells is reduced compared to Figure 7.21. By using (7.96) as an excitation source, the surface of the coaxial aperture is assumed to be conducting ground plane ($E_\rho^n = 0$). The advantage is that the monopole driven from coaxial line can be computed without any FDTD cell modeling of the coaxial feed line.

7.3.3 Boundary Conditions for the 2-D FDTD

A concept similar to the one described in Section 7.2.4 can be implemented in the 2-D FDTD. However, in the 2-D case, only two directions of UPML (radial, ρ-axis, and axial, z-axis) should be considered. Hence, the electromagnetic field components in the UPML are expressed as:

$$
-j\omega\left(1+\frac{\sigma_\rho}{j\omega\varepsilon_o}\right)\left(1+\frac{\sigma_z}{j\omega\varepsilon_o}\right)\mu_o\mu_r H_\phi = \frac{\partial E_\rho}{\partial z}-\frac{\partial E_z}{\partial \rho} \tag{7.101a}
$$

$$
j\omega\left(1+\frac{\sigma_\rho}{j\omega\varepsilon_o}\right)^{-1}\left(1+\frac{\sigma_z}{j\omega\varepsilon_o}\right)\varepsilon_o\varepsilon_r E_\rho = -\frac{\partial H_\phi}{\partial z} \tag{7.101b}
$$

$$
j\omega\left(1+\frac{\sigma_x}{j\omega\varepsilon_o}\right)\left(1+\frac{\sigma_z}{j\omega\varepsilon_o}\right)^{-1}\varepsilon_o\varepsilon_r E_z = \frac{1}{\rho}\frac{\partial\left(\rho H_\phi\right)}{\partial \rho} \tag{7.101c}
$$

The updated equations in UPML can be derived from (7.101a) through (7.101c), which are listed next.

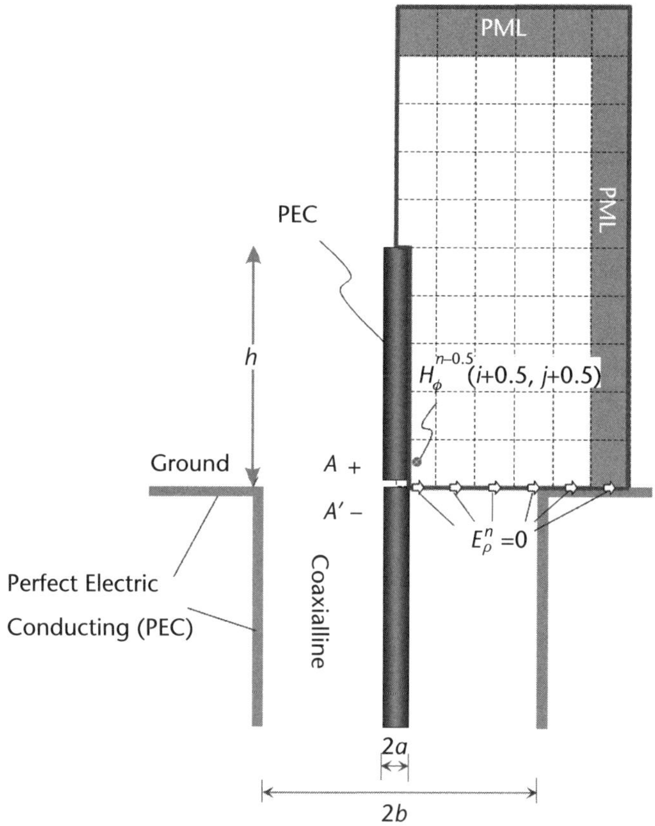

Figure 7.23 Combination feed model (combining the infinitesimal-gap source model and the equivalent load model of the feed line).

For the updated equations in a UPML:

$$\begin{aligned}H_\phi^{n+0.5}(i,j) &= g_3(i)g_3(j)H_\phi^{n-0.5}(i,j)\\&+ g_2(i)g_2(j)\frac{\Delta t}{\Delta\rho}\left[E_z^n(i+0.5,j) - E_z^n(i-0.5,j)\right]\\&- g_2(i)g_2(j)\frac{\Delta t}{\Delta z}\left[E_\rho^n(i,j+0.5) - E_\rho^n(i,j-0.5)\right]\end{aligned} \tag{7.102a}$$

$$\begin{aligned}D_\rho^{n+1}(i,j-0.5) &= f_3(j-0.5)D_\rho^n(i,j-0.5)\\&- f_2(j-0.5)\frac{\Delta t}{\Delta z}\left[H_\phi^{n+0.5}(i,j) - H_\phi^{n+0.5}(i,j-1)\right]\\&- f_2(j-0.5)\frac{\Delta t}{\Delta z}I_{D_\rho}^{n+1}(i,j-0.5)\end{aligned} \tag{7.102b}$$

where $D_\rho = \varepsilon_o\varepsilon_r E_\rho$ and

$$I_{D_\rho}^{n+1}(i,j-0.5) = I_{D_\rho}^n(i,j-0.5) + f_1(i)\frac{\Delta t}{\Delta z}\left[H_\phi^{n+0.5}(i,j) - H_\phi^{n+0.5}(i,j-1)\right]$$

$$\begin{aligned} D_z^{n+1}(i+0.5,j) &= f_3(i+0.5)D_z^n(i+0.5,j) \\ &+ f_2(i+0.5)\frac{\Delta t}{\Delta \rho}\frac{1}{\rho_{i+0.5}}\left[\rho_{i+1}H_\phi^{n+0.5}(i+1,j) - \rho_i H_\phi^{n+0.5}(i,j)\right] \\ &+ f_2(i+0.5)\frac{\Delta t}{\Delta \rho}\frac{1}{\rho_{i+0.5}} I_{D_z}^{n+1}(i+0.5,j) \end{aligned} \tag{7.102c}$$

where $D_z = \varepsilon_o \varepsilon_r E_z$ and

$$I_{D_z}^{n+1}(i+0.5,j) = I_{D_z}^{n}(i+0.5,j) + f_1(j)\frac{1}{\rho_{i+0.5}}\left[\rho_{i+1}H_\phi^{n+0.5}(i+1,j) - \rho_i H_\phi^{n+0.5}(i,j)\right]$$

where coefficients in (7.102a) through (7.102c) are given as [5]:

$$g_2(i) = \frac{1}{1+\sigma_x} \tag{7.103a}$$

$$g_3(i) = g_3(j) = \frac{1-\sigma_x}{1+\sigma_x} \tag{7.103b}$$

$$f_1(i) = f_1(j) = 0.25\left(\frac{L-i-0.5}{L}\right)^3 \tag{7.103c}$$

$$f_2(i+0.5) = f_2(j-0.5) = \frac{1}{1+\sigma_x'} \tag{7.103d}$$

$$f_3(i+0.5) = f_3(j-0.5) = \frac{1-\sigma_x'}{1+\sigma_x'} \tag{7.103e}$$

$$\sigma_x = 0.33\left(\frac{L-i}{L}\right)^3, \quad i = j = 0,1,2,\ldots,L \tag{7.104a}$$

$$\sigma_x' = 0.25\left(\frac{L-i-0.5}{L}\right)^3 \tag{7.104b}$$

7.4 Dielectric-Coated Dipole and Monopole

Boonzaaier and Pistonius [19] extended the 3-D subcell FDTD (in Section 7.2) to a dielectric-coated dipole as shown in Figure 7.24. Later, Nadobny et al. [20] implemented a simple concept of effective permittivity ε_{eff} in the 3-D subcell FDTD in order to compute the input impedance of the dielectric coated monopole. In the

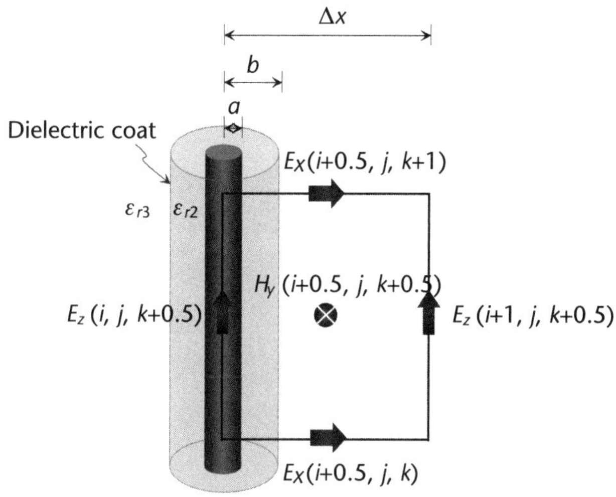

Figure 7.24 FDTD subcell model for a dielectric-coated dipole or monopole.

real implementation, the scattering field from the conductor dipole or monopole is radiated into its dielectric coating and partly into an exterior ambient medium, where both mediums have different values of relative permittivity. In Nadobny's technique, both mediums are assumed to be a single inhomogeneous medium and their dielectric properties are represented by the individual effective permittivity ε_{eff} as:

$$\varepsilon_{eff} = \frac{\ln\left(\frac{\Delta x}{a}\right)}{\frac{1}{\varepsilon_{r2}}\ln\left(\frac{b}{a}\right) + \frac{1}{\varepsilon_{r3}}\ln\left(\frac{\Delta x}{b}\right)} \tag{7.105}$$

where ε_{r2} and ε_{r3} are the relative permittivity of the low-loss dielectric coating and the outer ambient medium surrounding the coating, respectively. Thus, the FDTD routine in Section 7.3 can also be implemented for the coated monopole by assuming that $\varepsilon_r = \varepsilon_{eff}$ in the FDTD updated equations. The MATLAB program on the accompanying CD can be used for an open-ended coaxial probe and dielectric-coated monopole with minor changes of a few lines in the program codes. Table 7.2 shows the code lines on the accompanying CD that need to be changed according to the three kinds of configuration (assume that the monopole and the coated monopole have 14.32 mm of length).

Figure 7.25 shows the simulated half-symmetry radiation fields using the MATLAB program on the accompanying CD and Table 7.2. The radiation fields are excited from an open-ended coaxial probe, a monopole, and a dielectric-coated monopole. The open-ended coaxial probe has an inner radius, $a = 0.65$ mm and $b = 2.05$ mm. Both monopoles are 14.32 mm long and are driven from a Teflon-filled coaxial line into an exterior medium, which has dielectric constant, $\varepsilon_r' = 42$ and conductivity, $\sigma = 0.88$ S/m. In this work, up to 2,400 time steps N_t with an interval size of $\Delta t = 0.33$ ps are implemented in the FDTD computation.

Table 7.2 Modification of MATLAB Codes on the Accompanying CD Based on the Kinds of Probes

	1. Monopole Dimensions and Constant Values:	*2. Dielectric Properties of Mediums:*	*3. Updating Coefficients:*

Monopole driven from coaxial line	h=0.01432;	eps_exterior =42;
		sig_exterior =0.88
			gar(ia:ib,js:js+jl)=ca(1);
			gbr(ia:ib,js:js+jl)=cb(1);
			gaz(ia:ib,js:js+jl)=ca(1);
			gbz(ia:ib,js:js+jl)=cb(1);

Open-ended coaxial probe	h=0;	eps_exterior =42;
		sig_exterior =0.88
			gar(ia:ib,js:js+jl)=ca(1);
			gbr(ia:ib,js:js+jl)=cb(1);
			gaz(ia:ib,js:js+jl)=ca(1);
			gbz(ia:ib,js:js+jl)=cb(1);

Dielectric-coated monopole driven from coaxial line	h=0.01432;	eps_exterior =42;
		sig_exterior =0.88
			gar(ia:ib,js:js+jl+jh)=ca(1);
			gbr(ia:ib,js:js+jl+jh)=cb(1);
			gaz(ia:ib,js:js+jl+jh)=ca(1);
			gbz(ia:ib,js:js+jl+jh)=cb(1);

Dielectric-coated monopole driven from coaxial line [20]	h=0.01432;	er2 =eps_Teflon;
		er3=42+ 0.88*dt/ eps_0;
		eps_eff=log(dx./a)./ ((1./ er2).*log(b./a)+(1./ er3).*log(dx./b));
		eps_exterior =eps_eff;
			gar(ia:ib,js:js+jl)=ca(1);
			gbr(ia:ib,js:js+jl)=cb(1);
			gaz(ia:ib,js:js+jl)=ca(1);
			gbz(ia:ib,js:js+jl)=cb(1);

7.5 Heat Transfer from Dipole and Monopole

Recently, the heat diffusion from the monopole as shown in Figure 7.26 is applied to the hyperthermia treatment (treating tumors with heat). The advantage of a monopole compared to other applicators is that the size of the monopole is small. Particularly, for internal heating in the cranial cavity, thin or small-sized antennas are required in invasive treatments, such as brain tumor treatment. In this treatment, the tumor is normally heated up to the therapeutic temperature between 42°C and 45°C for 60 minutes without overheating the surrounding normal tissues. With the help of numerical calculations, the size of monopole and the power source parameters can be estimated for a given location and size of tumor before surgery is performed.

Several particular diffusion equations called bio-heat equations have been proposed to predict the heat distribution of the monopole (or dielectric-coated monopole), which is surrounded by a biological medium including the obtained specific absorption rate (SAR) distribution using the FDTD. Among the bio-heat equations, Pennes's equation is the pioneer and a simple model, which is given as [21]:

$$pC_p \frac{\partial T}{\partial t} = \kappa \left(\frac{\partial^2 T}{\partial \rho^2} + \frac{1}{\rho} \frac{\partial T}{\partial \rho} + \frac{\partial^2 T}{\partial z^2} \right) - \omega_b C_b p_b \left(T - T_b \right) + Q_{met} + Q_{ext} \tag{7.106}$$

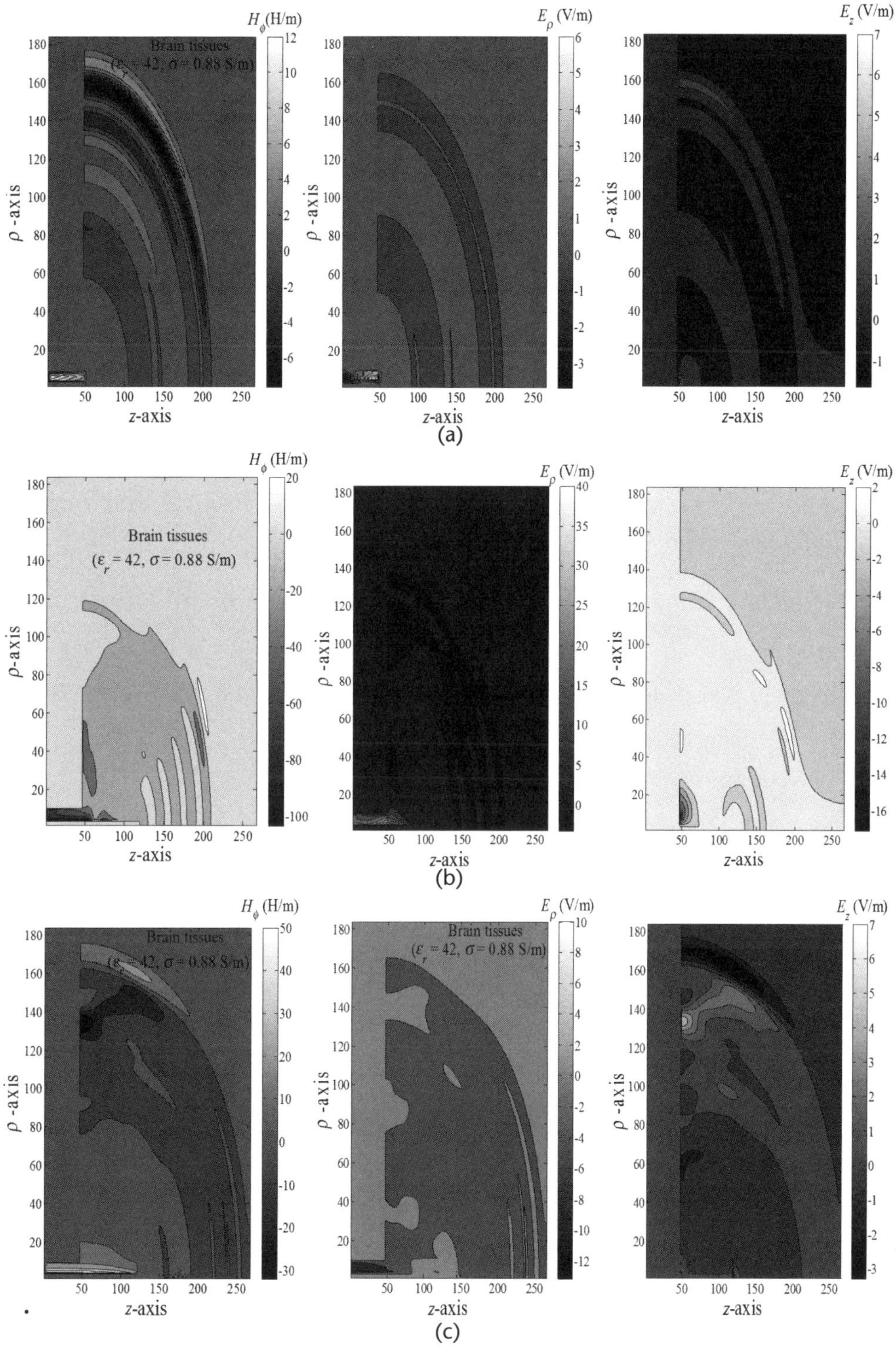

Figure 7.25 The radiation field components (H_ϕ, E_ρ, E_z) in brain tissues ($\varepsilon_r = 4.2$, $\sigma = 0.88$ S/m) excited from (a) open-ended coaxial probe, (b) monopole, and (c) dielectric-coated monopole.

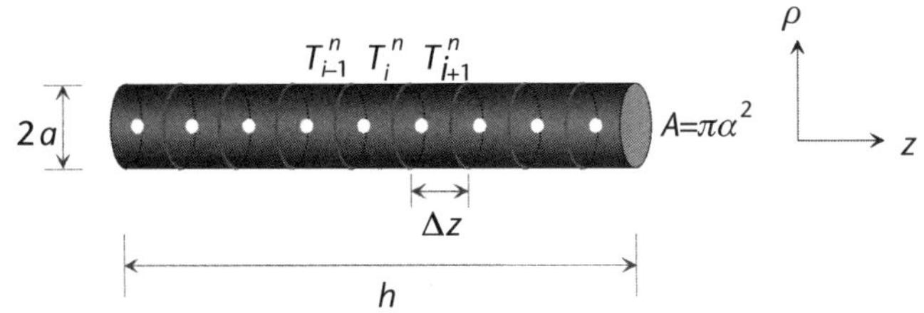

Figure 7.26 Heat diffusion in a dipole.

where p, C_p, and κ in (7.106) are the density (kg/m^3), specific heat capacity [J/(kg °C)], and thermal conductivity [W/(m °C)] of the biological medium, respectively. Parameters T_b, p_b, and C_b are the temperature (°C), density (kg/m^3), and specific heat capacity [J/(kg °C)] of blood, respectively; ω_b is the rate of blood flow [m^3/(kg.s)], which is assumed to be constant; Q_{met} is the metabolic heat generation rate (ignored in this work); and Q_{ext} is the external heat source that is related to the electric field E as:

$$\text{SAR} = \frac{\sigma \left| E_{\text{norm}} \right|^2}{2p} = \frac{Q_{ext}}{p} \tag{7.107}$$

where $E_{\text{norm}} = (E_\rho^2 + E_z^2)^{0.5}$ and SAR is the specific absorption rate (W/kg) of biological medium and σ is the conductivity of the biological medium (S/m). For FDM, (7.106) is discretized as:

$$\begin{aligned} \frac{\partial T}{\partial t} = \frac{\kappa}{pC_b} & \left[\frac{T(i+1,j) - 2T(i,j) + T(i-1,j)}{\Delta\rho^2} + \frac{1}{\rho}\frac{T(i+1,j) - T(i,j)}{2\Delta\rho} \right. \\ & \left. + \frac{T(i,j+1) - 2T(i,j) + T(i,j-1)}{\Delta z^2} \right] \\ & - \frac{\omega_b c_b p_b}{pC_b}\left[T(i,j) - T_b\right] + \frac{\text{SAR}}{C_b} \end{aligned} \tag{7.108}$$

The flowchart of the computational procedure is shown in Figure 7.27. Before SAR is calculated, the normal electric field E_{norm} distribution and some parameters (thermal and dielectric parameters) must be specified. The distribution of the E_{norm} field surrounding the applicator is determined using FDTD method and the E_{norm} field data at each location is stored in a file called "result.mat" (see the accompanying CD). From the known parameters and the E_{norm} field data, the SAR distribution can be calculated using (7.107) and its values are replaced in the bio-heat equation of (7.106). Later, the FDM is applied to (7.106) and the final result is interpreted in terms of temperature. In fact, the FDM routine can be found in MATLAB *ode45* function. Thus, in this work, (7.106) is solved explicitly by available MATLAB *ode45* function. A clear and extensive explanation of the heat transfer computation using MATLAB can be found in Schiesser and Griffiths [22]. The MATLAB

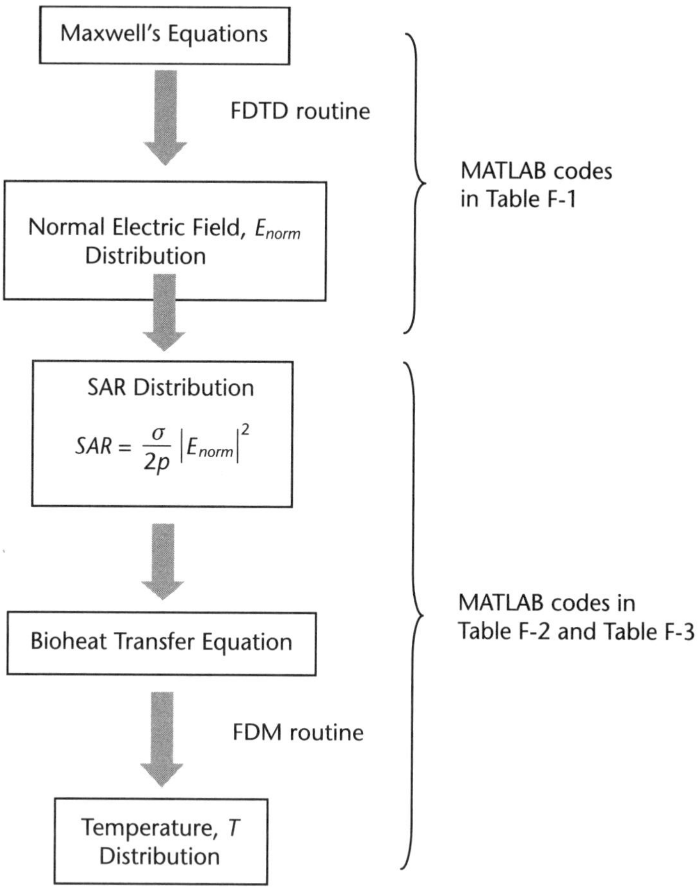

Figure 7.27 Flowchart of the temperature T distribution calculations.

code for calculating the heat distribution on the coated monopole in Lee et al. [23] is listed on the accompanying CD. In this FDM solution, an initial value in time, t (at $t = 0$ seconds) is set as:

$$T_{ko} = T_k^{n=0}(i, j) = 36.7\ ^\circ\text{C} \tag{7.109}$$

The specific implementations of (7.106) for two boundary conditions in ρ-axis and one boundary condition in z-axis are summarized next. Figure 7.28(a) shows the results for the simulated E_{norm} field of a dielectric-coated monopole with 14.32-mm length using MATLAB codes on the accompanying CD. The corresponding simulated temperature distribution of the monopole using MATLAB codes on the accompanying CD is shown in Figure 7.28(b).

For the symmetry axis, $\rho = 0$ ($i = 1$),

$$\frac{1}{\rho}\frac{\partial T}{\partial \rho}$$

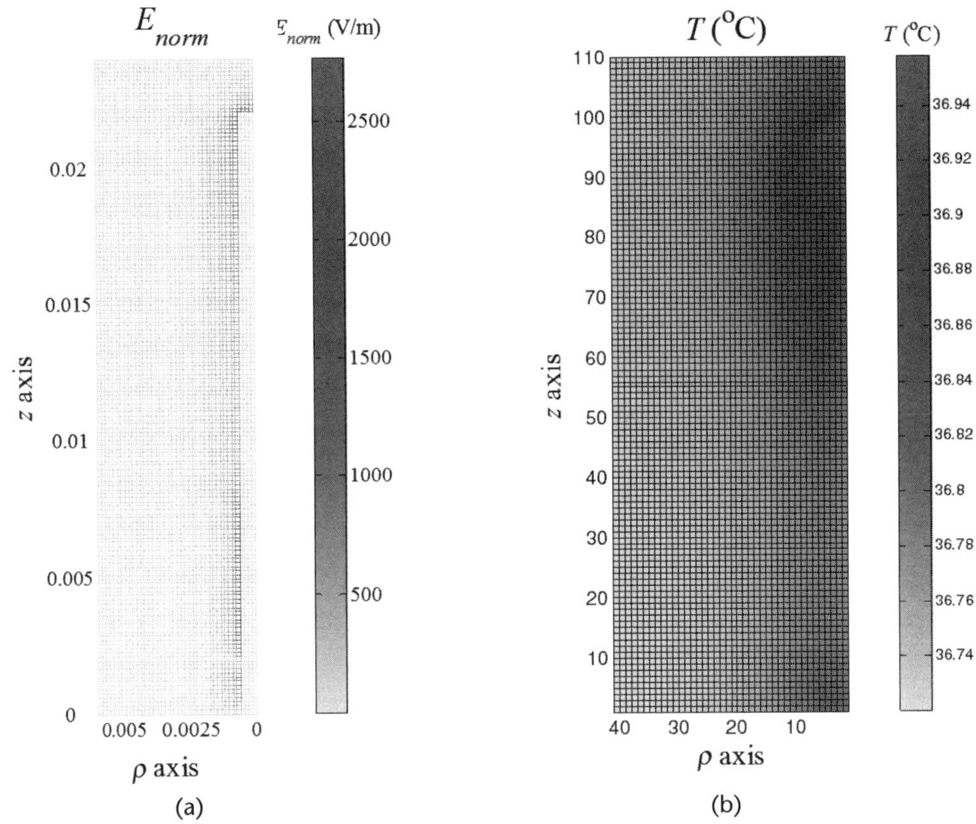

Figure 7.28 (a) The simulated E_{norm} field radiated from the dielectric-coated monopole into brain tissue. (b) The corresponding temperature T distribution in the brain tissue.

The first derivative radial term in (7.106) is equal to zero:

$$\frac{1}{\rho}\frac{\partial T}{\partial \rho} = 0 \tag{7.110}$$

By applying L'Hospital's rule, (7.110) becomes:

$$\lim_{\rho \to 0} \frac{1}{\rho}\frac{\partial T}{\partial \rho} = \frac{\partial^2 T}{\partial \rho^2} \tag{7.111}$$

Thus, the discretization of (7.111) is given as:

$$\begin{aligned} \lim_{\rho \to 0} \frac{1}{\rho}\frac{\partial T}{\partial \rho} &\approx \frac{T(0+\Delta\rho, z) - 2T(0,z) + T(0-\Delta\rho, z)}{\Delta\rho^2} \\ &= \frac{T(2,j) - 2T(1,j) + T(2,j)}{\Delta\rho^2} \\ &= \frac{2T(2,j) - 2T(1,j)}{\Delta\rho^2} \end{aligned} \tag{7.112}$$

The second derivative of the radial term in (7.106) is the same as (7.112):

$$\frac{\partial^2 T}{\partial \rho^2} = \frac{2T(2,j) - 2T(1,j)}{\Delta\rho^2} \tag{7.113}$$

For the vertical edge boundary, $\rho = N_\rho \Delta\rho$ ($i = N_\rho$). The first derivative radial term in (7.106) is modified as:

$$\begin{aligned}\frac{1}{\rho}\frac{\partial T}{\partial \rho} &\approx \frac{1}{N_\rho \Delta\rho}\frac{T\left(N_\rho \Delta\rho + \Delta\rho, z\right) - T\left(N_\rho \Delta\rho - \Delta\rho, z\right)}{2\Delta\rho} \\ &= \frac{1}{\rho}\frac{T\left(N_\rho + 1, j\right) - T\left(N_\rho - 1, j\right)}{2\Delta\rho}\end{aligned} \tag{7.114}$$

$T(N_\rho + 1, j)$ in (7.114) can be assumed as:

$$T\left(N_\rho + 1, j\right) = T\left(N_\rho - 1, j\right) + \left(\frac{2\Delta\rho}{pC_b}\right)\left[T_{\text{wall}} - T\left(N_\rho, j\right)\right] \tag{7.115}$$

Substituting (7.115) into (7.114) yields

$$\begin{aligned}&\frac{1}{\rho}\frac{\partial T}{\partial \rho} \\ &= \frac{1}{\rho}\frac{T\left(N_\rho - 1, j\right) + \left(\frac{2\Delta\rho}{pC_b}\right)\left[T_{\text{wall}} - T\left(N_\rho, j\right)\right] - T\left(N_\rho - 1, j\right)}{2\Delta\rho} \\ &= \frac{1}{\rho}\left(\frac{1}{pC_b}\right)\left[T_{\text{wall}} - T\left(N_\rho, j\right)\right]\end{aligned} \tag{7.116}$$

The second derivative radial term in (7.106) is given as:

$$\frac{\partial^2 T}{\partial \rho^2} = \frac{T\left(N_\rho + 1, j\right) - 2T\left(N_\rho, j\right) + T\left(N_\rho - 1, j\right)}{\Delta\rho^2} \tag{7.117}$$

where $T(N_\rho + 1, j)$ in (7.117) is equal to (7.115).

For the horizontal boundary, $z = 0$ ($j = 1$), the second derivative axial term in (7.106) is given as:

$$\frac{\partial^2 T}{\partial z^2} = \frac{T(i,2) - 2T(i,1) + T(i,0)}{\Delta z^2} \tag{7.118}$$

where $T(i, 0)$ is the source and set to be constant value of 37°C.

References

[1] Yee, K. S., "Numerical Solution of Initial Boundary Value Problems Involving Maxwell's Equations in Isotropic Media," *IEEE Trans. on Antennas Propagat.*, Vol. 14, 1966, pp. 302–307.

[2] Gedney, S. D., *Introduction to the Finite-Difference Time-Domain (FDTD) Method for Electromagnetics (Synthesis Lectures on Computational Electromagnetics)*, Morgan & Claypool, 2011.

[3] Inan, U. S., and R. A. Marshall, *Numerical Electromagnetics: The FDTD Method*, Cambridge, U.K: Cambridge University Press, 2011.

[4] Kunz, K. S., and R. J. Luebbers, *The Finite Difference Time Domain Method for Electromagnetics*, Boca Raton, FL: CRC Press, 1993.

[5] Sullivan, D. M., *Electromagnetic Simulation Using the FDTD Method*, New York: IEEE Press, 2000.

[6] Taflove, A., and S. C. Hagness, *Computational Electrodynamics: The Finite-Difference Time Domain Method*, 3rd ed., Norwood, MA: Artech House, 2005.

[7] Umashankar, K. R., A. Taflove, and B. Beker, "Calculation and Experimental Validation of Induced Currents on Coupled Wires in an Arbitrary Shaped Cavity," *IEEE Trans. on Antennas Propagat.*, Vol. 35, 1987, pp. 1248–1257.

[8] Maloney, J. G., G. S. Smith, and W. R. Scott, "Accurate Computation of the Radiation from Simple Antennas Using the Finite-Difference Time-Domain Method," *IEEE Trans. on Antennas Propagat.*, Vol. 38, 1990, pp. 1059–1068.

[9] Elsherbeni, A. Z., and V. Demir, *The Finite-Difference Time-Domain Method for Electromagnetics with MATLAB Simulations*, SciTech Publishing, 2009.

[10] Sullivan, D. M., "An Unsplit Step 3-D PML for Use with the FDTD Method," *IEEE Microwave Guided Wave Lett.*, Vol. 7, 1997, pp. 184–186.

[11] Mäkinen, R. M., J. S. Juntunen, and M. A. Kivikoski, "An Improved Thin-Wire Model for FDTD," *IEEE Trans. on Microwave Theory Tech.*, Vol. 50, 2002, pp. 1245–1255.

[12] Hyun, S. Y., and S. Y. Kim, "3-D Thin-Wire FDTD Analysis of Coaxial Probe Fed in Asymmetric Microwave Components," *IEEE Trans. on Microwave Theory Tech.*, Vol. 59, 2011, pp. 2808–2815.

[13] Watanabe, S., and M. Taki, "An Improved FDTD Model for the Feeding Gap of a Thin-Wire Antenna," *IEEE Microwave and Guided Wave Letters*, Vol. 8 (4), pp. 152–154, 1998.

[14] Mur, G., "Absorbing Boundary Conditions for the Finite-Difference Approximation of the Time Domain Electromagnetic Field Equations," *IEEE Trans. on Electromagn. Compat.*, Vol. 23, 1981, pp. 377–384.

[15] Berenger, J. P., "A Perfectly Matched Layer for the Absorption of Electromagnetic Waves," *J. Comput. Phys.*, Vol. 114, 1994, pp. 185–200.

[16] Sacks, Z. S., et al., "A Perfectly Matched Anisotropic Absorber for Use as an Absorbing Boundary Condition," *IEEE Trans. on Antennas Propagat.*, Vol. 43, 1995, pp. 1460–1463.

[17] Maloney, J. G., K. L. Shlager, and G. S. Smith, "A Simple FDTD Model for Transient Excitation of Antennas by Transmission Lines," *IEEE Trans. on Antennas Propagat.*, Vol. 42, 1994, pp. 289–292.

[18] Hyun, S. Y., S. Y. Kim, and Y. S. Kim, "Thin-Wire Finite-Difference Time-Domain Model for Insulated and Resistively Loaded Cylindrical Antennas Driven by Coaxial Lines," *IET Microw. Antennas Propag.*, Vol. 3, 2009, pp. 507–513.

[19] Boonzaaier, J. J., and C. W. I. Pistonius, "Finite-Difference Time-Domain Field Approximations for Thin Wires with a Lossy Coating," *IEE Proc. Microw. Antennas Propag.*, Vol. 141, 1994, pp. 107–113.

[20] Nadobny, J., et al., "A Thin-Rod Approximation for the Improved Modeling of Bare and Insulated Cylindrical Antennas Using the FDTD Method," *IEEE Trans. on Antennas Propagat.*, Vol. 51, 2003, pp. 1780–1796.

[21] Pennes, H. H., "Analysis of Tissue and Arterial Blood Temperature in the Resting Human Forearm," *J. Appl. Physiol.*, Vol. 1, 1948, pp. 93–122.

[22] Schiesser, W. E., and G. W. Griffiths, *A Compendium of Partial Differential Equation Models: Method of Lines Analysis with Matlab*, Cambridge, U.K.: Cambridge University Press, 2009.

[23] Lee, C. W., K. Y. You, and C. Y. Lee, "FDTD-Based CAD Simulator for Coaxial Applicator—Biomedical Application," *Progress in Electromagnetics Research Symposium*, Guangzhou, August 25–28, 2014.

Selected Bibliography

Balanis, C. A., *Advanced Engineering Electromagnetic*, New York: John Wiley & Sons, 1989.

Davidson, D. B., *Computational Electromagnetics for RF and Microwave Engineering*, Cambridge, U.K: Cambridge University Press, 2005.

DiStasio, M., and W. C. McHarris, "Electrostatic Problems? Relax!" *Am. J. Phys.*, Vol. 47, 1979, pp. 440–444.

Garg, R., *Analytical and Computational Methods in Electromagnetic*, Norwood, MA: Artech House, 2008.

Hallén, E., "Theoretical Investigations into Transmitting and Receiving Antennae," *Nova Acta Regiae Sco. Sci. Upsaliensis*, Ser. IV, Vol. 4, No. 2, 1938, pp. 1–44.

Hertel, T. W., and G. S. Smith, "On the Convergence of Common FDTD Feed Models for Antennas," *IEEE Trans. on Antennas Propagat.*, Vol. 51, 2003, pp. 1771–1779.

Sadiku, M. N. O., *Numerical Techniques in Electromagnetics*, Boca Raton, FL: CRC Press, 2000.

White, R. E., *Computational Mathematics: Models, Methods, and Analysis with MATLAB and MPI*, Boca Raton, FL: CRC Press, 2004.

CHAPTER 8

Calibration and Measurement Techniques for Coaxial Slot Radiators

This chapter reviews calibration and measurement techniques for coaxial slot radiators, including some instruments used in the measurement. In this chapter, principles and mathematical calculations for calibration are described in detail. Up to four kinds of one-port calibration techniques are described in Section 8.2. The description part of the measurement is mainly focused on the broadband reflection measurement using the coaxial radiators. The reflection wave is caused by the impedance mismatch either due to the coaxial aperture terminated with the exterior medium or the medium partly filled into the coaxial line.

Based on the measured reflection coefficient, Γ, the efficiency of RF power transfer (transmitted or received) from the coaxial radiators can be identified. Most of the communication applications require a good impedance match along the coaxial line so that the maximum power can be received or transmitted by the radiator. Conversely, for material processing applications, the radiator is required to have a highly sensitive reflection response to the impedance mismatch due to the exterior medium or material under test. From the measured reflection signal, the macroscopic properties of the material can be predicted.

8.1 Basic RF System to Measure the Reflection Coefficient

8.1.1 Time-Domain Reflectometry

The time-domain reflectometry (TDR) technique was primarily used to detect the fault and breaks in cables during the 1930s. Later, the same TDR principle was applied to radar technology in the early 1960s. The basic TDR system is shown in Figure 8.1. The pulse generator produces a repetitive pulse into a 50-Ω coaxial cable. The incident pulse is detected by oscilloscope. The pulse is transmitted to the coaxial radiator aperture and an incident pulse signal will partially reflect and travel back due to the discontinuity impedance at the radiator aperture. Then the reflected signal is received by the oscilloscope. The incident and reflected pulse signals' response are separated in time, *t*.

In order to understand the concept of TDR measurements, the measurement can be observed by using a 50-Ω, air-filled coaxial line with mismatched sections, a stepped impedance airline as shown in Figure 8.2(a). In TDR, normally, two kinds of test signals are used, the step response and impulse response signals, as shown in Figure 8.2(b, c), respectively. According to the amplitude of the test signal, the

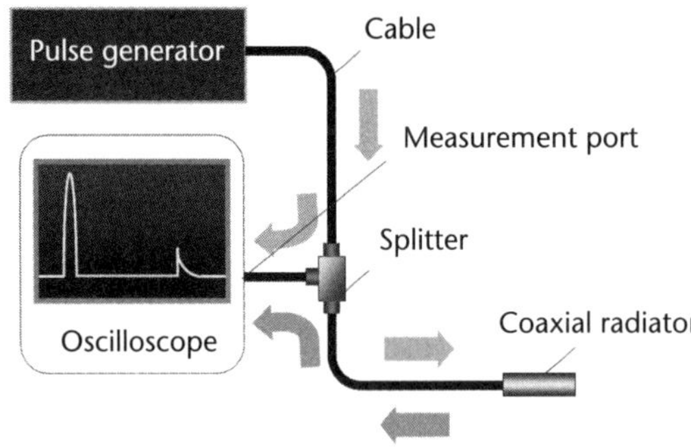

Figure 8.1 Basic time-domain reflectometry system.

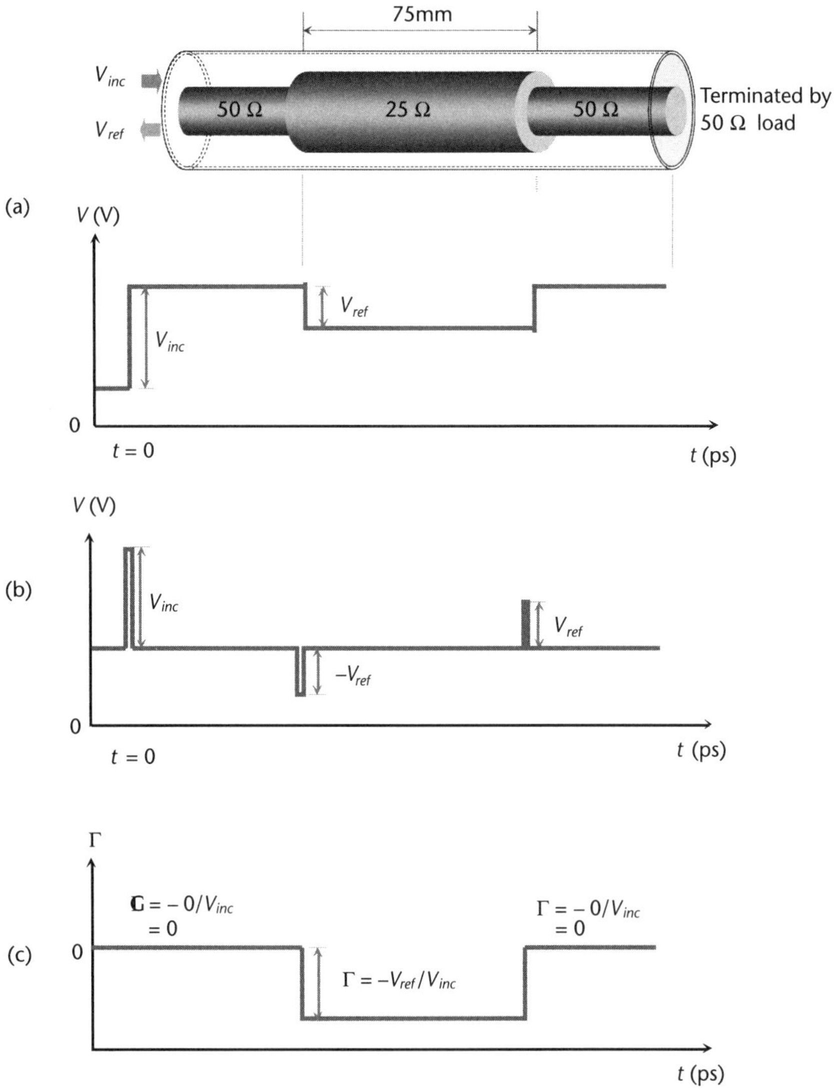

Figure 8.2 (a) Stepped impedance air line (from 50Ω to 25Ω). (b) Ideal step response. (c) Ideal impulse response. (d) Ideal reflection coefficient Γ.

reflection coefficient, Γ, due to the mismatches impedance in the transmission line can be obtained by:

$$\Gamma = \frac{V_{\text{ref}}}{V_{\text{inc}}} \tag{8.1}$$

where V_{ref} and V_{inc} are the reflected and incident pulse amplitudes (in volts), respectively. The V_{ref} can be either a positive or a negative value, which is based on the impedance Z_L at the discontinuity section in the transmission line. Thus, the impedance Z_L can be calculated from the reflection coefficient Γ as:

$$Z_L = Z_o \frac{1+\Gamma}{1-\Gamma} \tag{8.2}$$

where Z_o (equal to 50Ω) is the characteristic impedance of the coaxial line. Figure 8.2(d) shows the ideal reflection response in time for each discontinuity along the transmission coaxial line.

For instance, in Figure 8.2(a), if the amplitudes of V_{inc} and V_{ref} shown in the oscilloscope are 5 mV and –1.67 mV, then the magnitude of the reflection coefficient $|\Gamma|$ and impedance Z_L can be calculated using (8.1) and (8.2) as:

$$|\Gamma| = \left| -\frac{1.67 \text{ mV}}{5 \text{ mV}} \right| = |-0.334| = 0.334$$

$$Z_L = 50\Omega \left(\frac{1 + (-0.334)}{1 - (-0.334)} \right) \approx 25\Omega$$

However, for the case in Figure 8.3(a), the value V_{ref} is expected to be a positive value. When the amplitudes of V_{inc} and V_{ref} are given as 5 mV and 1 mV then, the magnitude of the reflection coefficient $|\Gamma|$ and impedance Z_L is given as:

$$|\Gamma| = \left| \frac{1 \text{ mV}}{5 \text{ mV}} \right| = 0.2$$

$$Z_L = 50\Omega \left(\frac{1 + 0.2}{1 - 0.2} \right) \approx 75\Omega$$

It is clear that the polarities of reflected amplitude, V_{ref}, can be used to predict the value of Z_L either larger or smaller than $Z_o = 50\Omega$. The delay time between the incident and reflected signals can be used to determine the distance or position of discontinuity point from the pulse source. The corresponding signal travel distance d can be converted from the measured time t that the signal needs to travel along the coaxial line:

$$d = \frac{c}{2\sqrt{\varepsilon_c}} t \qquad \text{(one-way distance)} \tag{8.3}$$

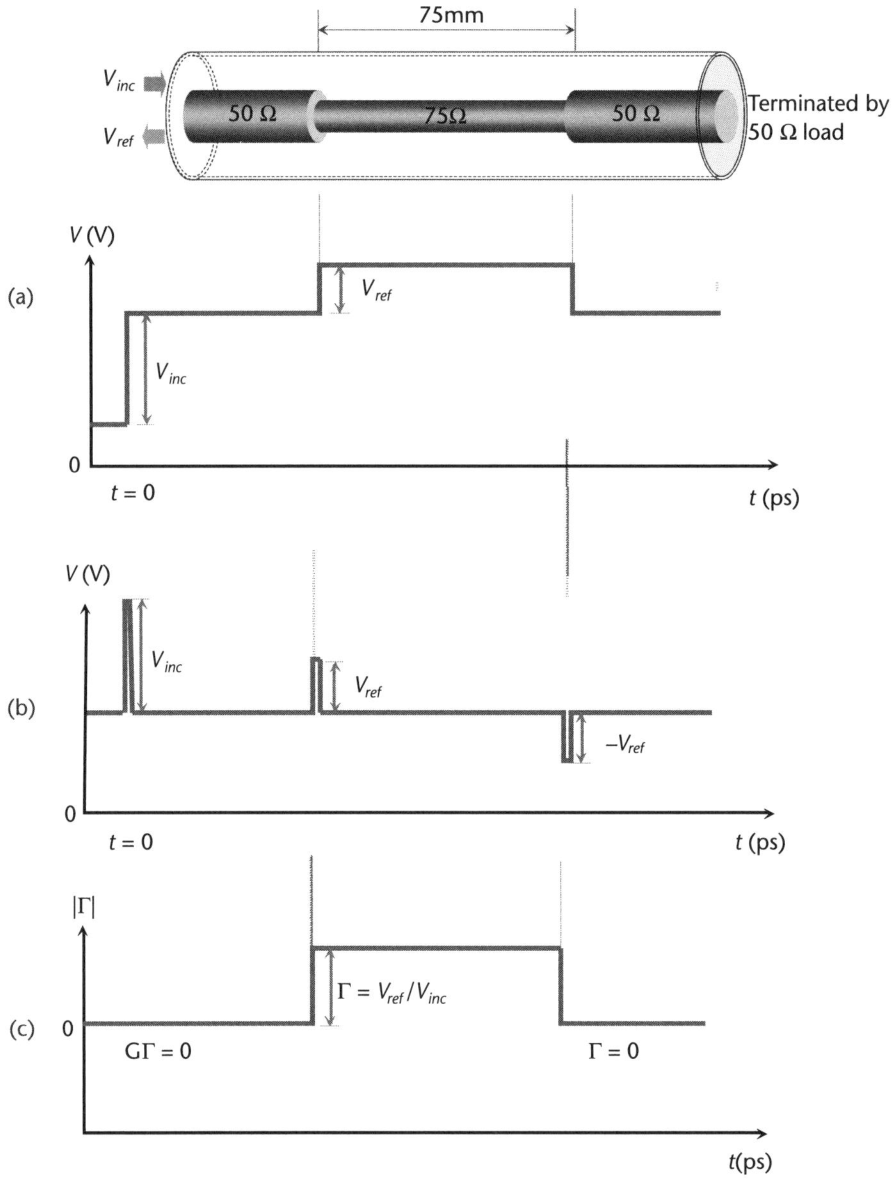

Figure 8.3 (a) Stepped impedance air line (from 50Ω to 75Ω). (b) Ideal step response. (c) Ideal impulse response. (d) Ideal reflection coefficient $|\Gamma|$.

If the round-trip distance d' (distance for signal transmitted and return back to the measurement port) is considered, the equation for d' is written as:

$$d' = \frac{c}{\sqrt{\varepsilon_c}}t \qquad \text{(round-trip distance)} \tag{8.4}$$

where c is the light velocity and ε_c is the complex relative permittivity of the medium filled in the coaxial line. Thus, by using (8.3), the measured incident and reflected voltage at a particular point along the transmission line can be monitored. For an advanced TDR system, the reflection coefficient Γ (or S_{11}) is first measured in the frequency domain, and then the measured reflection data is mathematically calculated to give the time-domain response using an inverse Fourier transform. Hence, the time-domain measurement range (in seconds) is determined by the sampling points N and the span frequency Δf as:

$$t = \frac{N - 1}{\Delta f} \tag{8.5}$$

The time-domain response is repeated every $1/\Delta f$ seconds and the corresponding one-way signal travel distance d is written as:

$$d = \frac{c}{2\Delta f\sqrt{\varepsilon_c}} \tag{8.6}$$

Figure 8.4 shows the measured time-domain reflection (Γ or S_{11}) of the stepped impedance airline shown in Figure 8.2(a). By comparing Figure 8.2(d) and Figure 8.4, clearly, the measured reflection pulse has a finite rise time, T_r, while the ideal reflection has zero rise time. The rise time T_r is the time taken by a signal to change from a specified low value to a specified high value. The corresponding measured impulse response is shown in Figure 8.5.

In fact, the measured pulse resolution is mainly limited by the rise time T_r of the TDR system. For a traditional TDR system (in Figure 8.1), the total rise time T_r of the TDR system is determined by the rise time of the pulse generator, T_{r_g}, and the oscilloscope, T_{r_o}:

$$T_r = \sqrt{T_{r\_g}^2 + T_{r\_o}^2} \tag{8.7}$$

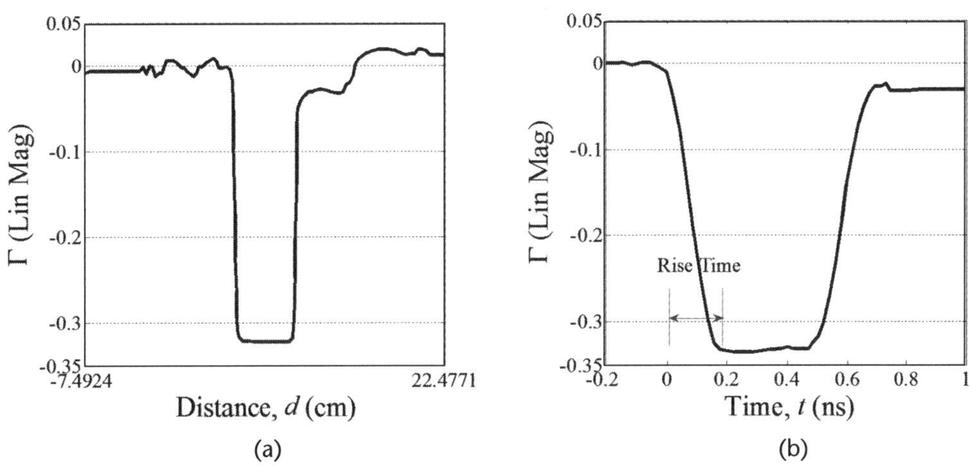

Figure 8.4 Measured step response of a stepped air line [in Figure 8.2(a)] extracted from (a) Grace [1] and (b) Rohde & Schwarz [2].

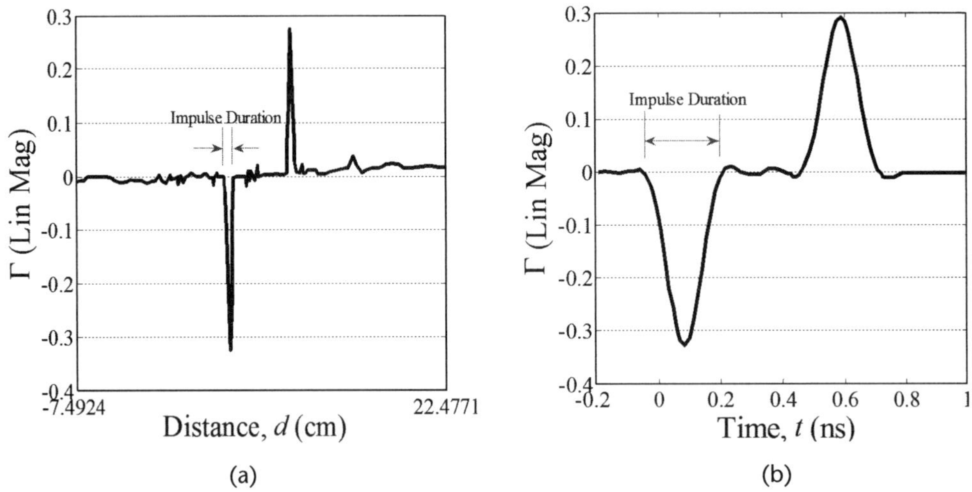

Figure 8.5 Measured impulse response of a stepped air line [in Figure 8.2(a)] extracted from (a) Grace [1] and (b) Rohde & Schwarz [2].

For instance, the effect of the pulse rise time in mismatch measurement can be observed using 50Ω of the air-filled coaxial line, which has three mismatch sections (75Ω) with lengths of 3, 6, and 12 mm, respectively [3], as shown in Figure 8.6.

Figure 8.6(b) shows the ideal reflection response in time for each discontinuity along the transmission coaxial line. However, in actual measurements, two neighboring discontinuities may be distinguishable by the measurement instrument if the signal travel time (t_1, t_2, and t_3) for the distance between them is more than half of the rise time T_r of the TDR instrument:

$$t_1 \geq \frac{T_r}{2}\text{s}, \; t_2 \geq \frac{T_r}{2}\text{s}, \text{ and } t_3 \geq \frac{T_r}{2}\text{s}$$

Therefore, for the system with $T_r = 28$ ps (or $T_r/2 = 14$ ps), the individual discontinuities are clearly distinguished ($t_1 = 20$ ps, $t_2 = 40$ ps, and $t_3 = 80$ ps) as shown in Figure 8.6(c). If the TDR system has $T_r = 150$ ps (or $T_r/2 = 75$ ps), then the two closely spaced discontinuities may be smoothed together into a single aberration in the waveform as shown in Figure 8.6(d). In practice, the accuracy of TDR measurement is affected by several factors, such as rise time, settling time, and pulse aberrations. Normally, filter features are available in the advanced TDR system, which can reduce the pulse aberrations.

8.1.2 Frequency-Domain Reflectometry

Similar to TDR, the frequency-domain reflectometry (FDR) measurement technique is also based on the measured reflection of waves due to discontinuity impedance at the interface of interest. The basic FDR system is shown in Figure 8.7, which consists of the signal generator, dual directional coupler, diode detector, and voltmeter. Sometimes, the FDR system is also described as a reflectometer.

The generator produces an incident signal in the dual-directional coupler; subsequently, the signal is coupled by the coupler. The coupled incident signal is detected

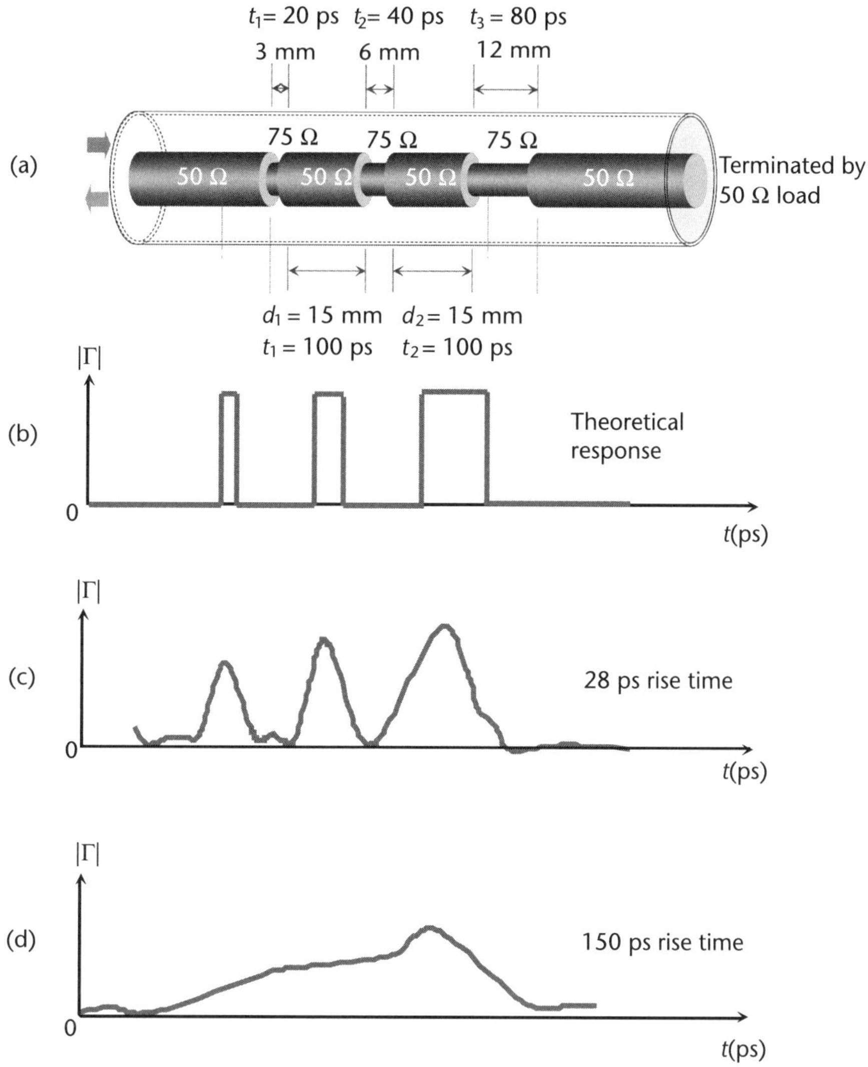

Figure 8.6 (a) A 50-Ω air-filled coaxial line that has three mismatch sections. (b) Ideal reflection coefficient, |Γ|. Reflection coefficient measurement using TDR with (c) 28-ps rise time and (d) 150-ps rise time [3].

by the detector, and its signal value, V_{inc} (volt) is displayed by the voltmeter. Simultaneously, the original incident signal is directly transmitted through the coupler and reaches the coaxial radiator aperture. At the aperture, a partial incident signal will reflect and travel back due to the impedance discontinuity. The reflected signal is coupled by other port in the directional coupler, and the reflected voltage, V_{ref} (volt), value is displayed by the voltmeter. Thus, the scalar measurement of reflection coefficient Γ can be obtained as:

$$\Gamma = \frac{V_{ref}}{V_{inc}} \tag{8.8}$$

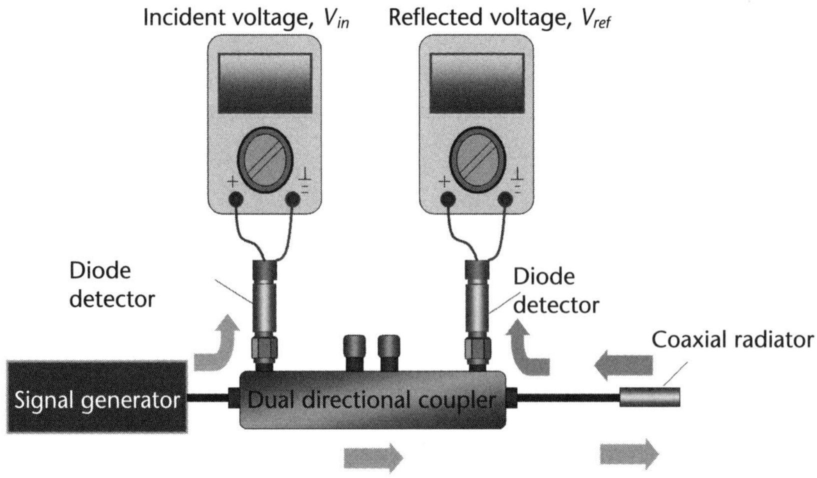

Figure 8.7 Basic scalar frequency-domain reflectometry system.

The vector reflectometer has become increasingly important at present because of measurements involving high-frequency signals. The basic vector FDR system is shown in Figure 8.8, which is able to measure the magnitude $|\Gamma|$ and the phase shift ϕ of the reflection coefficient simultaneously.

Using trigonometry, the relationship between incident voltage, V_{inc}, reflected voltage, V_{ref}, and combined voltage, V_{com}, in Figure 8.8 can be written as:

$$V_{com}^2 = V_{inc}^2 + V_{ref}^2 - 2V_{inc}V_{ref}\cos\phi \tag{8.9}$$

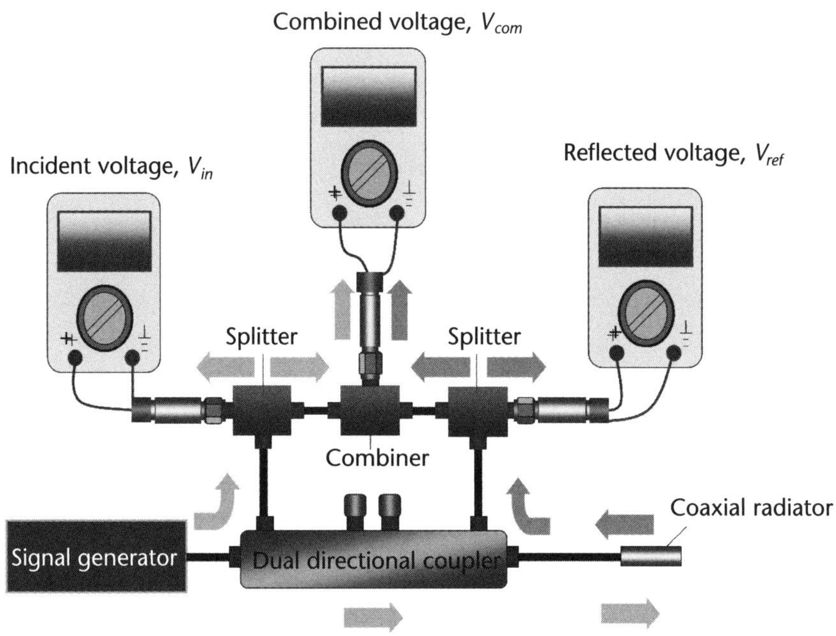

Figure 8.8 Basic vector frequency-domain reflectometry system.

Rearranging (8.9), the phase shift ϕ between V_{inc} and V_{ref} can be calculated as:

$$\phi = \cos^{-1}\left(\frac{V_{inc}^2 + V_{ref}^2 - V_{com}^2}{2V_{inc}V_{ref}}\right) \tag{8.10}$$

The vector diagram of the ideal relationship between V_{inc}, V_{ref}, and V_{com} is shown in Figure 8.9. However, in practice, the combination between the vector $V_{ref,}$ V_{inc}, and V_{com} are usually not in the same plane as shown in Figure 8.10.

Thus, (8.9) is modified as:

$$V_{com}^2 = V_{inc}^2 + V_{ref}^2 - 2\alpha V_{inc}V_{ref}\cos\phi + \beta \tag{8.11}$$

where

$$\alpha = \frac{\mathrm{Max}(\cos\phi) - \mathrm{Min}(\cos\phi)}{2} \tag{8.12}$$

$$\beta = \frac{\mathrm{Max}(\cos\phi) + \mathrm{Min}(\cos\phi)}{2} \tag{8.13}$$

From (8.11), the phase shift ϕ is rewritten as:

$$\phi = \cos^{-1}\left\{\frac{1}{\alpha}\left(\frac{V_{inc}^2 + V_{ref}^2 - V_{com}^2}{2V_{inc}V_{ref}} - \beta\right)\right\} \tag{8.14}$$

The derivation of formulation (8.11) is described in Section 8.1.3.1.

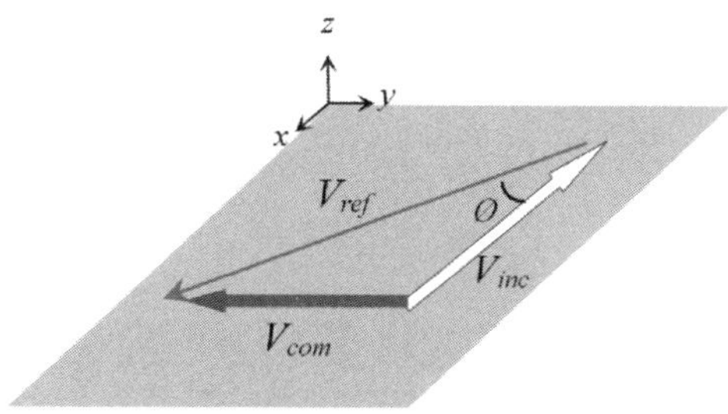

Figure 8.9 Ideal relationship between V_{inc}, V_{ref}, and V_{com}.

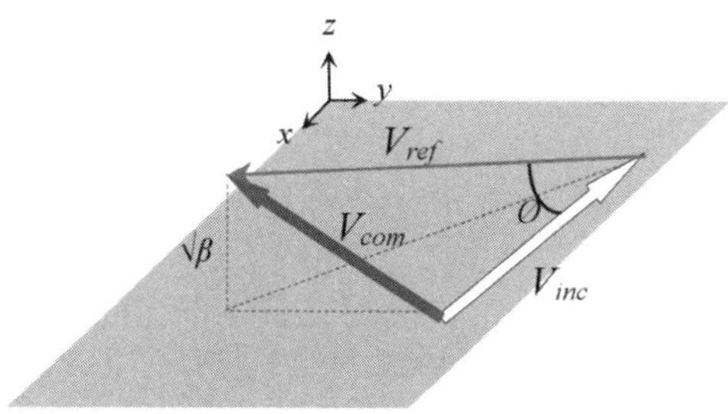

Figure 8.10 Actual relationship between V_{inc}, V_{ref}, and V_{com}.

8.1.3 Operating Frequency

Besides the difference in operating frequencies, the applications and the technologies involved with coaxial radiators are different as well. These frequencies are grouped into several smaller bands. The most commonly used frequency spectrum classification today is created by the Institute of Electrical and Electronic Engineers (IEEE), which is listed in Table 8.1.

8.1.3.1 Derivation of (8.11)

In the ideal case, the expected maximum value of the $\cos\phi$ is equal to 1 and the minimum value is 0. However, in the actual measurement, the maximum $\cos\phi$ is found to be equal to 1.0065 and the minimum value is equal to 0.22, using (8.15), as shown in Figure 8.11.

$$\cos\phi = \frac{V_{\text{inc}}^2 + V_{\text{ref}}^2 - V_{\text{com}}^2}{2V_{\text{inc}}V_{\text{ref}}} \tag{8.15}$$

Thus, (8.15) must be calibrated by adding a correction term as:

$$\begin{aligned}\cos\phi &= \frac{V_{\text{inc}}^2 + V_{\text{ref}}^2 - V_{\text{com}}^2}{2V_{\text{inc}}V_{\text{ref}}} - \frac{1}{2}\left[\max(\cos\phi) + \min(\cos\phi)\right] \\ &= \frac{V_{\text{inc}}^2 + V_{\text{ref}}^2 - V_{\text{com}}^2}{2V_{\text{inc}}V_{\text{ref}}} - \frac{1}{2}[1.0065 + 0.2232]\end{aligned} \tag{8.16}$$

The results of calculations using (8.16) are shown in Figure 8.12. Then, the maximum and minimum values of $\cos\phi$ are adjusted in order to achieve 1 and 0, respectively, by multiplying (8.16) by the coefficient term as:

$$\begin{aligned}\cos\phi &= \frac{2}{\max(\cos\phi) - \min(\cos\phi)} \\ &\quad \times \left\{\frac{V_{\text{inc}}^2 + V_{\text{ref}}^2 - V_{\text{com}}^2}{2V_{\text{inc}}V_{\text{ref}}} - \frac{1}{2}\left[\max(\cos\phi) + \min(\cos\phi)\right]\right\} \\ &= \frac{2}{1.0065 - 0.2232} \times \left\{\frac{V_{\text{inc}}^2 + V_{\text{ref}}^2 - V_{\text{com}}^2}{2V_{\text{inc}}V_{\text{ref}}} - \frac{1}{2}[1.0065 + 0.2232]\right\}\end{aligned} \tag{8.17}$$

The calculated result from (8.17) is shown in Figure 8.13. The inverse cosine of the phase shift ϕ (in radians):

$$\phi = \cos^{-1}\left\{\begin{aligned}&\frac{2}{1.0065 - 0.2232} \\ &\times \left\{\frac{V_{\text{inc}}^2 + V_{\text{ref}}^2 - V_{\text{com}}^2}{2V_{\text{inc}}V_{\text{ref}}} - \frac{1}{2}[1.0065 + 0.2232]\right\}\end{aligned}\right\} \tag{8.18}$$

The phase shift ϕ is shown in Figure 8.14. Clearly, there is an inherent ambiguity of the phase shift, ϕ, since there is no distinction between positive and negative phase shifts. To solve the ambiguity of the phase shift, ϕ an algorithm is developed to compare one phase data point, n to the next at a given frequency. If the phase voltage data has a positive slope, the phase must be positive. However, if the slope is negative, the phase must be negative. Finally, the phase shift will always alternate between $-\pi$ to π when the operating frequencies are sweeping as shown in Figure 8.15.

The limitation on the frequency range of measurement using a coaxial radiator is dependent on its coaxial line cross-sectional dimensions. It is well known that wavelength λ decreases as the frequency f increases. When the wavelength λ is smaller than the diameter of the probe, the polarizations of electric and magnetic field components in the coaxial line are no longer perpendicular to the other

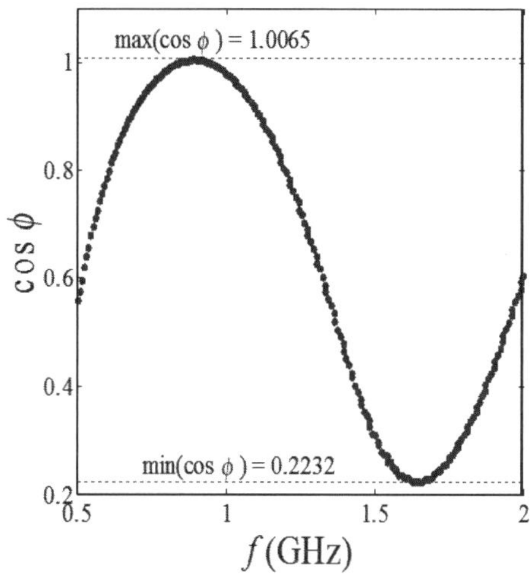

Figure 8.11 Calculated result from (8.15).

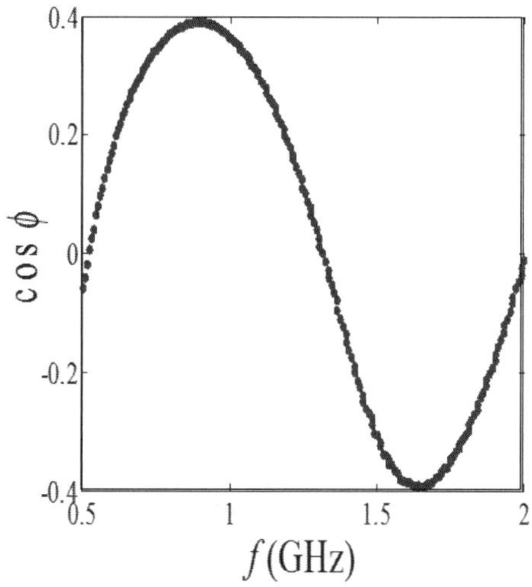

Figure 8.12 Calculated result from (8.16).

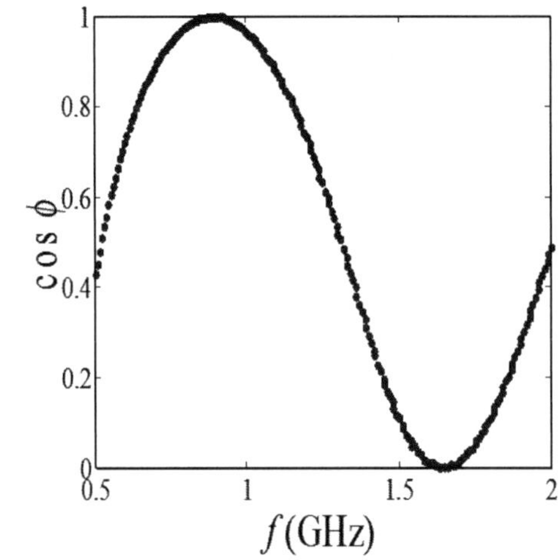

Figure 8.13 Calculated result from (8.17).

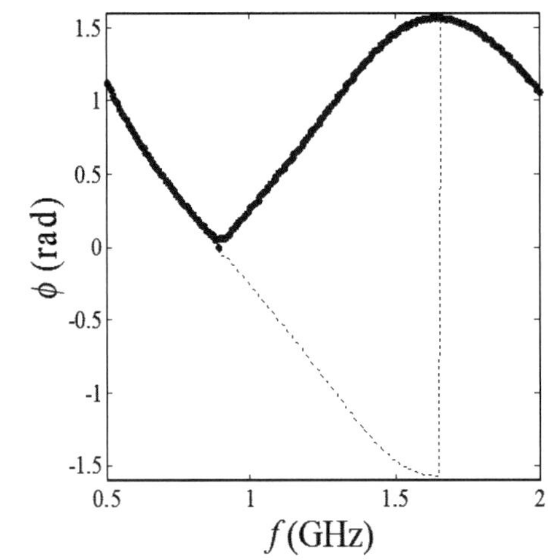

Figure 8.14 Calculated result from (8.18).

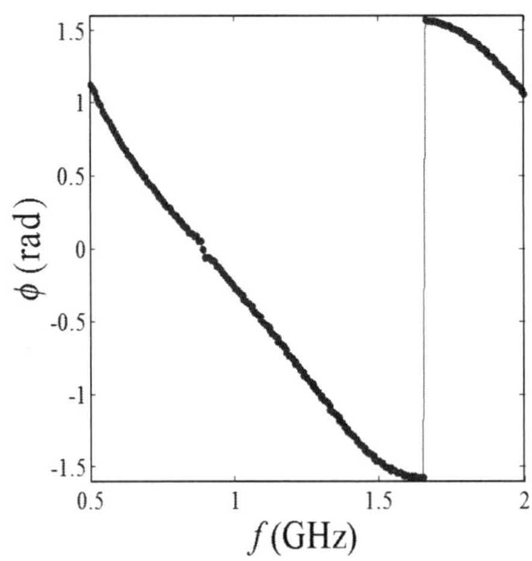

If $\phi_{n+1} - \phi_n > 0$, Output: $+\phi_n$

else $\phi_{n+1} - \phi_n < 0$, Output: $-\phi_n$

Figure 8.15 Calculated result from (8.19).

Table 8.1 Institute of Electrical and Electronic Engineers (IEEE) Frequency Spectrum

Electromagnetic Wave Spectrum	*Frequency Band*	*Wavelength*
Radio waves	Very high frequency (VHF) (30–300 MHz)	10m–1m
Microwaves	Ultrahigh frequency (UHF) (300–3,000 MHz)	(100 cm–10 cm)
	P-band (230 MHz–1 GHz)	130 cm–30 cm
	L-band (1 GHz–2 GHz)	30 cm–15 cm
	S-band (2 GHz–4 GHz)	15 cm–7.5 cm
	Superhigh frequency (SHF) (3–30 GHz)	(10 cm–1 cm)
	S-band (2 GHz–4 GHz)	15 cm–7.5 cm
	C-band (4 GHz–8 GHz)	7.5 cm–3.75 cm
	X-band (8 GHz–12.5 GHz)	3.75 cm–2.4 cm
	Ku-band (12.5 GHz–18 GHz)	2.4 cm–1.67 cm
	K-band (18 GHz–26.5 GHz)	1.67 cm–1.13 cm
	Ka-band (26.5 GHz–40 GHz)	1.13 cm–0.75 cm
	Extremely high frequency (EHF) (30–300 GHz)	(10 mm–1 mm)
Millimeter waves	Ka-band (26.5 GHz–40 GHz)	1.13 cm–0.75 cm
	V-band (40 GHz–75 GHz)	7.5 mm–4 mm
	W-band (75 GHz–110 GHz)	4 mm–2.73 mm
	Millimeter band (110 GHz–300 GHz)	2.73 mm–1 mm

and the higher-order mode, TE_{11}, is excited. Recently, the available dimension of coaxial line has been able to support an operating frequency up to 145 GHz. Some air-filled coaxial lines and their manufacturer's specified operating frequency, *f*, are tabulated in Table 8.2.

Table 8.2 Manufacturer's Specified Operating Frequency for a 50-Ω Air-Filled Coaxial Line

	Manufacturer's Operating Frequency (GHz)								
Air-filled coaxial line (50Ω)	*8.5*	*18*	*33*	*40*	*50*	*65*	*110*	*125/145*	*To be defined (TBD)*
14-mm coaxial line	X								
7-mm coaxial line	X	X							
3.5-mm coaxial line	X	X	X						
2.92-mm coaxial line	X	X	X	X					
2.4-mm coaxial line	X	X	X	X	X				
1.85-mm coaxial line	X	X	X	X	X	X			
1-mm coaxial line	X	X	X	X	X	X	X		
0.8-mm coaxial line	X	X	X	X	X	X	X	X	
0.6-mm coaxial line, b = 0.3 mm, a = 0.13 mm [5]	X	X	X	X	X	X	X	X	X
0.4-mm coaxial line, b = 0.2 mm, a = 0.087 mm [5]	X	X	X	X	X	X	X	X	X

8.1.4 Commercial Test Instruments

For some time now, robust RF reflectometer or vector network analyzer (VNA) products have been available in the market, commercialized by several manufacturers, such as Keysight Technologies (formerly Agilent Technologies), Anritsu, Rohde & Schwarz, and Advantest. Several RF instruments from various brands are shown in Figure 8.16. Currently, these devices are becoming common and popular due to the rapid development of RF and microwave technologies that have created a high demand for this test equipment. Thus, within the past 10 years, a lot of cost-efficient and compact PC-based VNAs (with a normal operating frequency up to 3 GHz or 6 GHz) have been commercialized by manufacturers, such as Copper Mountain Technologies, LA Techniques Ltd., AKELA, Inc., and Mini Radio Solutions.

Besides that, recently, the high-performance VNA has been more likely to be built in the form of the open industry-standard PXI (PCI eXtensions for Instrumentation). The PXI VNA has the benefit of low cost, performance, space, and flexibility of the latest PC technology. In addition, the PXI compatibility also allows the VNA to be integrated with other test and automation modules in a single chassis.

8.2 One-Port Calibration

8.2.1 Principles of One-Port Calibration

In this section, the procedures for reflection coefficient measurements at the coaxial radiator aperture (or driving point) using the VNA will be discussed in detail. The measurement setup is shown in Figure 8.17.

Before the VNA is connected to the coaxial radiator, the VNA is routinely calibrated at the end of the cable using commercial calibration kits, such as the

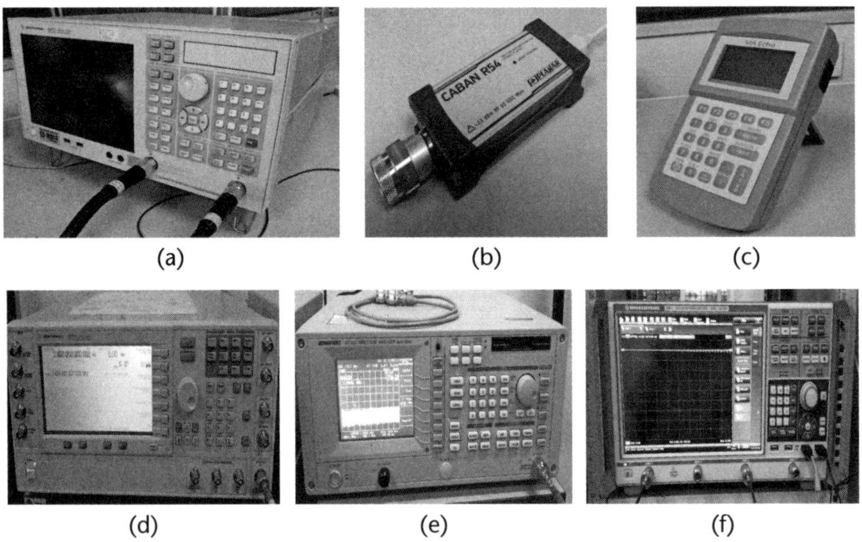

Figure 8.16 (a) Keysight E5071C network analyzer. (b) Planar R54 vector reflectometer. (c) VIA Echo vector impedance analyzer. (d) Keysight E8267D PSG vector signal generator. (e) Advantest R3131A spectrum analyzer. (f) Rohde & Schwarz ZNB 8 vector network analyzer.

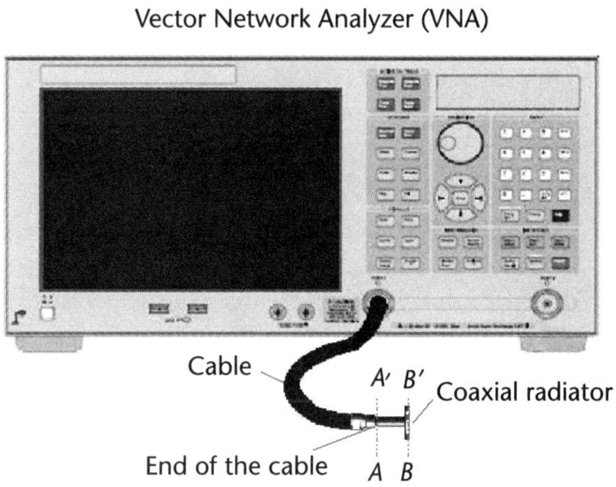

Figure 8.17 The open-ended coaxial radiator is connected to the VNA via cable.

Keysight (formerly Agilent) 85052D calibration kit. This means that all the reflection coefficient measurements are referred to the end of the cable even after any coaxial radiator is connected to it.

In fact, the reflection coefficient Γ_a at the radiator aperture is should be measured as shown in Figure 8.17. Hence, the radiator measurement must be calibrated due to the reflection at the end of the cable (at the AA' plane), which is separated from the radiator aperture (at the BB' plane) by a coaxial line. Thus, a de-embedding process should be done to remove the effects of the coaxial line. However, there are no commercial calibration kits for arbitrary coaxial radiators. Traditionally, the measured reflection coefficient Γ_{meas} at the plane AA' is de-embedded to the end of the probe aperture (plane BB') by adding the error network to the measured Γ_{meas} as shown in Figure 8.18. The relationship between the actual reflection coefficient Γ_a at the radiator aperture and the measured Γ_{meas} is expressed in a bilinear equation as:

$$\Gamma_{\text{meas}} = e_{11} + \frac{e_{21}e_{12}}{1 - e_{22}\Gamma_a}\Gamma_a \tag{8.19}$$

where e_{11}, e_{12}, e_{21}, and e_{22} are the scattering parameters of the error network. Rearranging (8.19), Γ_a at the probe aperture (plane BB') can be determined by:

$$\Gamma_a = \frac{\Gamma_{\text{meas}} - e_{11}}{e_{22}\Gamma_{\text{meas}} + \left(e_{12}e_{21} - e_{11}e_{22}\right)} \tag{8.20}$$

The derivation process of the formulation in (8.19) can be found in Figure 8.19. In fact, the $e_{12}e_{21}$ terms in (8.19) and (8.20) represent the values of the tracking error, and e_{11} and e_{22} are the values of the directivity error and the source match error,

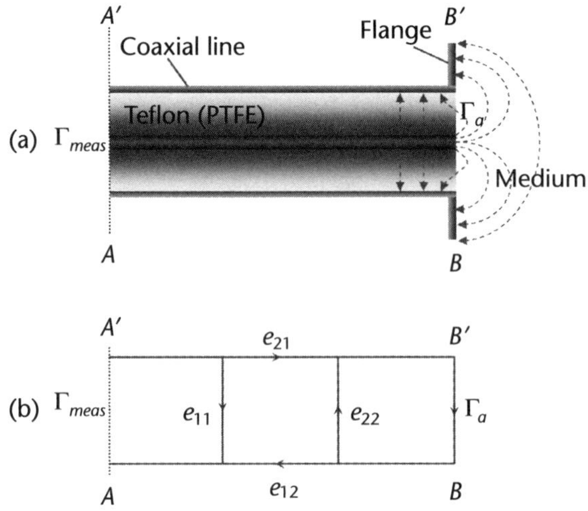

Figure 8.18 Signal flow graph representation of an error network.

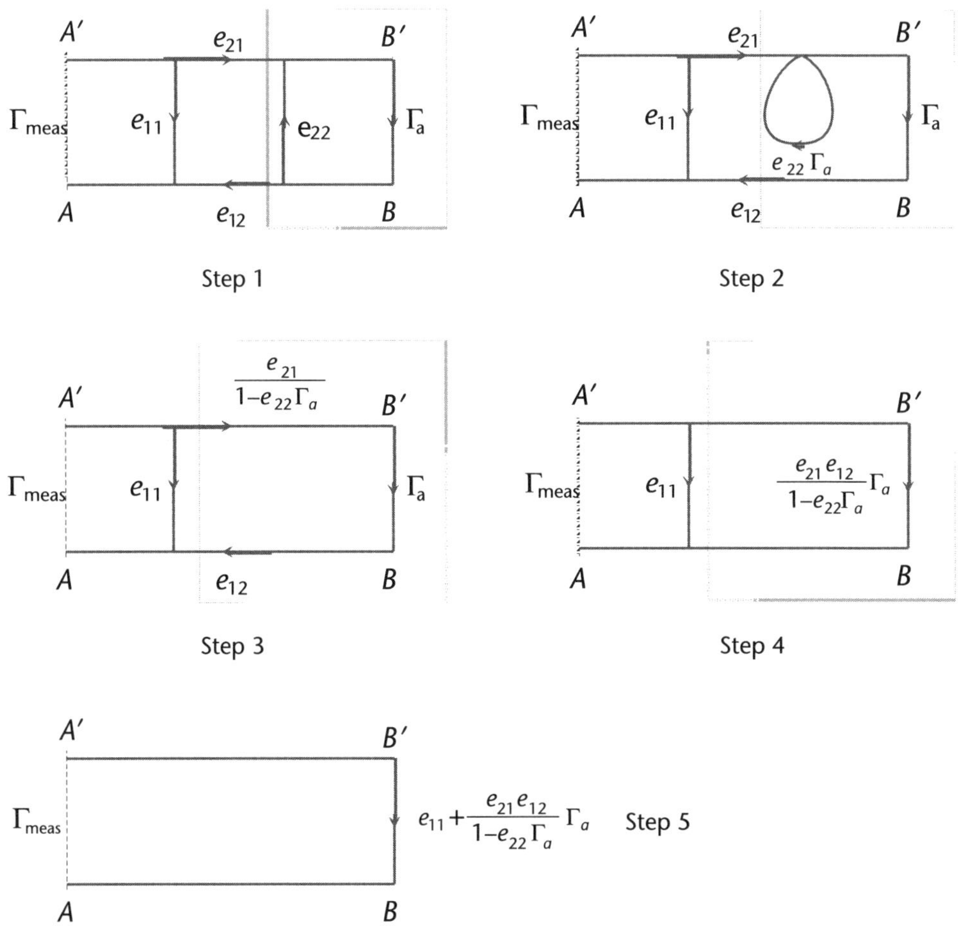

Figure 8.19 The derivation steps of the calibration equation.

respectively. These errors are contributed to by a shifted phase and attenuation loss in the coaxial line and from the radiating fringing effects at the radiator aperture, which are mainly caused by the occurrence of standing waves in the coaxial line.

8.2.2 Three-Standard Calibration

Normally, for a coaxial radiator, three independent calibration standards, the open-short-load (OSL) calibration techniques, are used in the de-embedding process. In this process, the e_{11}, e_{22}, and $(e_{12}e_{21} - e_{11}e_{22})$ in (8.20) are assumed to be unknown values that are desired to be determined. Let $c_1 = (e_{12}e_{21} - e_{11}e_{22})$, $c_2 = e_{11}$, and $c_3 = -e_{22}$, and (8.20) can be rewritten as:

$$\Gamma_a = \frac{\Gamma_{\text{meas}} - c_2}{-c_3\Gamma_{\text{meas}} + c_1} \tag{8.21}$$

Equation (8.21) can be rearranged and expressed in a linear form as:

$$c_1\Gamma_a + c_2 - \Gamma_a\Gamma_{\text{meas}}c_3 = \Gamma_{\text{meas}} \tag{8.22}$$

The three unknown complex coefficients (c_1, c_2, and c_3) values in (8.22) are found using three independent reflection coefficients, Γ_{meas}, of the open-short-load standards. Normally, a polar liquid, such as water, is used as a load standard for the calibration process, since it does not have any available match-load kit due to the difficulty in constructing a kit that can be connected to the aperture radiator [6]. Let Γ_{a_air}, Γ_{a_short}, and Γ_{a_water} represent the known reflection coefficients for the open, short, and load standards that are terminated at the aperture plane BB', while, $\Gamma_{\text{meas\_air}}$, $\Gamma_{\text{meas\_short}}$, and $\Gamma_{\text{meas\_water}}$ are the measured reflection coefficients for open, short, and load standards at plane AA'. $\Gamma_{\text{meas\_air}}$ is measured by leaving the open end of the probe in air as shown in Figure 8.20(a). Later, the measurement is repeated to obtain $\Gamma_{\text{meas\_short}}$ by terminating the probe aperture with metal plate as shown in Figure 8.20(b). Finally, $\Gamma_{\text{meas\_water}}$ is obtained by immersing the probe in the water as shown in Figure 8.20(c).

However, the values of the open, short, and water-load standards (Γ_{a_air}, Γ_{a_short}, and Γ_{a_water}) at the probe aperture are calculated using analytical formulations or

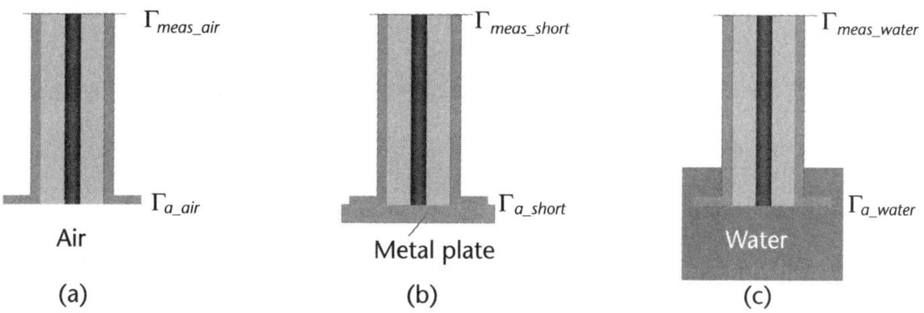

Figure 8.20 Finite coaxial line: (a) terminated by free space, (b) shorted by a metal plate, and (c) immersed into water.

numerical methods. Thus, by using (8.22), three sets of linear equations are created as:

$$c_1\Gamma_{a\_air} + c_2 - \Gamma_{a\_air}\Gamma_{meas\_air}c_3 = \Gamma_{meas\_air} \tag{8.23a}$$

$$c_1\Gamma_{a\_short} + c_2 - \Gamma_{a\_short}\Gamma_{meas\_short}c_3 = \Gamma_{meas\_short} \tag{8.23b}$$

$$c_1\Gamma_{a\_water} + c_2 - \Gamma_{a\_water}\Gamma_{meas\_water}c_3 = \Gamma_{meas\_water} \tag{8.23c}$$

Equation (8.23) can be written in a matrix form as:

$$\begin{bmatrix} \Gamma_{a\_air} & 1 & -\Gamma_{a\_air}\Gamma_{meas\_air} \\ \Gamma_{a\_short} & 1 & -\Gamma_{a\_short}\Gamma_{meas\_short} \\ \Gamma_{a\_water} & 1 & -\Gamma_{a\_water}\Gamma_{meas\_water} \end{bmatrix} \begin{bmatrix} c_1 \\ c_2 \\ c_3 \end{bmatrix} = \begin{bmatrix} \Gamma_{meas\_air} \\ \Gamma_{meas\_short} \\ \Gamma_{meas\_water} \end{bmatrix} \tag{8.24}$$

Finally, (8.24) is solved by using a Gaussian elimination routine. In fact, (8.24) can also be manually solved by Cramer's rule as:

$$c_1 = \frac{\Gamma_{a\_air}\Gamma_{meas\_air}\Delta_{w\_s} + \Gamma_{a\_water}\Gamma_{meas\_water}\Delta_{s\_a} + \Gamma_{a\_short}\Gamma_{meas\_short}\Delta_{a\_w}}{\Gamma_{a\_water}\Gamma_{a\_short}\Delta_{w\_s} + \Gamma_{a\_air}\Gamma_{a\_short}\Delta_{s\_a} + \Gamma_{a\_water}\Gamma_{a\_air}\Delta_{a\_w}} \tag{8.25a}$$

$$c_2 = \frac{\left\{\begin{array}{l} \Gamma_{a\_short}\Gamma_{a\_water}\Gamma_{meas\_air}\Delta_{w\_s} + \Gamma_{a\_open}\Gamma_{a\_short}\Gamma_{meas\_water}\Delta_{s\_a} \\ +\Gamma_{a\_air}\Gamma_{a\_water}\Gamma_{meas\_short}\Delta_{a\_w} \end{array}\right\}}{\Gamma_{a\_water}\Gamma_{a\_short}\Delta_{w\_s} + \Gamma_{a\_air}\Gamma_{a\_short}\Delta_{s\_a} + \Gamma_{a\_water}\Gamma_{a\_air}\Delta_{a\_w}} \tag{8.25b}$$

$$c_3 = \frac{\Gamma_{a\_air}\Delta_{w\_s} + \Gamma_{a\_water}\Delta_{s\_a} + \Gamma_{a\_short}\Delta_{a\_w}}{\Gamma_{a\_water}\Gamma_{a\_short}\Delta_{w\_s} + \Gamma_{a\_air}\Gamma_{a\_short}\Delta_{s\_a} + \Gamma_{a\_water}\Gamma_{a\_air}\Delta_{a\_w}} \tag{8.25c}$$

where

$$\Delta_{m\_a} = \Gamma_{meas} - \Gamma_{meas\_air}, \qquad \Delta_{m\_s} = \Gamma_{meas} - \Gamma_{meas\_short}$$
$$\Delta_{m\_w} = \Gamma_{meas} - \Gamma_{meas\_water}, \qquad \Delta_{a\_w} = \Gamma_{meas\_air} - \Gamma_{meas\_water}$$
$$\Delta_{s\_a} = \Gamma_{meas\_short} - \Gamma_{meas\_air}, \qquad \Delta_{w\_s} = \Gamma_{meas\_water} - \Gamma_{meas\_short}$$

Once the values of c_1, c_2, and c_3 are obtained, the calibrated reflection coefficient Γ_a of the exterior sample can be calculated by using (8.21).

The accuracy of the OSL calibration is affected significantly by the accuracy of the values of the calibration standard (Γ_{a_open}, Γ_{a_short}, and Γ_{a_load}). If the standard values are closer to the actual situation, the accuracy of the calibration will be higher. In the previous calibration study, the reflection coefficient of an open standard is assumed to be ideal ($\Gamma_{a\_open} = 1$). However, in an actual case, all terminated open-circuits have fringing field due to the fact that stray capacitance across the probe aperture, which is $\Gamma_{a\_open} \neq 1$. Thus, the fringing effect of the open-circuit standard is important to be taken into account in the calibration. For microwave broadband frequency measurement, the fringing effect can be modeled using a frequency-dependent equivalent capacitor, C, which is simple and applicable to provide the accuracy reference values for the open-circuit calibration. Normally, the equivalent capacitance C element is modeled as a cubic polynomial function of frequency f as:

$$C = C_o + C_1 f + C_2 f^2 + C_3 f^3 \tag{8.26}$$

with units of f (Hz), $C_o(10^{-15}$ Farad), C_1 (10^{-27} Farad/Hz), C_2 (10^{-36} Farad/Hz^2) and C_3 (10^{-45} Farad/Hz^3). At low frequencies, a single capacitance C_o element may be sufficient. Finally, the values of Γ_{a_open} can be obtained through the aperture admittance $\tilde{Y}_{air}$ formulation as [7]:

$$\Gamma_{a\_open} = \frac{1-\tilde{Y}_{air}}{1+\tilde{Y}_{air}} = \frac{1 - j(\omega/Y_o)(C_o + C_1 f + C_2 f^2 + C_3 f^3)}{1 + j(\omega/Y_o)(C_o + C_1 f + C_2 f^2 + C_3 f^3)} \tag{8.27}$$

For ultrabroadband modeling (from dc to more than 30 GHz), the reflection coefficient Γ_{a_open} is more suitable to be expressed as:

$$\Gamma_{a\_open} = \frac{1 - j(\omega/Y_o)(C_o f^{-1} + C_1 + C_2 f + C_3 f^2)}{1 + j(\omega/Y_o)(C_o f^{-1} + C_1 + C_2 f + C_3 f^2)} \tag{8.28}$$

The complex values of the C_o, C_1, C_2, and C_3 in (8.27) for various types and size of coaxial probes are listed on the accompanying CD.

8.2.3 Alternative Three-Standard Calibration

The alternative calibration formulation without any unknown complex coefficients (c_1, c_2, and c_3) can be derived by using the cross-ratio invariance concept [8]:

$$\frac{(\tilde{Y}_a - \tilde{Y}_{air})(\tilde{Y}_{water} - \tilde{Y}_{short})}{(\tilde{Y}_a - \tilde{Y}_{short})(\tilde{Y}_{air} - \tilde{Y}_{water})} = \frac{(\Gamma_{meas} - \Gamma_{meas\_air})(\Gamma_{meas\_water} - \Gamma_{meas\_short})}{(\Gamma_{meas} - \Gamma_{meas\_short})(\Gamma_{meas\_air} - \Gamma_{mean\_water})} \tag{8.29a}$$

or

$$\frac{(\Gamma_a - \Gamma_{a\_air})(\Gamma_{a\_water} - \Gamma_{a\_short})}{(\Gamma_a - \Gamma_{a\_short})(\Gamma_{a\_air} - \Gamma_{a\_water})} = \frac{(\Gamma_{meas} - \Gamma_{meas\_air})(\Gamma_{meas\_water} - \Gamma_{meas\_short})}{(\Gamma_{meas} - \Gamma_{meas\_short})(\Gamma_{meas\_air} - \Gamma_{meas\_water})} \tag{8.29b}$$

where Γ_{meas} is the measured reflection coefficient of the exterior medium at plane AA'. Γ_a and $\tilde{Y}_a$ are the aperture reflection coefficient and normalized input admittance of the exterior medium at the open end of coaxial line (at plane BB'), respectively. $\tilde{Y}_{air}$, $\tilde{Y}_{short}$, and $\tilde{Y}_{water}$ are the calculated normalized admittances for air, metallic plate, and water at plane BB', respectively. Marsland and Evans [9] rearranged the bilinear transform of (8.29a) into a convenient form as:

$$\tilde{Y}_a = -\left(\frac{\Delta_{m\_a}\Delta_{w\_s}\tilde{Y}_{water}\tilde{Y}_{short} + \Delta_{m\_s}\Delta_{a\_w}\tilde{Y}_{air}\tilde{Y}_{water} + \Delta_{m\_w}\Delta_{s\_a}\tilde{Y}_{short}\tilde{Y}_{air}}{\Delta_{m\_a}\Delta_{w\_s}\tilde{Y}_{air} + \Delta_{m\_s}\Delta_{a\_w}\tilde{Y}_{short} + \Delta_{m\_w}\Delta_{s\_a}\tilde{Y}_{water}} \right) \tag{8.30}$$

For the short-circuit standard, the normalized admittance, $\tilde{Y}_{short} = \infty$. Hence, the denominator term in (8.30) can be approximated as:

$$\Delta_{m\_a}\Delta_{w\_s}\tilde{Y}_{air} + \Delta_{m\_s}\Delta_{a\_w}\tilde{Y}_{short} + \Delta_{m\_w}\Delta_{s\_a}\tilde{Y}_{water} \approx \Delta_{m\_s}\Delta_{a\_w}\tilde{Y}_{short} \tag{8.31}$$

Replacing (8.31) into (8.30), the simplified equation (8.30) can be rewritten as:

$$\begin{aligned} \tilde{Y}_a &= -\left(\frac{\Delta_{m\_a}\Delta_{w\_s}\tilde{Y}_{water}\tilde{Y}_{short} + \Delta_{m\_s}\Delta_{a\_w}\tilde{Y}_{air}\tilde{Y}_{water} + \Delta_{m\_w}\Delta_{s\_a}\tilde{Y}_{short}\tilde{Y}_{air}}{\Delta_{m\_s}\Delta_{a\_w}\tilde{Y}_{short}} \right) \\ &= -\left(\frac{\Delta_{m\_a}\Delta_{w\_s}}{\Delta_{m\_s}\Delta_{a\_w}}\tilde{Y}_{water} + \frac{\Delta_{m\_w}\Delta_{s\_a}}{\Delta_{m\_s}\Delta_{a\_w}}\tilde{Y}_{air} \right) \end{aligned} \tag{8.32}$$

Because the cross-ratio invariance and admittance of the exterior medium are proportional to their relative permittivity ($\tilde{Y}_a \propto \varepsilon_r$), (8.32) can also be expressed as a function of relative permittivity, ε_r:

$$\varepsilon_r = -\left(\frac{\Delta_{m\_a}\Delta_{w\_s}}{\Delta_{m\_s}\Delta_{a\_w}}\varepsilon_{r\_water} + \frac{\Delta_{m\_w}\Delta_{s\_a}}{\Delta_{m\_s}\Delta_{a\_w}}\varepsilon_{r\_air} \right) \tag{8.33}$$

Through (8.33), the relative permittivity ε_r of the exterior medium can be calculated directly from the corresponding measured reflection coefficient Γ_{meas} without any calculation of the aperture admittance and matrix solution. The relative permittivity, ε_{r_air}, of air in (8.33) is unity. The relative permittivity, ε_{r_water}, of water up to 250 GHz can be computed with two relaxation terms of the Debye model as [10]:

$$\varepsilon_{r\_water} = \varepsilon_{\infty\_water} + \frac{\varepsilon_{s\_water} - \varepsilon_{1\_water}}{1 + j2\pi\tau_{1\_water}f} + \frac{\varepsilon_{1\_water} - \varepsilon_{\infty\_water}}{1 + j2\pi\tau_{2\_water}f} \tag{8.34}$$

where τ_{1_water} and τ_{2_water} are the relaxation time of the water for low-frequency and high-frequency processes, respectively. The term ε_{s_water} is the static relative permittivity at zero frequency and the term ε_{1_water} is the static relative permittivity for second relaxation process. The term $\varepsilon_{\infty\_water}$ is the relative permittivity at the high-frequency limit (maximum operating frequency), which is known as optical permittivity. Those parameters in (8.34) that vary as a function of water temperature T (in degrees Celsius) are tabulated on the accompanying CD.

However, the calibration that involves the short-circuit standard requires a coaxial line with a length exceeding $\lambda/4$. The reason is that the strong interference at the probe aperture terminated by short circuits will directly affect the measured Γ_{meas}, due to the shorter length of coaxial line (positions of plane AA' and BB' are too close). The above condition will cause the open (air) and short standards failed to maintain 180° of phase separate throughout the operation frequency range (especially at low operation frequencies). In addition, the measurement of the short-circuit standard is quite difficult to determine accurately and precisely since it depends on the density of pressing a coaxial probe to a metal plate. In addition, a small air gap in between the aperture probe and the metal plate will cause significant errors in measuring. To solve this problem, a perfect short calibration standard must be designed to suit the probe in use.

Thus, the short-circuit standard can be replaced by methanol liquids and it provides the calibration for coaxial radiators that do not have a flat surface aperture, such as a monopole driven from the coaxial line as shown in Figure 8.21.

When the air, methanol, and water are used as the calibration standards, (8.32) must be rewritten as:

$$\tilde{Y}_a = -\left(\frac{\Delta_{m\_a}\Delta_{w\_s}\tilde{Y}_{water}\tilde{Y}_{short} + \Delta_{m\_s}\Delta_{a\_w}\tilde{Y}_{air}\tilde{Y}_{water} + \Delta_{m\_w}\Delta_{s\_a}\tilde{Y}_{short}\tilde{Y}_{air}}{\Delta_{m\_a}\Delta_{w\_s}\tilde{Y}_{air} + \Delta_{m\_s}\Delta_{a\_w}\tilde{Y}_{short} + \Delta_{m\_w}\Delta_{s\_a}\tilde{Y}_{water}}\right) \tag{8.35}$$

Similarly, (8.33) can be rewritten as:

$$\varepsilon_r = -\left(\frac{P + Q + R}{\Delta_{m\_a}\Delta_{w\_me}\varepsilon_{r\_air} + \Delta_{m\_me}\Delta_{a\_w}\varepsilon_{r\_methanol} + \Delta_{m\_w}\Delta_{me\_a}\varepsilon_{r\_water}}\right) \tag{8.36}$$

where

$$P = \Delta_{m\_a}\Delta_{w\_me}\varepsilon_{r\_water}\varepsilon_{r\_methanol} \tag{8.37a}$$

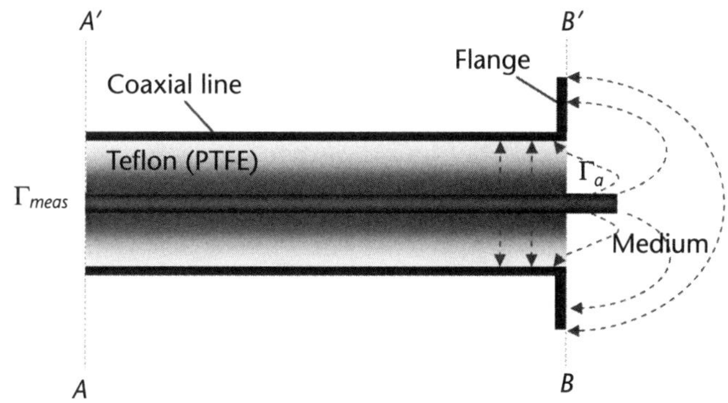

Figure 8.21 Short monopole driven from a coaxial line.

$$Q = \Delta_{m\_me}\Delta_{a\_w}\varepsilon_{r\_air}\varepsilon_{r\_water} \tag{8.37b}$$

$$R = \Delta_{m\_w}\Delta_{me\_a}\varepsilon_{r\_methanol}\varepsilon_{r\_air} \tag{8.37c}$$

and

$$\Delta_{m\_a} = \Gamma_{meas} - \Gamma_{meas\_air}, \qquad \Delta_{m\_me} = \Gamma_{meas} - \Gamma_{meas\_methanol},$$
$$\Delta_{m\_w} = \Gamma_{meas} - \Gamma_{meas\_water}, \qquad \Delta_{a\_w} = \Gamma_{meas\_air} - \Gamma_{meas\_water},$$
$$\Delta_{s\_a} = \Gamma_{meas\_short} - \Gamma_{meas\_air}, \qquad \Delta_{w\_me} = \Gamma_{meas\_water} - \Gamma_{meas\_methanol}$$

The relative permittivity $\varepsilon_{r\_methanol}$ of the methanol in (8.36) and (8.37) is obtained from three relaxation terms of Debye model (8.38) with parameters, which are listed on the accompanying CD.

$$\begin{aligned}\varepsilon_{r\_methanol} &= \varepsilon_{\infty\_methanol} + \frac{\varepsilon_{s\_methanol} - \varepsilon_{1\_methanol}}{1 + j2\pi\tau_1 f} \\ &+ \frac{\varepsilon_{1\_methanol} - \varepsilon_{2\_methanol}}{1 + j2\pi\tau_2 f} + \frac{\varepsilon_{2\_methanol} - \varepsilon_{\infty\_methanol}}{1 + j2\pi\tau_3 f}\end{aligned} \tag{8.38}$$

The parameters of (8.38) as a function of temperature T are tabulated on the accompanying CD.

8.2.4 One-Standard Calibration

Apart from the three standards, the calibration can be simplified by using only one standard, which is air (open-circuited). The de-embedding formulation is expressed as:

$$\Gamma_a = \Gamma_{meas}\left(\frac{\Gamma_{a\_air}}{\Gamma_{meas\_air}}\right) \tag{8.39}$$

where Γ_{meas_air} is the reflection coefficient measurement, for air at plane AA', and Γ_{a_air} is the standard value of the air reflection coefficient at plane BB', obtained using the analytical calculation or numerical simulation (commercial simulator). However, the error of a standing wave is not taken into account in the de-embedding calibration techniques. The standing-wave effect can be ignored when the operating quarter-wave length, $\lambda/4$, in the coaxial line is equal to or larger than the physical length, l, of the coaxial line.

8.2.5 De-Embedding Without Any Standard Value

This calibration routine assumed that coaxial line is perfect line ($e_{11} = 0$ and $e_{22} = 0$), as well as that the propagated wave along the line is represented by the transmission

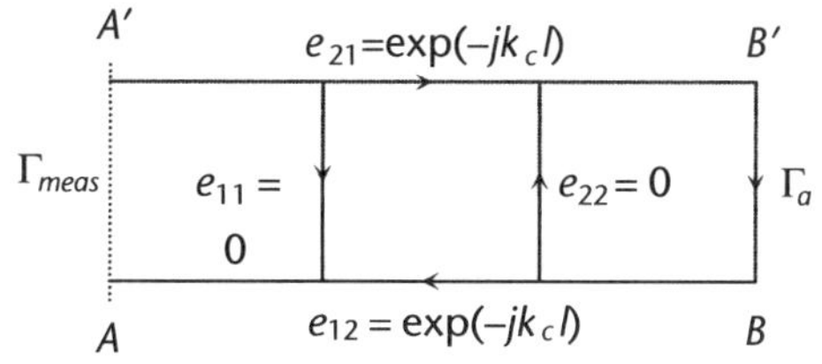

Figure 8.22 Signal flow along the transmission coaxial line.

factor, $\exp(-jk_c l)$ as shown in Figure 8.22. Under the assumption of quasi-TEM mode in the coaxial line, the measured reflection coefficient Γ_{meas} of the sample at the plane AA' can be de-embedded to the connector end of the probe, which is coincident with the calibration plane BB' to give a reflection coefficient Γ_a by:

$$\Gamma_a = \Gamma_{\text{meas}} e^{2jk_c l} \tag{8.40}$$

where l and $k_c = (\omega\sqrt{\varepsilon_c})/c$ are the physical length (in meters) and the propagation constant of the coaxial line, respectively.

In de-embedding, only the values of the physical length l and the propagation constant k_c of the coaxial line for the coaxial radiator are required to be inserted into (8.40). The advantage of this approach is that it is suitable for arbitrary size of the coaxial fed radiator, without knowing the radiated field (calibration standard value) at the open end of coaxial line; thus, the cost and measuring time will be reduced. However, this approach gives a relatively less repeatable and precise measurement result than three standard calibration methods.

8.3 Coaxial Radiator Performance Measurements

Generally, the performance of the coaxial radiators can be tested by measuring the reflection using the VNA (or the frequency/time-domain reflectometer). From the reflection measurements, radiator efficiency transmitting energy to the environment or receiving energy from the environment can be determined. Furthermore, measurements of reflection can be analyzed and interpreted in various ways, such as voltage standing-wave ratio (VSWR), gain, quality factor (*Q*-factor), resonant frequency, return loss, Smith chart, and matching complex impedance. Those measurements concept are further described on the accompanying CD.

8.4 Dielectric Measurements Using Coaxial Radiators

Material characterization using coaxial discontinuity methods can be modified into various kinds of measurement configurations and sample holder types. In macroscopic material characteristic investigations, three properties of the material are often measured: relative permittivity ε_r, relative permeability μ_r, and conductivity σ. Normally, many microwave measurements only focus on the properties of relative permittivity ε_r rather than the permeability μ_r and the conductivity σ. Various measurement techniques are explained in detail on the accompanying CD.

8.5 Leaky Coaxial Cable Characteristics Measurements

A leaky coaxial cable is similar to a normal coaxial cable [Figure 8.23(a)], except that the leaky coaxial cable has opening apertures on the outer conductor surface to allow a small amount of RF signal to radiate out into the surrounding environment from the cable as shown in Figure 8.23(b). The dielectric used in the cable is usually foamed polyethylene with relative permittivity, $\varepsilon_c \approx 1.25$.

Since the late 1970s, the leaky coaxial cable has been widely used as a radiator in underground tunnels, mines, subways, or underground parking spaces, due to the weakly propagated radio signal underground where the signal is screened by concrete walls or building. Recently, this kind of radiator has been implemented indoors in buildings, especially in the blind zone area where the discrete antennas fail to perform adequately. Further description of leaky coaxial cable applications is available in Chapter 9.

8.5.1 Design Rules of Leaky Cable

The leaky cable only produces radiating waves when the radiated wave is in the first harmonic [11]. Parameters such as size, composition, and position of the slot aperture on the outer conductor of the coaxial cable are important factors in the cable design in order to produce only single mode radiation (only the first harmonic, $m = -1$) of the radiated wave as shown in Figure 8.24.

For a general leaky cable design, Delogne and Deryck [12] proposed that the optimum length d_o (in meters) of each leaky section [in Figure 8.24(a)] relative to the operating wavelength λ can be determined as:

$$d_o = \frac{\lambda}{2\left(\sqrt{\varepsilon_c} - 1\right)} \tag{8.41}$$

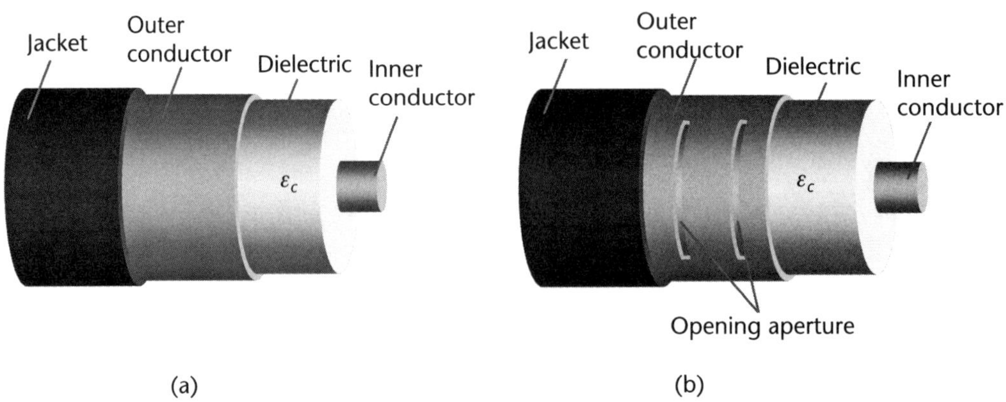

Figure 8.23 (a) Normal coaxial cable. (b) Leaky coaxial cable.

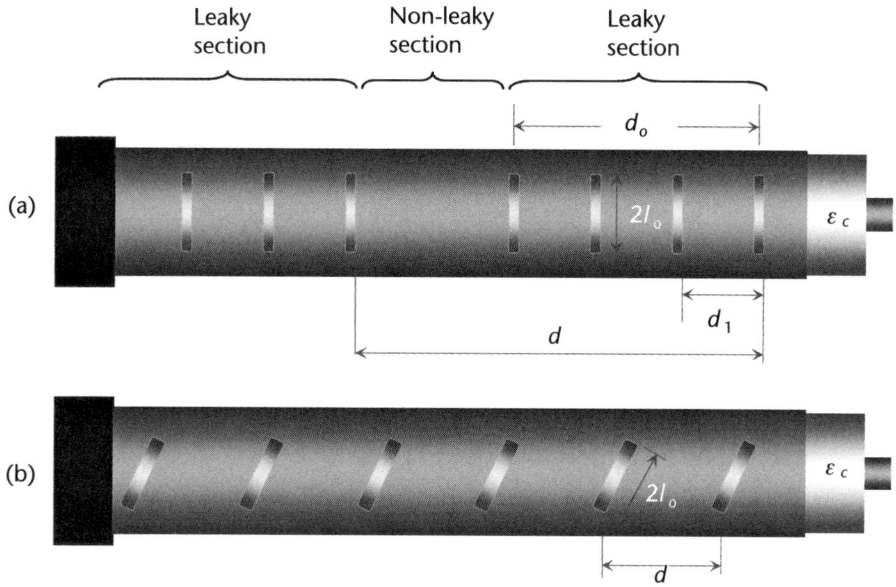

Figure 8.24 (a) Longitudinal and (b) zigzag slot apertures on a leaky cable.

However, a resonant frequency f_o within the single-mode radiation band is given as [11]:

$$f_o = \frac{c}{d\sqrt{\varepsilon_c}} \qquad \text{(in hertz)} \tag{8.42}$$

where d is the periodic distance of slot aperture and c is the velocity of light in free space, respectively. Recently, Li et al. [13] empirically expressed the resonant frequency f_o (in gigahertz) of the leaky cable relative to the length of its slot aperture l_o (in meters) as:

$$f_o = \frac{0.0428346}{l_o} + 0.50853 \qquad \text{(in gigahertz)} \tag{8.43}$$

When the operating frequency f increases, higher-mode waves (second harmonics, $m = -2$; third harmonics, $m = -3$; and so forth) will emerge and mix in the radiated wave. The multiple mode radiated wave may interfere the stability and uniformity of the transmitted electrical fields. The condition for radiation from the mth harmonic in the radial direction is given as [11]:

$$-mf_1 < f < -mf_2 \tag{8.44}$$

where

$$f_1 = \frac{c}{d\left(\sqrt{\varepsilon_c}+1\right)} \quad \text{and} \quad f_2 = \frac{c}{d\left(\sqrt{\varepsilon_c}-1\right)}$$

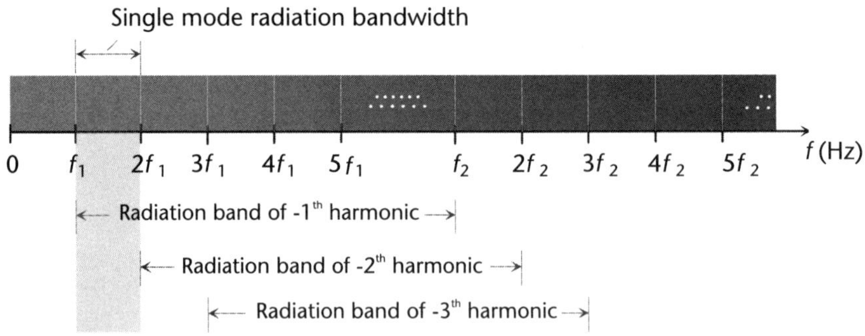

Figure 8.25 Operating frequency bands of the harmonics of a conventional leaky coaxial cable.

The subscript m is the order of higher-mode radiation harmonics. The operating frequency band of the harmonics is shown in Figure 8.25. In addition, the measured VSWR data versus normalized frequency, (f/f_o), for the leaky cable, obtained from Wang and Mei [11], is shown in Figure 8.26, which clearly shows that its useable single-mode radiation bandwidth of the first harmonic ($m = -1$) is only within f_1 to $2f_1$.

In order to increase the operating frequency band of the leaky cable, normally the slot aperture on the outer conductor is arranged as an array group in a predetermined cycle as shown in Figure 8.27. For the $f_1/f_o < 0.5$ region, only surface waves exist and no radiated wave is produced from the leaky cable. A background theory for suppression of the higher-order radiation harmonics ($m = -2$, $m = -3$, and so forth) using slot array grouping techniques is described in detail next.

With regard to the theoretical concept of the suppression of the higher-order radiation harmonics, it is known that the electric field E around the leaky cable with periodic slot aperture on the outer conductor as shown in Figure 8.27 can be expressed as [11]:

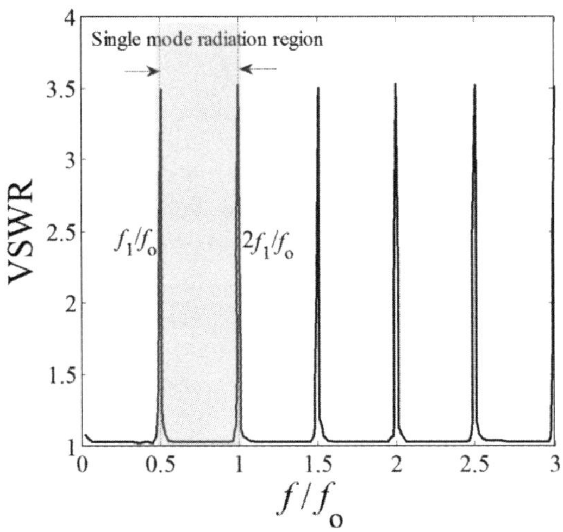

Figure 8.26 VSWR of a conventional leaky coaxial cable [11].

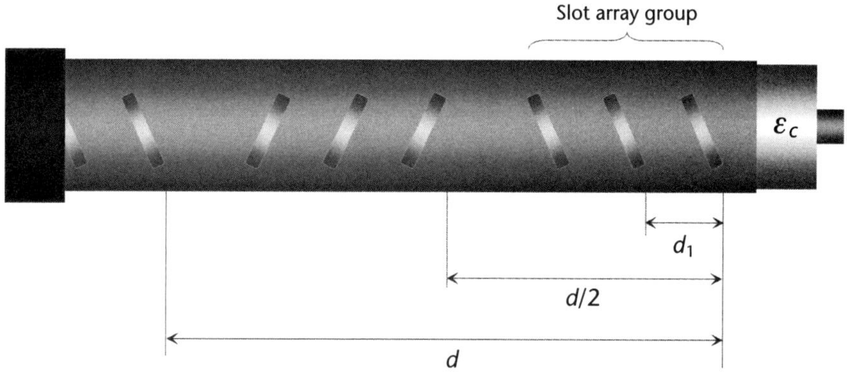

Figure 8.27 Configurations of leaky coaxial cable with array grouping of periodic zigzag slot aperture.

$$E = e^{-jk_c z} \sum_{m=-\infty}^{\infty} E_m e^{-j\frac{2m\pi}{d}z} \tag{8.45}$$

E_m in (8.45) is assumed to be expressed as:

$$E_m = \sum_{n=1}^{N} a_n e^{-j\frac{2m\pi}{d}z_n} \tag{8.46}$$

where the amplitude a_n in (8.46) is assumed to be a sine function as [14]:

$$a_n = a_o \sin\left(\frac{2\pi z_n}{d}\right) \tag{8.47}$$

where N is the total number of the slots within one period. The subscript n is the order of slot aperture in the period. The term z_n is the distance between the nth slot (within one period) and the middle point of a pitch. For a zigzag slot leaky cable with $N = 2$ (except the original one), $z_1 = -d/4$, and $z_2 = d/4$, as shown in Figure 8.28, (8.46) can be reexpressed as:

$$\begin{aligned} E_m &= \sum_{n=1}^{2} a_n e^{-j\frac{2m\pi}{d}z_n} \\ &= a_o \sin\left(\frac{2\pi z_1}{d}\right) e^{-j\frac{2m\pi}{d}z_1} + a_o \sin\left(\frac{2\pi z_2}{d}\right) e^{-j\frac{2m\pi}{d}z_2} \\ &= a_o \sin\left(\frac{2\pi}{d} \times \frac{-d}{4}\right) e^{-j\frac{2m\pi}{d}\times\frac{-d}{4}} + a_o \sin\left(\frac{2\pi}{d} \times \frac{d}{4}\right) e^{-j\frac{2m\pi}{d}\times\frac{d}{4}} \\ &= -a_o\left(e^{j\frac{m\pi}{2}} - e^{-j\frac{m\pi}{2}}\right) \end{aligned} \tag{8.48}$$

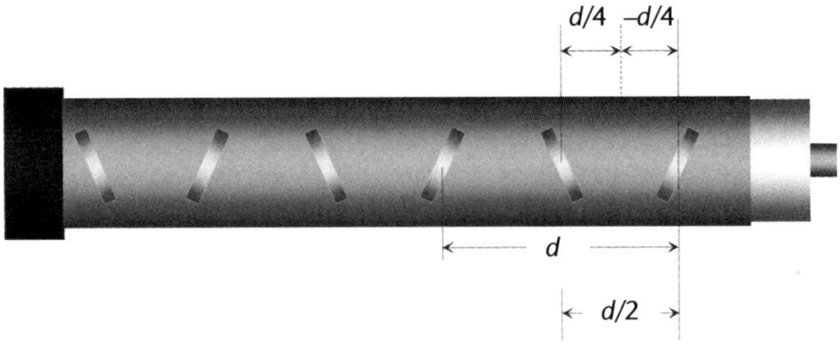

Figure 8.28 Leaky cable with one zigzag slot in a period.

Then each harmonic mode ($m = -1, -2, -3, \ldots, \infty$) is replaced, respectively, in (8.48), yielding

$$E_{-1} = -a_o\left(e^{-j\frac{\pi}{2}} - e^{j\frac{\pi}{2}}\right) = j2a_o \quad ; \quad E_{-2} = -a_o\left(e^{-j\pi} - e^{j\pi}\right) = 0$$

$$E_{-3} = -a_o\left(e^{-j\frac{3\pi}{2}} - e^{j\frac{3\pi}{2}}\right) = -j2a_o \quad ; \quad E_{-4} = -a_o\left(e^{-j2\pi} - e^{j2\pi}\right) = 0$$

From the above calculation results, clearly, the zigzag slot leaky cable with $N = 2$ does not have the radiation wave with even harmonic modes ($m = -2, -4, -6, \ldots$) but only with the odd harmonic modes ($m = -1, -5, -7, \ldots$). Figure 8.29 shows that the second harmonic is suppressed in the operating frequency band, in which the maximum single-mode radiation frequency is increased from $2f_1$ to $3f_1$. The corresponding VSWR data versus normalized frequency, (f/f_o), is shown in Figure 8.30.

The single mode radiation bandwidth can be further improved using leaky cable with $N = 4$, $z_1 = -d/6$, $z_2 = d/6$, $z_3 = -d/3$, and $z_4 = d/3$, as shown in Figure 8.31. Using a similar calculation routine, the calculated results from (8.46) are given as:

$$E_{-1} = j3a_o; \; E_{-2} = 0; \; E_{-3} = 0; \; E_{-4} = 0; \; E_{-5} = -j3a_o$$

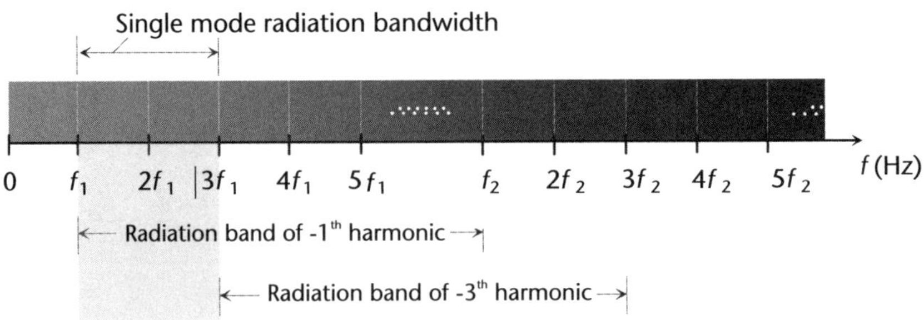

Figure 8.29 Operating frequency bands of the harmonics of leaky coaxial cable with one zigzag slot in a period.

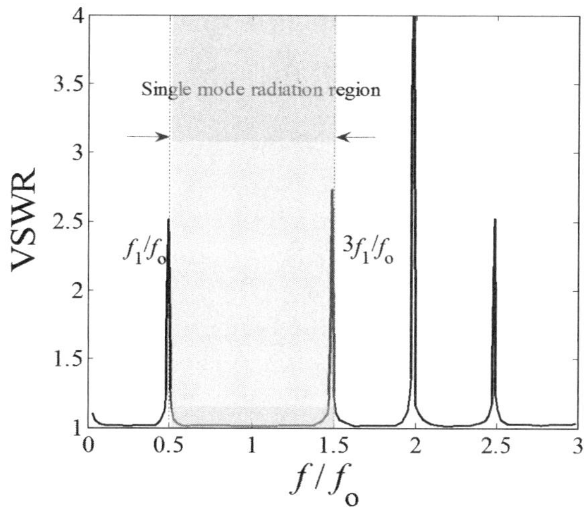

Figure 8.30 VSWR of leaky coaxial cable with one zigzag slot in a period [11].

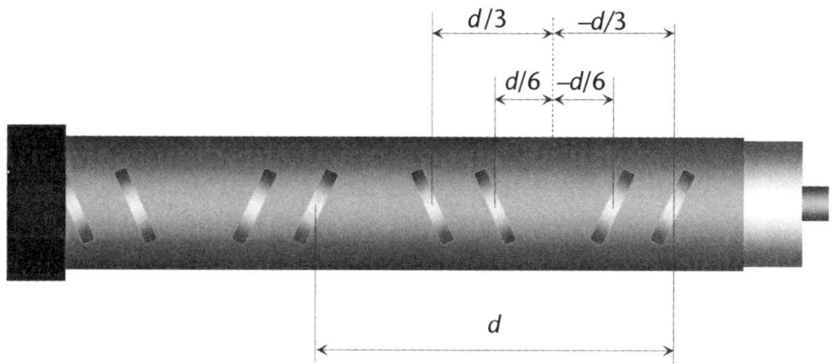

Figure 8.31 Leaky cable with two zigzag slots in a period.

In this case, not only the even harmonic modes ($m = -2, -4, -6, \ldots$) have been suppressed, but also $m = -3$. Hence, the limitation of single-mode radiation bandwidth of the first harmonic ($m = -1$) is expanded to $5f_1$ as shown in Figure 8.32.

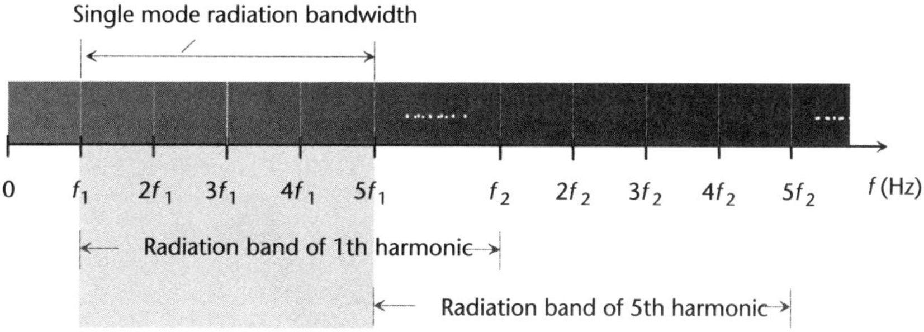

Figure 8.32 Operating frequency bands of the harmonics of leaky coaxial cable with two zigzag slots in a period.

8.5.2 Insertion Loss and Coupling Loss Measurements

In telecommunication applications, the performance of leaky coaxial cable is normally measured based on two principal parameters, which are the insertion loss, L_i, and the coupling loss, L_c, as shown in Figure 8.33. The insertion loss L_i is due to the signal's attenuation along the coaxial cable. However, the coupling loss L_c is contributed to by the signal loss between the transmitted power from the antenna into the leaky coaxial cable (or the transmitted power from the leaky cable into a test antenna) at a radial distance.

The insertion loss L_i and coupling loss L_c of the leaky coaxial cable can be measured by using a ground method or free space method [15] as shown in Figures 8.34 and 8.35, respectively. For the ground method, the test leaky cable is laid on nonmetallic spacers, which gives the test cable a distance from the floor of 10 cm to 12 cm. For the free-space method, the test cable is laid on nonmetallic posts at a height of 1.5m to 2m from the floor. Both methods require that the length of leaky cable be at least 10λ, where λ is the operating wavelength.

From the measurement in Figure 8.34, the insertion loss L_i (in dB/meter) of the leaky cable can be determined as:

$$L_i = \frac{P_{\text{in}} - P_{\text{out}}}{d} \tag{8.49}$$

where P_{in} and P_{out} in (8.49) are the power levels at the cable input (in dBm) and output (in dBm), respectively. The term d is the length of the leaky cable (in meters). The input power P_{in} is excited from a signal generator, while the output power P_{out} is detected using a spectrum analyzer or equivalent instrument.

From the measurement in Figure 8.35, the measured local coupling loss L_c (in decibels) can be found as:

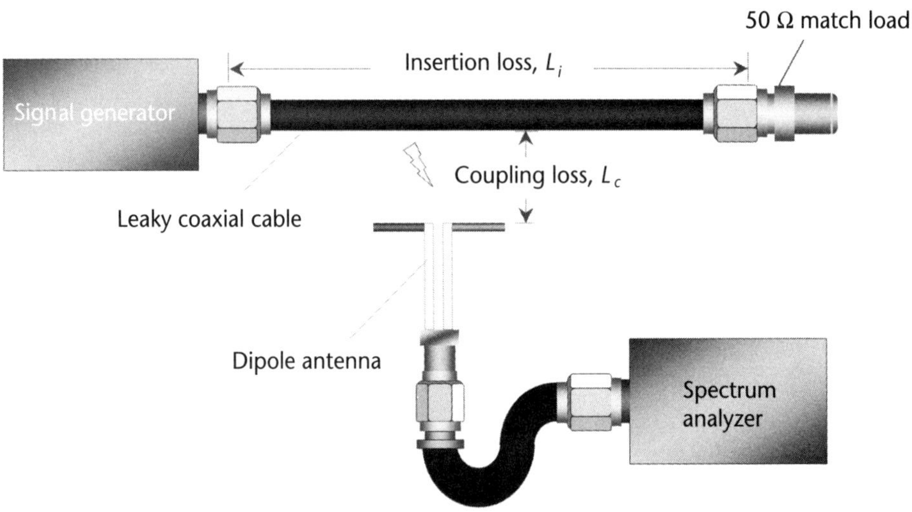

Figure 8.33 The insertion loss L_i and coupling loss L_c of the leaky coaxial cable.

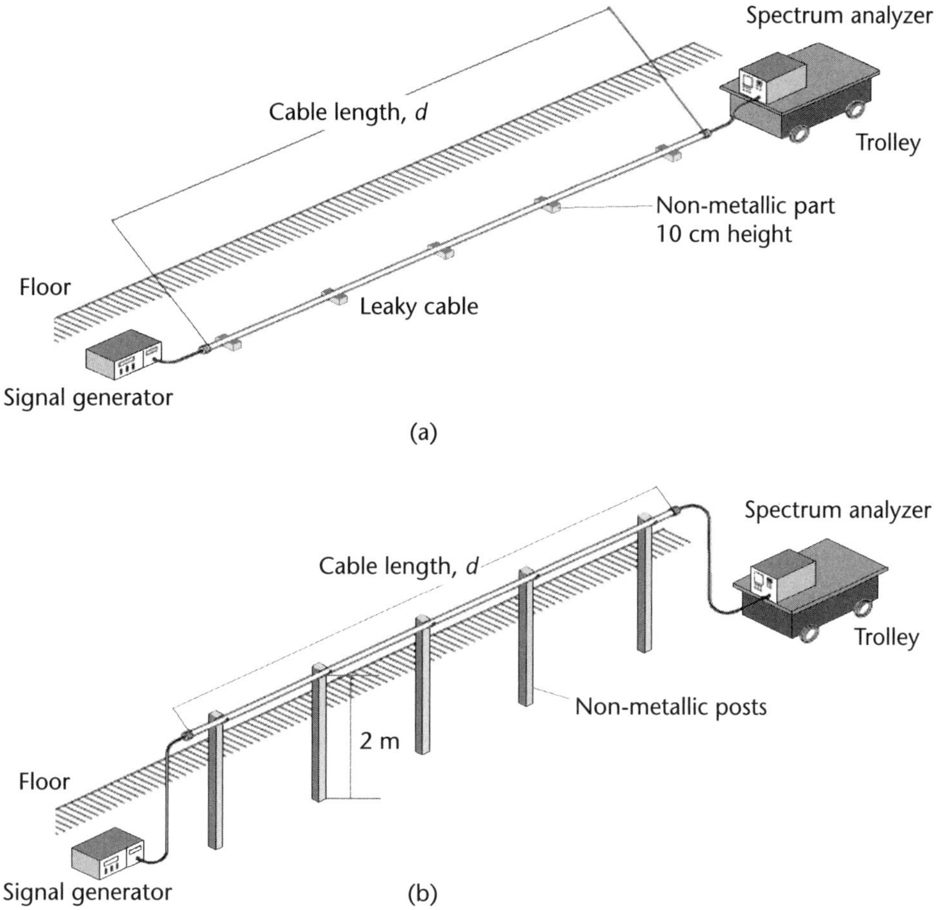

Figure 8.34 Insertion loss measurements using (a) ground method and (b) free-space method [15].

$$L_c = P_{in} - P'_{out} - (L_i \times d') \tag{8.50}$$

where P_{in} and P'_{out} in (8.50) are the power level at the cable input (in dBm) and output (in dBm), respectively. P'_{out} is the power received by spectrum analyzer or equivalent instrument. The term d' is the distance along the leaky cable from the cable input to the nearest point on the dipole antenna (in meters). In order to obtain more accurate measurement results, the coupling loss is measured with three kinds of dipole antenna orientations as shown in Figure 8.36.

Hence, the average value of the coupling loss L_{c_mean} (in decibels) can be calculated as:

$$L_{c\_mean} = -10\log_{10}\left\{\frac{1}{3}\left(10^{-\frac{L'_c}{10}} + 10^{-\frac{L''_c}{10}} + 10^{-\frac{L'''_c}{10}}\right)\right\} \tag{8.51}$$

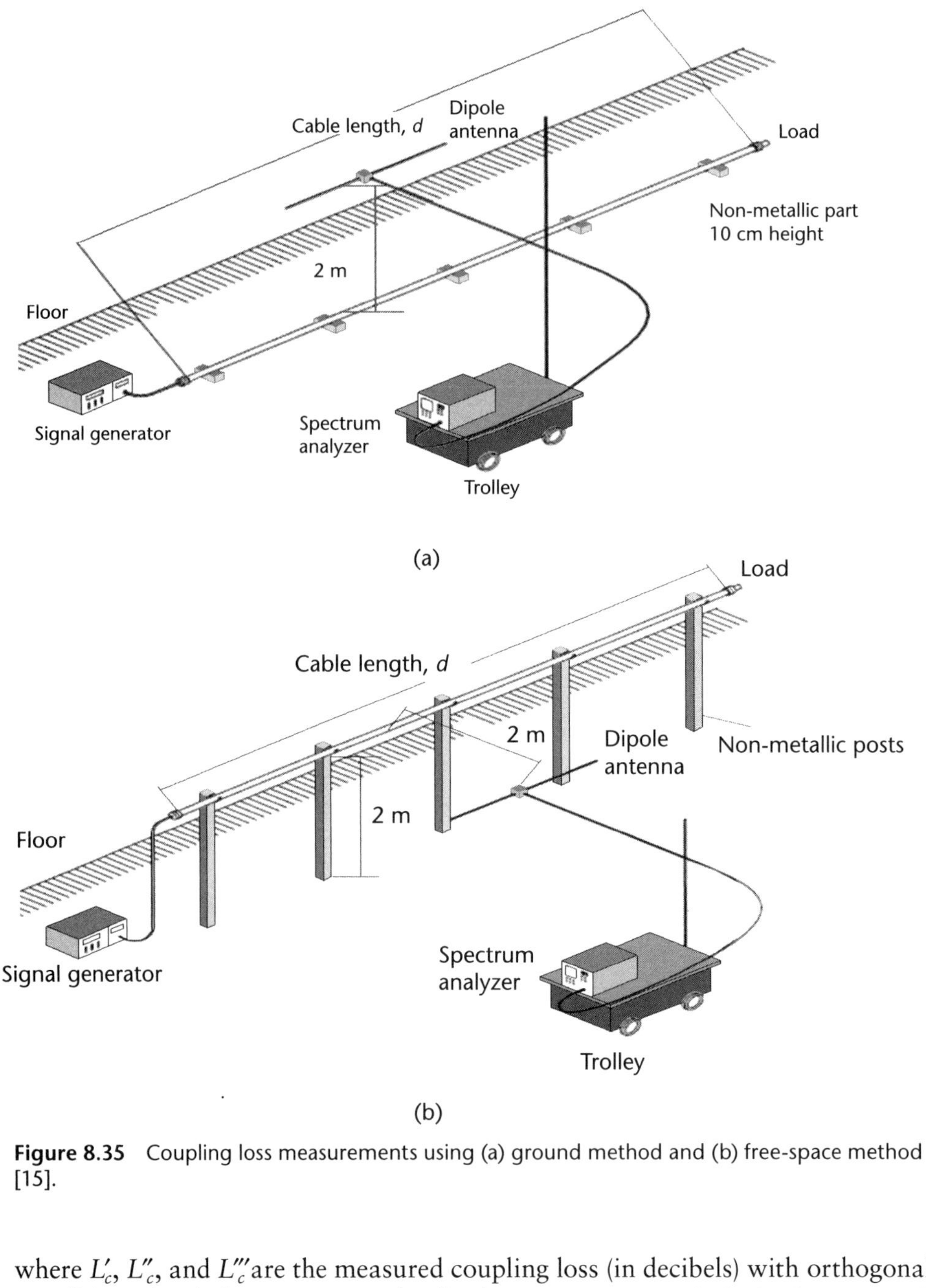

Figure 8.35 Coupling loss measurements using (a) ground method and (b) free-space method [15].

where L'_c, L''_c, and L'''_c are the measured coupling loss (in decibels) with orthogonal, parallel, and radial orientations of a dipole antenna, respectively. In fact, an early use of leaky feeder cable was made by Halstead in 1940. Recently, in the market, most of the leaky coaxial cables are manufactured by RADIAX. There are various types of leaky cables for different applications as shown in Figure 8.37, and the corresponding features and their implementation are listed in Table 8.3. Each type of leaky cable has a different shape, size, and arrangement of the opening aperture, which may influence the operating wavelength, coupling loss, and beamwidth of the radiated field [16].

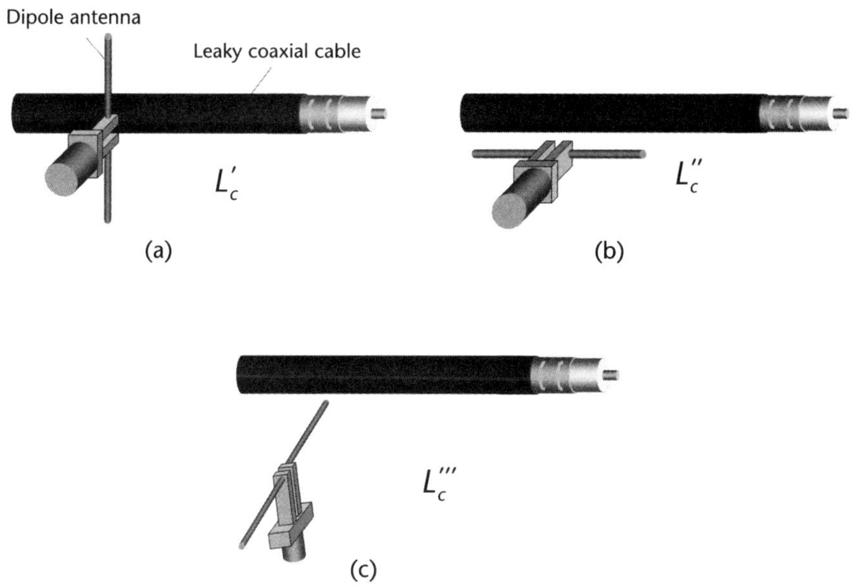

Figure 8.36 Spatial orientations of dipole antenna in coupling loss measurement using the free-space method: (a) orthogonal, (b) parallel, and (c) radial.

Table 8.3 Features and Applications of Leaky Coaxial Cable

Types	*Operation Frequency, f*	*Slot Group Intervals*	*Features*	*Use*
RLK	0.075–0.6 GHz, 0.075–0.96 GHz, 0.075–1.9 GHz, 0.075–2.7 GHz, 0.8–6 GHz	1m	Broadband, low coupling loss at 0.9 GHz, 1.8 GHz, and 2.2 GHz.	In-building, tunnel for areas under 80m.
RLF	0.075–0.96 GHz, 0.075–2.4 GHz	22m	Robust; low loss; broadband; negligible influence of dust, salt, and moisture accumulation on cable loss.	Single-band and multiband systems, long tunnels, aggressive environments.
RAY	0.075–0.96 GHz, 0.075–2.4 GHz, 0.075–2.7 GHz, 2.4–2.5 GHz	≤50 cm	Very low coupling loss, optimized for high frequencies and digital transmission. Optimized for vertical antennas.	In-building, dry tunnels.
RSF/RCF	0.075–2.6 GHz, 0.075–6 GHz	Milled slots on corrugated outer conductor	Robust, low bending radius.	Heavy-duty applications, in-building, mines.
RLV	0.15–0.96 GHz, 0.075–2.4 GHz, 0.075–2.7 GHz	Graded groups of apertures	Constant attenuation loss along the cable.	Long tunnels.

Source: [16].

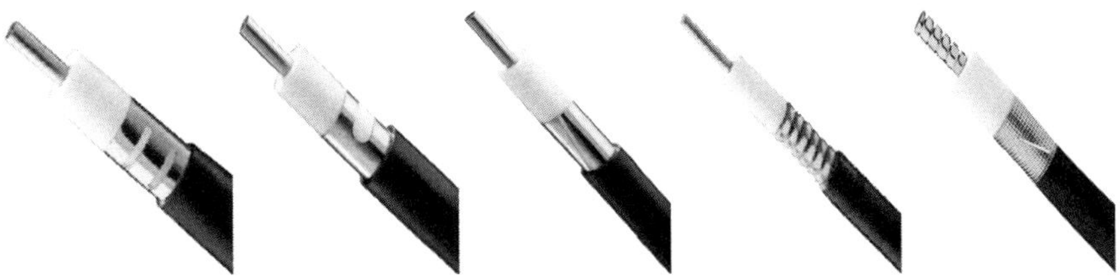

Figure 8.37 Various kinds of leaky coaxial cables [16]: (a) RLK, (b) RLF, (c) RAY, (d) RCF, and (e) RLV.

References

[1] Grace, M. I., "Time Domain Analysis Using Vector Network Analyzers," Anritsu, Inc. U.S., 2010.

[2] Rohde & Schwarz, "Time Domain Measurements Using Vector Network Analyzer ZVR," Application Note 1EZ44_0E, 1998.

[3] Steele, J. M., "Time Domain Reflectometry," Ch. 18 in *Microwave Measurements*, 2nd ed., A. E. Bailey, (ed.), London, U.K.: Peter Peregrinus, 1989.

[4] Barthel, J., et al., "Dielectric Spectra of Some Common Solvents in the Microwave Region, Water and Lower Alcohols," *Chem. Phys. Lett.*, Vol. 165, 1990, pp. 369–373.

[5] Oldfield, W. W., "Reaching Beyond 100 GHz with Coaxial Connectors," *Microwaves & RF*, 2014, pp. 64–68.

[6] Kraszewski, A., M. A. Stuchly, and S. S. Stuchly, "ANA Calibration Method for Measurements of Dielectric Properties," *IEEE Trans. on Instrum. Meas.*, Vol. 32, No. 2, 1983, pp. 385–386.

[7] Agilent Technologies, "Specifying Calibration Standards and Kits for Agilent Vector Network Analyzers," Application Note 1287-11, 2010.

[8] Beatty, R. W., "Invariance of the Cross Ratio Applied to Microwave Network Analysis," NBS Tech. Note 623, U.S. National Bureau of Standards, 1972.

[9] Marsland, T. P., and S. Evans, "Dielectric Measurements with an Open-Ended Coaxial Probe," *IEE Proceedings H*, Vol. 134, No. 4, 1987, pp. 341–349.

[10] Stogryn, P. A., et al., *The Microwave Dielectric Properties of Sea and Fresh Water*, GenCorp Aerojet, Azusa, CA, Tech. Rep., 1995.

[11] Wang, J. H., and K. K. Mei, "Theory and Analysis of Leaky Coaxial Cables with Periodic Slots," *IEEE Trans. on Antennas Propag.*, Vol. 49, No. 12, 2001, pp. 1723–1732.

[12] Delogne, P. P., and L. Deryck, "Underground Use of a Coaxial Cable with Leaky Sections," *IEEE Trans. on Antennas Propag.*, Vol. 28, No. 6, 1980, pp. 875–883.

[13] Li, Y, J. H. Wang, Z. Zhang, and M. Chen, "Empirical Formula for the Electric Field Calculation in Rectangular Slots on Leaky Coaxial Cable," *IEEE Trans. Antennas Propag.*, Vol. 60 (12), pp. 5825-5833, 2012.

[14] Yang, X. D., et al., "Research of Single Mode Radiation Leaky Coaxial Cable Used for Ultra Wide-Band Communication," *Proc. of 2006 International Conference on Communications, Circuits, and Systems, Vol. 2*, Guilin, China, June 2006, pp. 1261–1264.

[15] International Electrotechnical Commission Standard, *Coaxial Communication Cables—Part 4: Sectional Specification for Radiating Cables*, International Electrotechnical Commission (IEC) 61196-4, Switzerland, 2004.

[16] RADIAFLEX, The Optimal Confined-Coverage Solution, Radio Frequency Systems (RFS), Germany, 2013.

Selected Bibliography

Agilent Technologies, "Time Domain Analysis Using a Network Analyzer," Application Note 1287-12, 2012.

Baker-Jarvis, J., et al., "Transmission/Reflection and Short-Circuit Line Methods for Measuring Permittivity and Permeability," Nat. Inst. of Standards and Technol., Washington, D.C., Tech. Note 1355-R, 1993.

Belhadj-Tahar, N. E., and A. Fourrier-Lamer, "Broad-Band Analysis of a Coaxial Discontinuity Used for Dielectric Measurements," *IEEE Trans. on Microwave Theory Tech.*, Vol. 34, No. 3, 1986, pp. 346–350.

Blakney, T. L., "Comments on 'Automatic Measurement of Complex Dielectric Constant and Permeability at Microwave Frequencies,'" *Proceedings of the IEEE*, Vol. 63, No. 1, 1975, pp. 203–204.

Buckley, F., and A. A. Maryott, "Tables of Dielectric Dispersion Data for Pure Liquids and Dilute Solutions," *National Bureau of Standards Circular 589*, Nov. 1, 1958.

Burdette, E., F. Cain, and J. Seals, "In Vivo Probe Measurement Technique for Determining Dielectric Properties at VHF Through Microwave Frequencies," *IEEE Trans. on Microwave Theory Tech.*, Vol. 28, No. 4, 1980, pp. 414–427.

Courtney, C. C., and W. Motil, "One-Port Time-Domain Measurement Approximate Permittivity and Permeability of Materials," *IEEE Trans. on Microwave Theory Tech.*, Vol. 47, No. 5, 1999, pp. 551–555.

Da Silva, E. F., and M. K. McPhun, "Calibration Techniques for One-Port Measurement," *Microwave Journal*, Vol. 21, No. 6, 1978, pp. 97–100.

Fitzpatrick, J., "Error Models for Systems Measurement," *Microwave Journal*, Vol. 21, No. 5, 1978, pp. 63–66.

Hwang, H. S., et al., "110 GHz Broadband Measurement of Permittivity on Human Epidermis Using 1 mm Coaxial Probe," *Proc. IEEE Microwave Theory and Techniques Society International Microwave Symposium*, Philadelphia, PA, June 2003.

Ishizone, K., "Poynting Vector Flow in the Vicinity of a Receiving Dipole Antenna," *IEICE Technical Report AP-86-38*, 1986, pp. 35–41.

Iskander, M. F., and J. B. DuBow, "Time- and Frequency-Domain Techniques for Measuring the Dielectric Properties of Rocks: A Review," *Journal of Microwave*, Vol. 18, No. 1, 1983, pp. 55–74.

Iskander, M. F., and S. S. Stuchly, "A Time-Domain Technique for Measurement of the Dielectric Properties of Biological Substances," *IEEE Trans. on Instrum. Meas.*, Vol. 21, No. 4, 1972, pp. 425–429.

Ito, K., and K. Saito, "Antenna Issues in Microwave Thermal Therapies," Ch. 5 in *Antennas for Portable Devices*, Z. N. Chen, (ed.), New York: John Wiley & Sons, 2007.

Jepsen, P. U., et al., "Terahertz Time-Domain Spectroscopy of Crystalline and Aqueous Systems," in *Terahertz Frequency Detection and Identification of Materials and Objects*, X. C. Zhang et al., (eds.), Berlin, Germany: Springer, 2007, pp. 147–165.

Kajfez, D., "Q-Factor Measurement with a Scalar Network Analyzer," *IEE Proc. Microw. Antennas Propag.*, Vol. 142, No. 5, 1995, pp. 369–372.

Kwok R. S., and J. F. Liang, "Characterization of High-Q Resonators for Microwave-Filter Applications," *IEEE Trans. on Microwave Theory Tech.*, Vol. 47, No. 1, 1999, pp. 111–114.

Ludwig, R., and P. Bretchko, *RF Circuit Design: Theory and Applications*, Upper Saddle River, NJ: Prentice Hall, 2000, pp. 178–184.

Marcuvitz, N., *Waveguide Handbook*, Boston, MA: Boston Technical Publishers, 1964.

Narayana Rao, D. A. A. S., "Dielectric Measurement on Liquids with a Coaxial Line," *Brit. J. Appl. Phys.*, Vol. 17, 1966, pp. 109–112.

Nicolson, A. M., and G. F. Ross, "Measurement of the Intrinsic Properties of Materials by Time-Domain Technique," *IEEE Trans. on Instrum. Meas.*, Vol. 19, No. 4, 1970, pp. 377–382.

Nishikawa, K. I., et al., "A New Position Detection Method Using Leaky Coaxial Cable," *IEICE Electronics Express*, Vol. 5, No. 8, 2008, pp. 285–290.

Hwang, H. S., et al., "110 GHz Broadband Measurement of Permittivity on Human Epidermis Using 1 mm Coaxial Probe," *Proc. IEEE Microwave Theory and Techniques Society International Microwave Symposium*, Philadelphia, PA, 2003.

Papanikolaou, N., et al., "Radio Coverage Measurements into Tunnel with Leaky Section Radiating Cable," *Proc. IEEE Ninth International Symposium on Computers and Communications (ISCC 2004)*, Alexandria, Egypt, June 2004.

Robert, S., and A. Von Hippel, "A New Method of Measuring Dielectric Constant and Loss in Range of Centimetre Waves," *J. Appl. Phys.*, Vol. 17, 1946, pp. 610–616.

Sato, T., A. Chiba, and R. Nozaki, "Hydrophobic Hydration and Molecular Association in Methanol-Water Mixtures Studied by Microwave Dielectric Analysis," *J. Chem. Phys.*, Vol. 112, 2000, pp. 2924–2932.

Schroeder, A., "Overcoming the Challenges of 4G LTE and MIMO for In-Tunnel Wireless Connectivity," *Radio Frequency Systems (RFS)*, 2014, pp. 1–8.

Seseña-Osorio, J. A., et al., "Indoor Propagation Measurements for Radiating Cable in the UHF Band 800 MHz–2500 MHz," *21st International Conference on Electric Communications and Computers* (*CONIELECOMP 2011*), San Andres Cholula, Mexico, March 2011.

Siggins, A. F., J. Gunning, and M. Josh, "A Hybrid Waveguide Cell for the Dielectric Properties of Reservoir Rocks," *Meas. Sci. Technol.*, Vol. 22, 2011, pp. 1–9.

Stuchly, M. A., and S. S. Stuchly, "Coaxial Line Reflection Methods for Measuring Dielectric Properties of Biological Substances at Radio and Microwave Frequencies—A Review," *IEEE Trans. on Instrum. Meas.*, Vol. 29, 1980, pp. 176–183.

Suzuki, T., T. Hanazawa, and S. Kozono, "Design of a Tunnel Relay System with a Leak Coaxial Cable in an 800-MHz Band Land Mobile Telephone System," *IEEE Trans. on Vehicular Technology*, Vol. 29, No. 3, 1980, pp. 305–316.

Tanabe, E., and W. T. Joines, "A Nondestructive Method for Measuring the Complex Permittivity of Dielectric Materials at Microwave Frequencies Using an Open Transmission Line Resonator," *IEEE Trans. on Instrum. Meas.*, Vol. 25, No. 3, 1976, pp. 222–226.

Tektronix, *TDR Impedance Measurements: A Foundation for Signal Integrity: Application Note*, 2008.

Wang, P., et al., "Improve Calculation of Reflection Coefficient for Coaxial Antennas with Feed Gap Effect," *IEEE Trans. on Antennas Propag.*, Vol. 57, No. 2, 2009, pp. 559–563.

Weir, W. B., "Automatic Measurement of Complex Dielectric Constant and Permeability at Microwave Frequencies," *Proceedings of the IEEE*, Vol. 62, No. 1, 1974, pp. 33–36.

CHAPTER 9

Applications of Coaxial Slot Radiators

This chapter briefly reviews research and measurements of coaxial slot radiators based on the reflection method for industrial, scientific, medical and telecommunication applications. Due to different configurations of coaxial slot radiators, the operating frequencies and the applications involved with radiators are different as well. There are two microwave frequencies allocated by the U.S. Federal Communications Commission (FCC) for industrial, scientific, and medical (ISM) use, which are 915 MHz and 2.45 GHz. Currently, most of the applications are devoted to 2.45 GHz, since it provides a suitable compromise between power deposition and penetration depth. The ISM bands defined by the International Telecommunication Union (ITU) are listed in Table 9.1. Currently, these radiators are widely used in various wireless communications, material processing, biomedical engineering, and remote sensing of vegetation. In this chapter, only a selected scope of radiator applications is discussed as listed in Table 9.2.

In the discussion of communication applications, some common types of antenna will be revisited, such as Yagi-Uda dipole array, log-periodic dipole array, helical monopole and conical monopole. Some simple and reliable formulas that are used to design those antennas will also be described. In practice, the radiator used as an antenna is normally fed by a coaxial line (antenna is driven from the coaxial line), since the characteristic impedance, Z_o of the coaxial line is almost 50Ω and the impedance matching can be achieved easily when the antenna is connected to the RF test device through the 50-Ω cable.

Most of the applications that will be discussed, except communication applications, involve the interactions between the microwave and material containing water. The interaction between agri-foods materials with microwave can be described by the complex relative permittivity, ε_r:

$$\varepsilon_r = \varepsilon_r' - j\varepsilon_r''$$

where the real part, ε_r' is the dielectric constant and imaginary part, ε_r'' is the dielectric loss factor. The dielectric constant, ε_r' influences the electric field distribution and the phase of waves traveling through the material. In contrast, the loss factor, ε_r'' influences the energy absorption or attenuation by the material.

Both ε_r' and ε_r'' are highly correlated with moisture content, since the electromagnetic energy is mainly absorbed by water [1] at microwave frequencies and the volume of moisture in the total volume of material influences the effective relative permittivity greatly. This is due to the relative permittivity of water (ε_r = 80 at dc) normally being much greater than the other constituents in the materials, such as

Table 9.1 ISM Operating Frequency Band Defined by the ITU

Frequency Range, f	*Bandwidth, Δf*	*Center Frequency, f_o*
6.765–6.795 MHz*	30 kHz*	6.780 MHz*
13.553–13.567 MHz	14 kHz	13.560 MHz
26.957–27.283 MHz	326 kHz	27.120 MHz
40.660–40.700 MHz	40 kHz	40.680 MHz
433.050–434.790 MHz**	1.74 MHz*	433.920 MHz*
902.000–928.000 MHz***	26 MHz*	915.000 MHz*
2.400–2.500 GHz	100 MHz	2.450 GHz
5.725–5.875 GHz	150 MHz	5.800 GHz
24.000–24.250 GHz	250 MHz	24.125 GHz
26.975–27.283 GHz*	308 MHz*	27.129 GHz*
61.000–61.500 GHz*	500 MHz*	61.250 GHz*
122.000–123.000 GHz*	1 GHz*	122.500 GHz*
244.000–246.000 GHz*	2 GHz*	245.000 GHz*

*Subject to local acceptance. **Only Europe, Africa, the Middle East (west of the Persian Gulf), the former Soviet Union, and Mongolia. ***Only Americas, Greenland, and some of the eastern Pacific Islands.

soils (mineral soil: 4, organic matter: 4, air: 1) and oil palm fruits (fiber: 2.2, oil: 2.5). If the value of the effective relative permittivity changes in the agricultural products, the coaxial radiator will measure a change in reflection/transmission coefficient or resonant frequency that can be directly correlated with a change in moisture content, *m.c*, of the product, which is obtained from the oven drying method. Thus, the dielectric measurements can be used to monitor the moisture content, *m.c.* in the material under test.

9.1 Communication Applications

Dipole and monopole wire antennas are the most basic types of radiators due to their simple structure and omnidirectional radiation capability. Until the present, dipole and monopole antennas remain popular in telecommunication applications, such as usage in satellite communication and radio astronomy. For instance, an array design of monopole antennas is used in the GPS III satellites in order to carry a new civil signal. The satellites can interoperate with other international global navigation satellite systems [2]. In addition, Canada's RADARSAT Constellation Mission (RCM) used 13 self-deploying monopole antennas in their Earth Observation satellites, which are used for maritime surveillance, disaster management, and ecosystem monitoring [3]. Due to different operating frequencies, the applications and the technologies involved with coaxial radiators are varied as shown in Table 9.3.

The individual monopole or dipole normally has low radiation efficiency and an omnidirectional radiation pattern [4]. Thus, in some applications, two or more monopoles or dipoles are arranged closely together to create a superposition of radiation fields in order to improve the radiation efficiency (gain). This close spacing yields a resultant field that fulfils certain desirable directional properties [8]. Typically,

Table 9.2 Selected Scopes of Applications for the Coaxial Radiators

Applications	*Coaxial Slot Radiators*	*Descriptions*
Wireless communications	Dipole/monopole array, dielectric-coated monopole, resistively loaded dipole/monopole, coaxial leaky cable	As a signal transmitter or receiver for communication systems, such as television antennas, walkie-talkie antennas, and WiFi antennas. Typically, the conventional monopole antenna has been modified to reduce its size, length, and the resonant condition of the modified monopole remains the same as the conventional monopole. In addition, monopole or dipole antennas are also modified to increase their radiation gain and operating bandwidth.
Biomedical	Open-ended coaxial probe, modified dielectric-coated monopole, resistively loaded dipole array	As a dielectric probe for measurement of the dielectric properties of biological tissues in order to provide reference dielectric databases. The dielectric database of the biological tissues can be used for radiation safety analysis, medical imaging processing, abnormal tissues detection, specific absorption rate (SAR) simulation.
		As a hyperthermia applicator for tumor or cancer treatments. The electromagnetic radiation distributed along the monopole is converted to heat, which can kill the cancer cells.
		As a detector in cancer imaging applications. The reflected signals detected by each individual dipole are different. The overall reflected signal data is processed and displayed in the form of contour images.
Agricultural	Open-ended coaxial probe, monopole	As a moisture detector, moisture content in fruits normally is correlated with the maturity and sweetness of the fruits. The moisture content can be used to determine the storage period and the quality of the agricultural products after processing by an industrial factory.
		As a soil moisture detector, the thin monopole structure can be easily thrust into the ground to detect underground water content.
Foods processing	Open-ended coaxial probe	As a dielectric probe for measurement of the dielectric properties of foods in order to provide reference dielectric databases for microwave food drying process.
		As a moisture detector, the moisture content in the food is correlated with the storage period and the crispness of the foods after processing by an industrial factory.
Industrial materials processing	Open-ended coaxial probe	As a metal crack detector. The reflection measurement using an open-ended coaxial probe is very sensitive to small air gap between the aperture probe and the metal surface.
		As a dielectric probe for measurement of the dielectric properties of thin-film materials, the performance of thin-film materials used in electronic or material science industry depends on their dielectric properties.

Table 9.3 Typical Communication Applications That Depend on the Operating Frequency Range

Frequency Band	*Typical Service*
3–30 kHz, very low frequency (VLF)	Maritime navigation
30–300 kHz, low frequency (LF)	Radio beacons, navigational aids
300–3,000 kHz, medium frequency (MF)	AM broadcast, maritime radio, Coast Guard communication, direction finding
3–30 MHz, high frequency (HF)	Telephone, telegraph, facsimile, short-wave radio, amateur radio, ship-to-coast and ship-to-aircraft communication
30–300 MHz, very high frequency (VHF)	Television, FM broadcast, air traffic control, navigational aids
300 MHz–3 GHz, ultrahigh frequency (UHF)	Television, satellite communication, Radiosonde, surveillance radar, navigational aids, mobile phone, Global Positioning System (GPS) and walkie-talkie
3–30 GHz, super high frequency (SHF)	Airborne radar, microwave links, satellite communication
30–300 GHz, extremely high frequency (EHF)	Radar landing system, radio astronomy

the distance between adjacent dipoles is less than half of a wavelength, $\lambda/2$. One such example of a dipole arrays is the Yagi-Uda antenna, which is normally used as a television antenna. The Yagi-Uda antenna is further described in Section 9.1.1.

In addition, the conventional monopole antenna does not exhibit a wideband characteristic, since its resonance exists only when the length h of the monopole is equal to a particular quarter-wavelength, $\lambda/4$.The antenna can be considered an ultrawideband antenna when it has a bandwidth greater than 500 MHz. In an ultrawideband (UWB) communication system, an antenna should have characteristics that are relatively independent of operating frequency over a large bandwidth. UWB is a wireless system for transmitting large amounts of digital data over a wide range of frequency bands with very low power (<0.5 mW) for a short distance (<70m). Recently, the UWB operating frequency-band standards for commercial communication applications have been shared with other unlicensed frequency-band standards, such as the ISM band. There are three categories of UWB systems that were categorized by the FCC in 2002 as described in Table 9.4. In Schantz [6], the UWB antennas were categorized into four types that depend on the shape, performance, and applications, as listed in Table 9.5. However, only two types of UWB antennas are described in this chapter: frequency-independent antennas and small element antennas. The dipole array antenna, the log-periodic antenna, is one of the frequency-independent antennas that has input impedance and radiation patterns that vary periodically with the logarithm of the operating frequency.

Besides that, there is another demand for the monopole antenna that is minimizing the size of the antenna, while the performance of the small antenna remains unchanged. Thus, different modification approaches have been used to reduce the length of the conventional monopole, such as a disk-loaded or helical structure monopole. Small antenna size is required for portable RF or MIMO equipment

Table 9.4 UWB Systems and Frequency Bands Defined by the FCC

UWB Systems	*Frequency Band, f*
1. **Image systems**	
Ground Penetrating Radars (GPRs): To detect the images of buried objects.	<960 MHz or 3.1–10.6 GHz
Wall: To detect the location of objects contained within a wall, such as a concrete structure, the side of a bridge, or the wall of a mine.	<960 MHz or 3.1–10.6 GHz
Through-wall: To detect the location or movement of persons or objects which are located on the other side of a wall.	1.99–10.6 GHz
Surveillance: To detect the intrusion of persons	1.99–10.6 GHz
Medical imaging devices: To detect a location or movement of objects within a human body or animal body.	3.1–10.6 GHz
2. **Vehicular radar systems:** To detect the location and movement of objects near a vehicle, enabling features such as near-collision avoidance, improved airbag activation, and suspension systems that better respond to road conditions	22–29 GHz
3. **Communications and measurement systems:** Wide variety of other UWB devices, such as high-speed home and business networking devices as well as storage tank measurement devices	3.1–10.6 GHz

Table 9.5 The Four Categories of the UWB Antennas

UWB Antenna	*Descriptions*
Frequency-independent antennas	Various sizes of antenna combined in array form. Smaller-scale geometry of the antenna elements in the array contributes to higher frequencies. However, larger-scale geometry of antenna elements contributes to lower frequencies. Example: log-periodic antenna.
Small element antennas	Antenna has smaller size and omnidirectional radiation pattern. Example: Biconical antenna or monoconical antenna.
Horn antennas	An antenna in the shape of a horn that has high gain and unidirectional radiation pattern. Example: Rectangular horn antenna.
Reflector antennas	A relatively large antenna which has high gain and uni-direction radiation pattern. Example: Hertz's parabolic cylinder reflector

since the space to install antennas is limited and the close proximity of antenna elements is unavoidable [4].

9.1.1 Frequency-Independent Antennas: Log-Periodic Antennas

The log-periodic antenna was first proposed by DuHamel and Isbell in 1957 [7] as shown in Figure 9.1. The log-periodic antenna looks very similar to the Yagi-Uda antenna, as both consist of a number of dipole elements spaced out along a support boom. In fact, the Yagi-Uda antenna was first proposed by Hidetsugu Yagi and Shintaro Uda in 1926 [8] and it has been commonly used as a television antenna due to its high value of gain as shown in Figure 9.2. In addition, both antennas have unidirectional radiation pattern. The distinction between the log-periodic antenna and the Yagi-Uda antenna is listed in Table 9.6. There are no simple formulas for designing Yagi-Uda antennas due to the complex relationships between the dipole array arrangement and its mutual coupling effect. However, a Yagi-Uda antenna

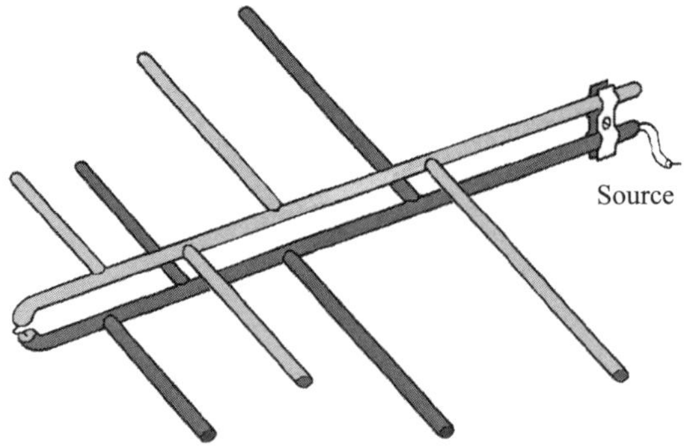

(a)

(b)

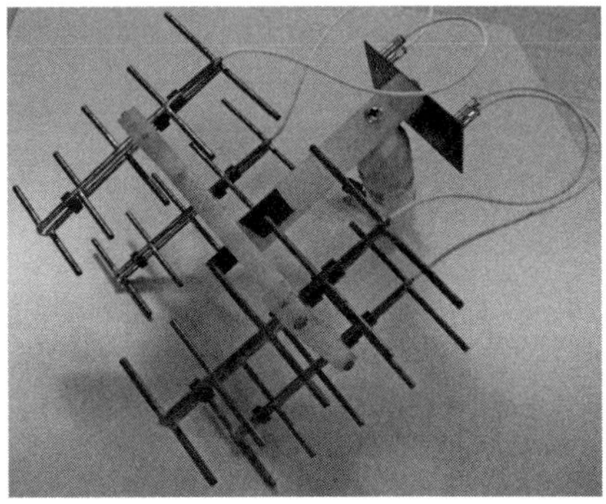

(c)

Figure 9.1 (a) Standard log-periodic antenna [10]. (b) The actual configuration of a commercial log-periodic antenna (Bandwidth: 0.29–2.0 GHz, Gain = 6.5 dBi). (c) Four-array log-periodic antenna.

can be designed to obtain optimum gain by adjusting the dipole element length and spacing between those elements based on some guidance rules as described on the accompanying CD. However, explicit formulas and the steps for designing the log-periodic antenna are listed on the accompanying CD.

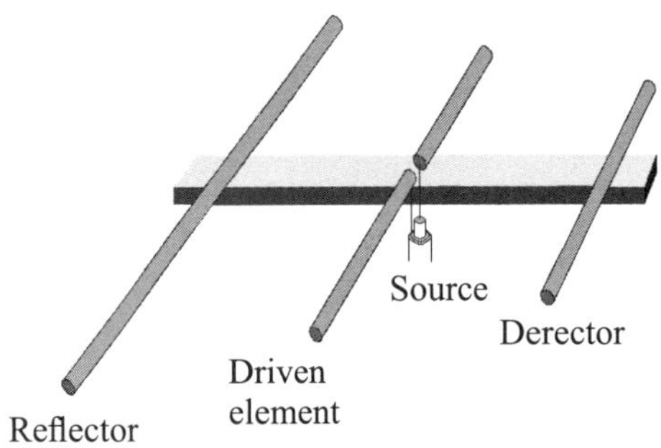

(a)

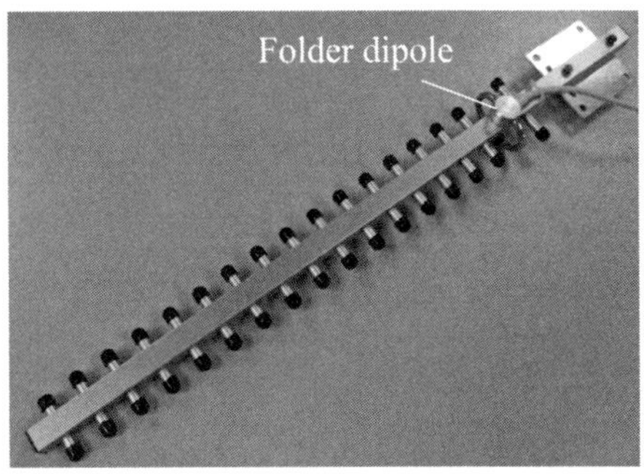

(b)

Figure 9.2 (a) Basic three-element Yagi-Uda array antenna. (b) The actual configuration of a commercial Yagi-Uda antenna (Bandwidth: 2.4–2.483 GHz, Gain = 18 dBi).

Table 9.6 Comparison Between a Yagi-Uda Antenna and a Log-Periodic Antenna

<table>
<tr><th>Yagi-Uda Antenna</th><th>Log-Periodic Antenna</th></tr>
<tr><td>1. All dipole elements have almost the same length in the array</td><td>1. The dipoles have different lengths in the array. The shorter length of the dipole element in the array contributes to higher frequencies. However, the greater length of the dipole element contributes to lower frequencies.</td></tr>
<tr><td>2. Only one dipole element is connected to the feed line (RF source). The antenna consists of one driven element, one reflector element (behind the driven element), and at least one director element (in front of the driven element). A number of dipole elements in the array are used as parasitic elements. The mutual coupling effect among the parasitic elements is based on the length $2h$ of each dipole and the spacing distance S between the dipole elements.</td><td>2. All of the dipole elements are connected to the feed line (RF source). The lengths of the dipole elements are increased by a constant ratio when the elements are located farther away from the feed line.</td></tr>
<tr><td>3. The antenna has a high gain with fewer dipole elements in the array. The parasitic elements will induce the current from the driven element and increase the gain of the antenna. The gain of the antenna is increased when the number of the dipole elements is increased [11]:
<table>
<tr><th>Number of the dipole elements, N</th><th>Typical gain (dBi)</th></tr>
<tr><td>3</td><td>7</td></tr>
<tr><td>4</td><td>9</td></tr>
<tr><td>6</td><td>10.5</td></tr>
<tr><td>8</td><td>12.5</td></tr>
<tr><td>12</td><td>14.5</td></tr>
<tr><td>15</td><td>15.5</td></tr>
<tr><td>18</td><td>16.5</td></tr>
</table>
</td><td>3. Require many dipole elements in the array to increase the gain value of the antenna.</td></tr>
<tr><td>4. Narrow bandwidth: The lengths of the reflector and director in the array are determined by the phase of the element's current with respect to its excitation from the driven element. The length of the reflector $2h$ is slightly longer than the length of the driven element (resonant length), which has inductive reactance. However, the length of director $2h$ is slightly shorter than $\lambda/2$ (resonant length), which has capacitive reactance. Hence, the phase of current induced on the reflector lags behind the phase of voltages. However, the phase of current induced on director leads the phase of the voltage. This will cause a continuous phase distribution across the array of elements and form a unidirectional radiation pattern.</td><td>4. Wider bandwidth: Each dipole element in the array has very closely spaced resonant frequency and they repeat periodically as the logarithm of the frequency. Thus, the dipole array will cause the continuity of resonant condition and give overall wider bandwidth as shown below:

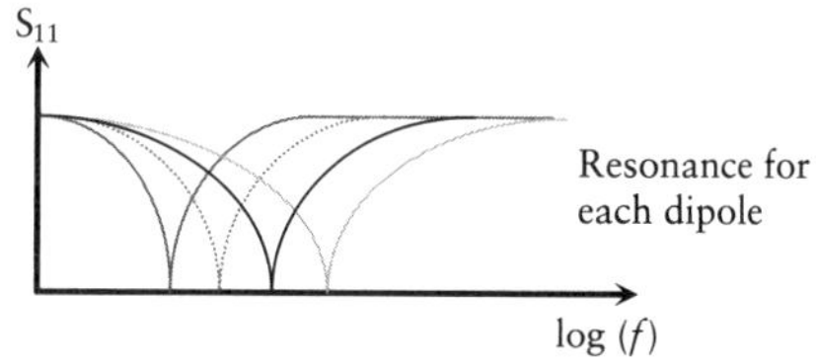

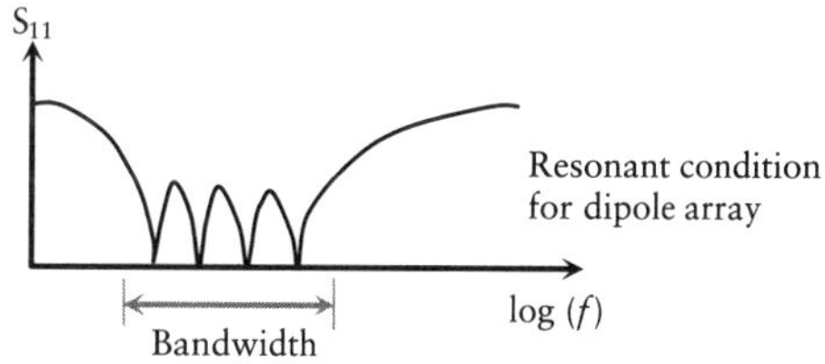

Each different length of dipole element in the array has different resonant frequency, (f_1, f_2, f_3 and f_4) which relationship is given as:

$$f_n = \frac{f_1}{\tau^{n-1}} \text{ for } n = 1,2,3,\ldots$$

where τ is the design scale factor.</td></tr>
</table>

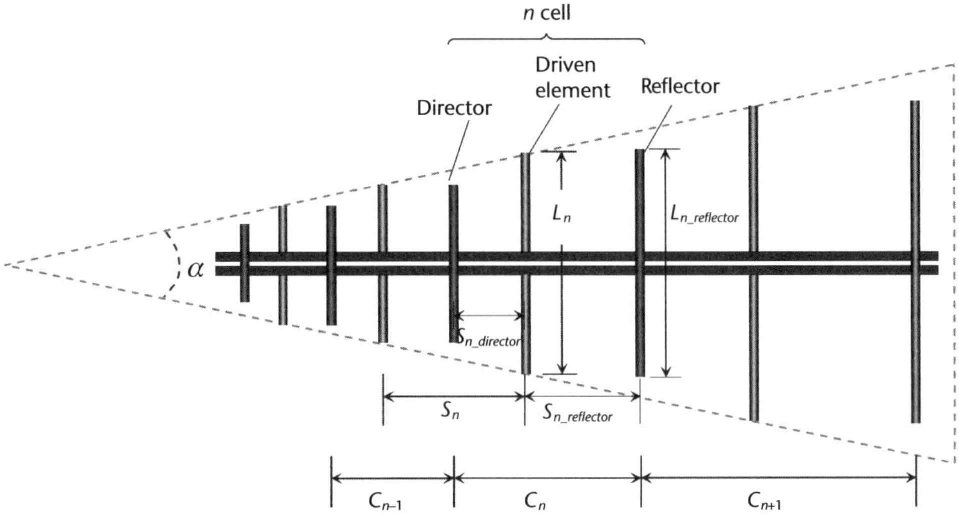

Figure 9.3 Log-periodic Yagi-Uda dipole array [9].

After the Yagi-Uda antenna has been designed based on the guidelines on the accompanying CD, the designed antenna needs to be recomputed using numerical simulation, such as the Method of Moments (MoM). A minor adjustment to the dimensions and geometry of the designed antenna may be needed in order to obtain the optimum gain. The gain of a log-periodic antenna can be improved by increasing the number of the dipole elements N in the array. However, the bandwidth of the antenna will decrease as the gain increases. In 1966, Barbano [9] applied the log-periodic techniques to the Yagi-Uda antenna in order to improve the operating bandwidth and the gain simultaneously. Thus, the improved antenna is called the log-periodic Yagi-Uda array antenna, as shown in Figure 9.3.

9.1.2 Sleeve Monopole Antennas

Some simple, small, and wideband monopoles driven from the coaxial line were invented by Lindenblad in 1941 [12] as shown in Figures 9.4 and 9.5. The monopole structure in Figure 9.4 permits a smoother transition into the ellipsoid shape of the extended center conductor. Furthermore, the impedance aperture at $z = 0$ can be maintained at 50Ω if the size ratio of $b_1/a_1 = b/a$.

The structure of monopole, as shown in Figure 9.5, is called a sleeve monopole. The outer conductor of the coaxial line extends a length H above the ground plate.

The inner conductor (or monopole) further extends beyond the coaxial line a length h. The inner conductor part contributes a series capacitive reactance. The outer conductor part has a parallel inductive reactance characteristic. The combination of both parts is equivalent to the circuit, which has an improved bandwidth characteristic [12]. The optimum design ratio $h/H = 2.25$ for the sleeve monopole was experimentally found by Poggio and Mayes [13]. In addition, the optimum total length of the monopole is found to be $(h + H) \approx \lambda/4$, which makes it possible to achieve optimum radiation patterns over a 4:1 band.

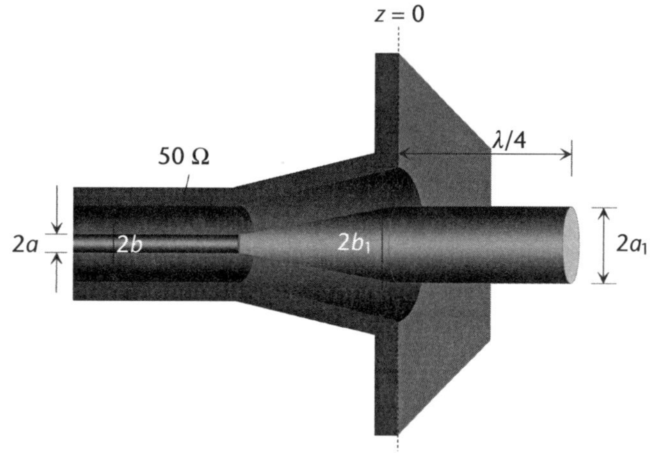

Figure 9.4 Expanding size diameter of monopole.

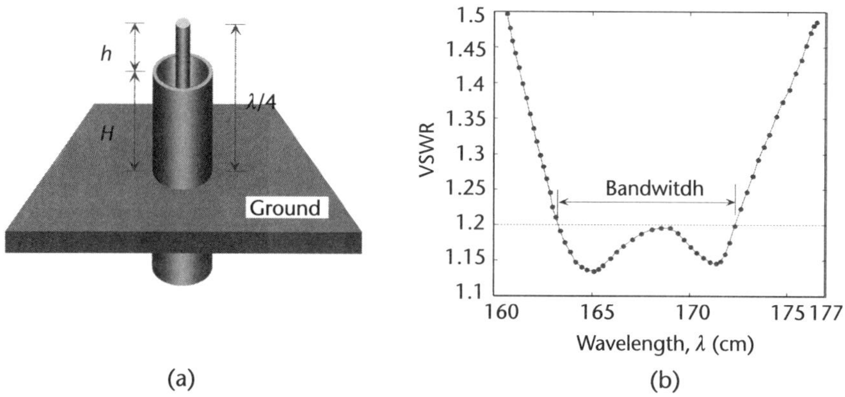

Figure 9.5 (a) Basic configuration of a sleeve monopole. (b) The corresponding performance of the sleeve monopole [12].

9.1.3 Conical Monopole Antennas

A modified monopole, the conical monopole antenna, is one of the members of the small-element antennas category with wideband applications. The conical monopole was first invented by Carter in 1939 [14]. The conical monopole has two primary design parameters, the cone angle θ and the cone length l. The conical monopole shows broadband characteristics as the cone angle is increased as shown in Figure 9.6.

The performance of the conical monopole antenna relates to its input impedance, Z_{in}, which was presented by Papas and King in 1949 [15]. The theoretical formulation of the input impedance Z_{in} is expressed as:

$$Z_{in} = Z_c \frac{1 - \dfrac{\beta}{\alpha}}{1 + \dfrac{\beta}{\alpha}} \tag{9.1}$$

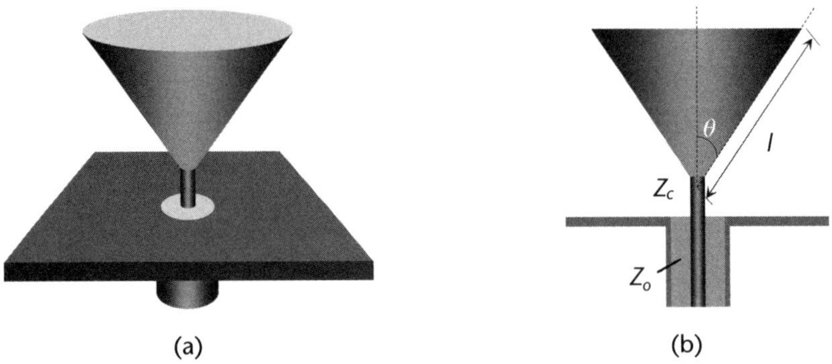

Figure 9.6 (a) Conical monopole antenna. (b) Dimensions of the conical antenna.

where Z_c is the characteristic impedance of the cone above the ground plane, which is given as:

$$Z_c = \frac{\eta_o}{2\pi}\ln\left[\cot\left(\frac{\theta}{2}\right)\right] \tag{9.2}$$

where η_o ($\approx$ 376.73Ω) is the intrinsic impedance of free space. The ratio of the amplitudes of the reflected and outwardly radiating TEM waves β/α in (9.1) is expressed as:

$$\frac{\beta}{\alpha} = e^{-2jkl}\frac{1 + j\frac{60}{Z_c}\sum_{n=1}^{\infty}\frac{2n+1}{n(n+1)}\left[P_n(\cos\theta)\right]^2\zeta_n}{-1 + j\frac{60}{Z_c}\sum_{n=1}^{\infty}\frac{2n+1}{n(n+1)}\left[P_n(\cos\theta)\right]^2\zeta_n} \quad \text{for } n = 1,3,5,\ldots \tag{9.3}$$

and

$$\zeta_n = \frac{H_n^{(2)}(kl)}{H_{n-1}^{(2)}(kl) - \frac{n}{kl}H_n^{(2)}(kl)} \quad \text{for } n = 1,3,5,\ldots \tag{9.4}$$

where $H_n^{(2)}(kl)$ is the nth-order spherical Hankel function of the second kind and $P_n(\cos\theta)$ is the Legendre polynomial of order n. Equation (9.2) accurately predicts the equivalent input impedance of a conical monopole if the characteristic impedance Z_c is equal to Z_o ($\approx$ 50Ω) of the coaxial feed line as [15]:

$$Z_c = Z_o = 50\Omega \text{ (impedence matching)} \tag{9.5}$$

Thus, the high accuracy of input impedance Z_{in} calculation can be achieved when $\theta = 47°$. The optimum cone angle θ is normally varied between 30° to 50°. When the value of θ is too large ($\theta > 50°$), the impedance mismatch between the coaxial feed line and the cone shape of monopole will become significant. The MATLAB

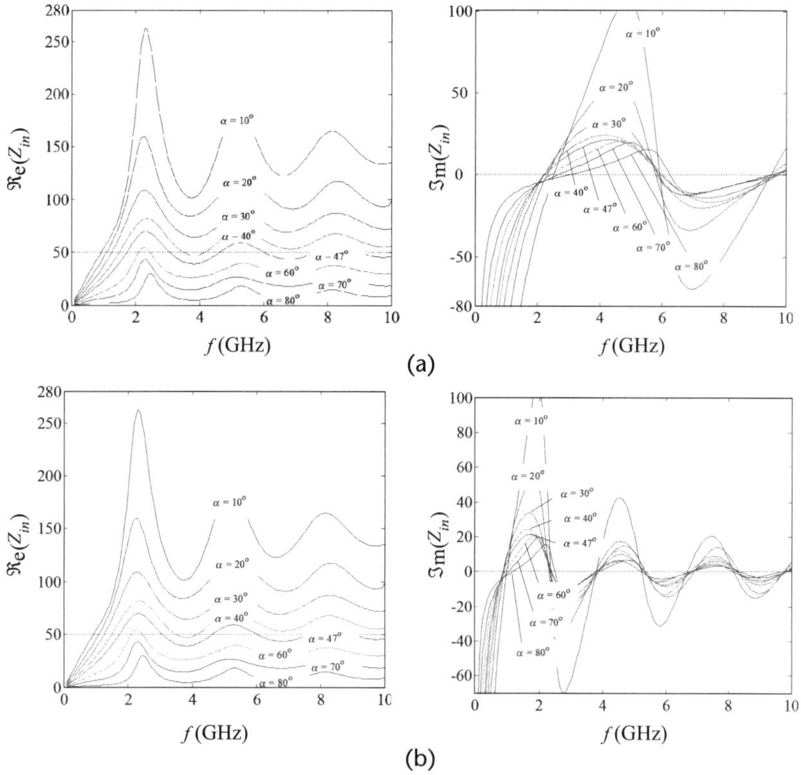

Figure 9.7 The resistance, $\mathfrak{Re}(Z_{in})$, and reactance, $\mathfrak{Im}(Z_{in})$, of conical monopole with (a) cone length, $l = 2$ cm, and (b) $l = 5$ cm.

calculation program of (9.13) is shown on the accompanying CD. The calculated input impedances Z_{in} (= $\mathfrak{Re}(Z_{in}) + j\mathfrak{Im}(Z_{in})$) of the conical monopole using (9.2), which varies with operating frequency f, are shown in Figure 9.7. Clearly, the first resonant condition of the conical monopole can be shifted to a lower frequency when the cone length l is increased. For $\theta = 47°$, overall, the value of $\mathfrak{Re}(Z_{in})$ is close to 50Ω. The values of $\mathfrak{Im}(Z_{in})$ are near zero and the values of $\mathfrak{Re}(Z_{in})$ are approximated to be constant over a range of frequencies for $l = 5$ cm rather than $l = 2$ cm. This implies that the conical monopole could be impedance-matched to a 50-Ω coaxial feed line over a broad range of frequencies with $\theta = 47°$ and $l \geq 5$ cm.

9.1.4 Disk-Loaded Monopole Antennas

Besides the conical monopole, there are various shapes of the end-loaded monopoles that are designed to provide wide operating frequency bands, as well as to reduce the size of the conventional monopole structure. For instance, one of the modified monopoles is the disk-loaded monopole, which has a radius a_1 of the horizontal disk attached on the top of length h of the monopole as shown in Figure 9.8.

This kind of antenna will generate a conically shaped radiation pattern. When $h \ll \lambda$ (operating wavelength), the optimum angle θ at which the maximum radiation is achieved can be predicted using [16]:

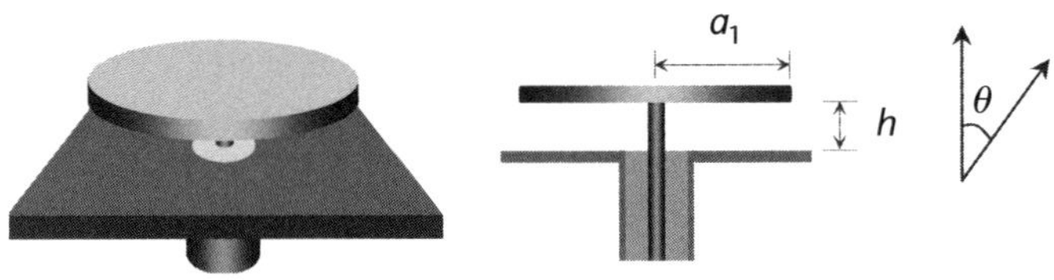

Figure 9.8 The disk-loaded monopole over a ground plane driven from a coaxial transmission line.

$$\theta = \sin^{-1}\frac{1.841}{ka_1} \tag{9.6}$$

where k (= $2\pi/\lambda$) is the propagation constant of free space. The electrical performance of the small disk-loaded monopole can be modeled by simple lumped-element circuit formulas [17]:

1. Capacitance:

$$C = \varepsilon_o \frac{A_{\text{eff}}}{h_{\text{eff}}} \tag{9.7}$$

2. Radiation series resistance:

$$R_{\text{eff}} = \frac{Z_o}{3\pi}\left(\frac{h_{\text{eff}}}{\ell}\right)^2 = 39.97\left(\frac{h_{\text{eff}}}{\ell}\right)^2 \tag{9.8}$$

3. Radiation shunt conductance:

$$G_{\text{eff}} = \frac{Y_o}{3\pi}\left(\frac{A_{\text{eff}}}{\ell^2}\right)^2 = 2.8164\times10^{-4}\left(\frac{A_{\text{eff}}}{\ell^2}\right)^2 \tag{9.9}$$

4. Radiation power factor:

$$p_{\text{eff}} = \frac{G_{\text{eff}}}{\omega C} = \frac{1}{3\pi}\frac{h_{\text{eff}}A_{\text{eff}}}{\ell^3} \tag{9.10}$$

where h_{eff} and A_{eff} are the effective monopole height and the top loading disk area, respectively. Symbol l (= $\lambda/2\pi$) is the radian length, and Z_o (= $1/Y_o$ = 376.73Ω) is the characteristic impedance of free space. In fact, the values of h_{eff} and A_{eff} in (9.7) to (9.10) are close to its actual physical dimension values (monopole height h and circular disk area A). However, accurate values of h_{eff} and A_{eff} can be found using the MoM [17]. From (9.7) and (9.10), the input admittance Y_{in} of the disk-loaded monopole can be modeled using simple equivalent lumped element circuit as:

$$Y_{in} = G_{\text{eff}} + j\omega C = 2.8164\times10^{-4}\left(\frac{A_{\text{eff}}}{\ell^2}\right)^2 + j\omega\varepsilon_o\frac{A_{\text{eff}}}{h_{\text{eff}}},\ \text{for } R_{\text{eff}} \ll \omega C \tag{9.11}$$

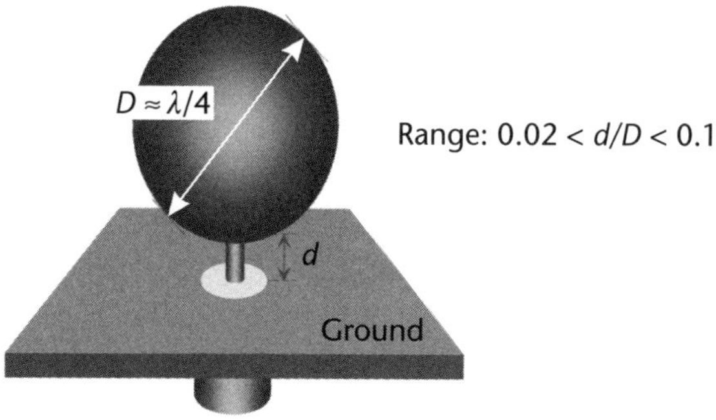

Figure 9.9 The ellipsoidal monopole [18].

where the admittance of the series circuit may be expressed as (9.11). The monopole end loaded by a hollow conducting ellipsoidal is another loaded monopole example, which was invented by Stöhr in 1968 [18], as shown in Figure 9.9.

9.1.5 Helical Monopole Antennas

The helical monopole antenna was first reported by Kraus in 1947 [19], as shown in Figure 9.10. Recently, this kind of antenna has been commonly used in walkie-talkie phones. The dimensions and structure of the helical monopole will strongly affect the gain and the operating frequency f. The structure of the helical monopole is formed as a wire spring with N turns. C in Figure 9.10 is the circumference of the turn. The D and S parameters are the diameter of the helix and the spacing distance between the turns, respectively. L and a are the total length of the helical monopole and the radius of the wire, respectively. In addition, the helical antenna may radiate either in axis mode (directional radiation pattern) or normal mode (omnidirectional radiation pattern) depending on the size of the antenna and operating wavelength, λ [19, 20].

The axis mode has a maximum radiation field in the direction of the helix axis and it is circularly polarized, like a horn antenna radiation pattern, as shown in Figure 9.11(a). The axis mode can be produced when the length of the helix circumference C is approximately equal to the operating wavelength λ (or the length of each helix turn, $l > 1.3\lambda$). The axis mode of radiation persists over a relatively wide operating frequency range [19, 20]. Thus, this kind of antenna is also widely used in broadband satellite communication systems as shown in Figure 9.12(a). However, the normal mode has a maximum radiation field in a normal direction to the helix axis, which was first studied by Wheeler in 1947 [21], as shown in Figure 9.11(b). The normal mode is produced when the dimensions of the helical monopole are significantly small compared to the wavelength λ (the length of each helix turn, $l \ll 0.8\lambda$). The helical monopole with normal mode has low radiation efficiency and narrow bandwidth like the small monopole performance [4]. Thus, this kind of antenna is widely used in mobile [in Figure 9.12(b)] or portable RF communication system since the helical wire structure can significantly reduce the length of the monopole antenna and maintains a high operating frequency performance simultaneously.

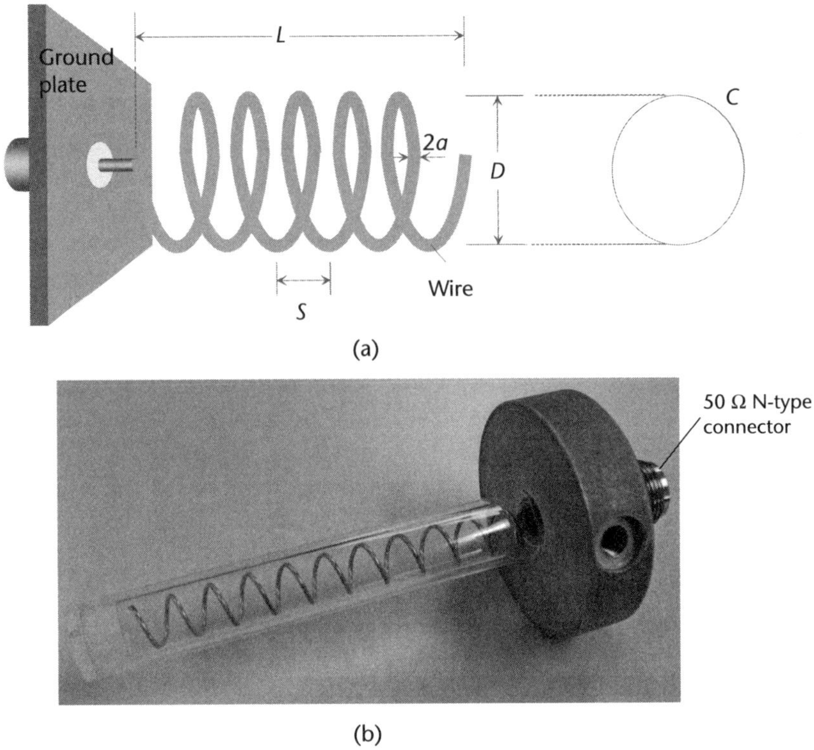

Figure 9.10 (a) Six-turn helical monopole antenna with labeled dimensions. (b) The actual configuration of commercial helical monopole antenna.

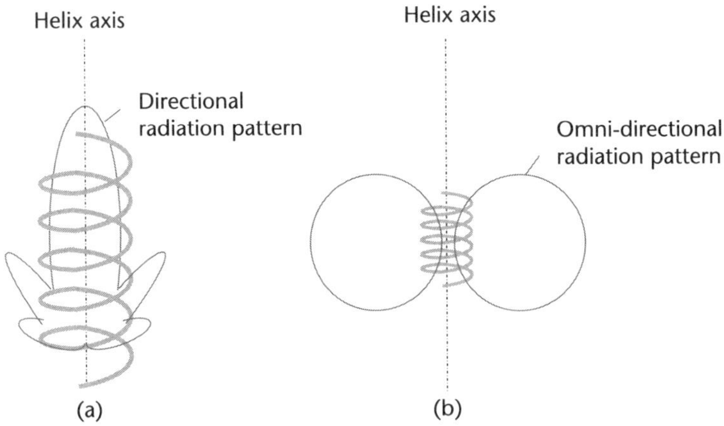

Figure 9.11 (a) Axis mode of radiation pattern. (b) Normal mode of radiation pattern.

The explicit formulas and steps for designing the helical monopole antenna in axis mode and normal mode are listed on the accompanying CD. Although the analytical formulas on the accompanying CD provide a tool for the optimal design of the helical monopole antenna, they do not guarantee actual performance for the practical antenna. Hence, the performance of the designed antenna is needs to be refined by numerical simulation or measurements.

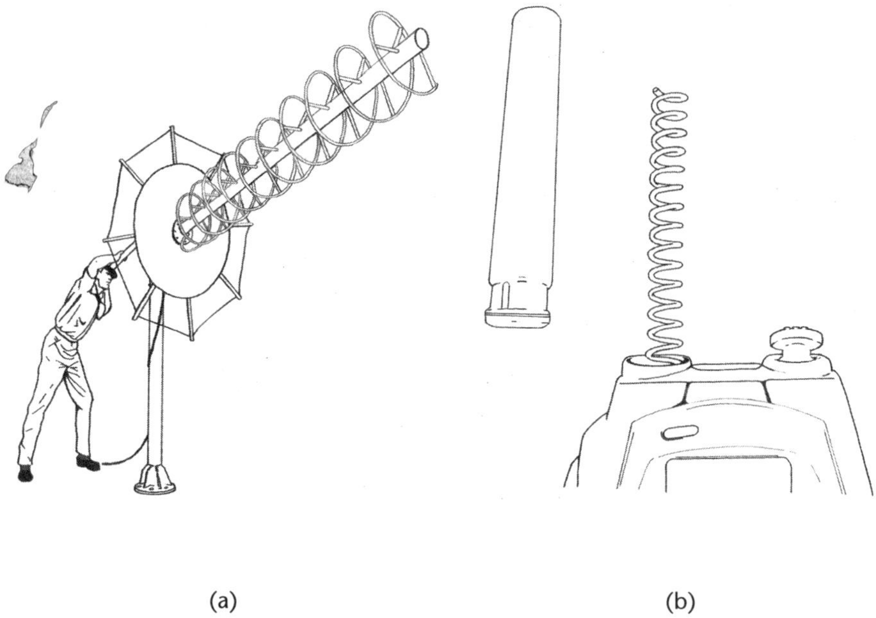

Figure 9.12 (a) A large helical monopole in axis mode is used in satellite communication application [22]. (b) A small helical monopole in normal mode is used in mobile and portable communication devices [22].

9.1.6 Insulated Monopole Antennas

In addition, the insulated monopole can be used as an antenna to replace the half-wave dipole antenna. However, the conventional insulated monopole in Figure 9.13 failed to produce a radiation pattern that is similar to half-wave dipole pattern. The two modified insulated monopoles are proposed by Drozd and Joines [23] in order to improve the radiation pattern more closely to that of the half-dipole pattern. The first type of modified insulated monopole has a small slot gap on the outer conductor, spaced a quarter-wavelength back from the end of the coaxial line as shown in

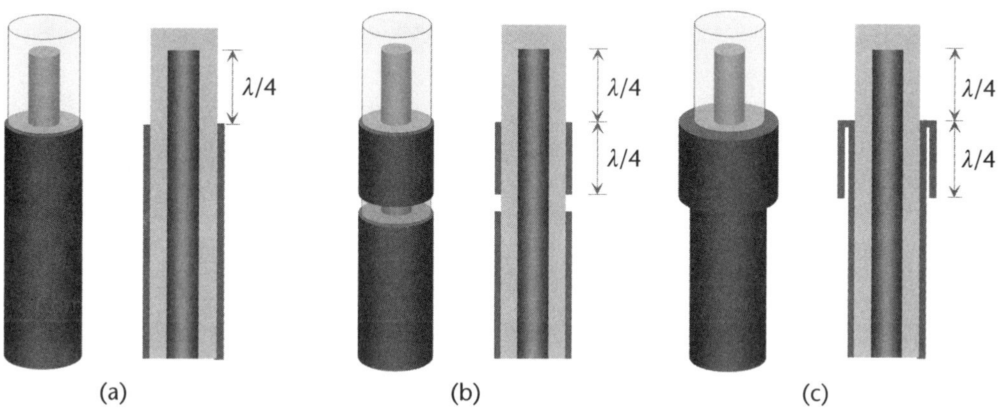

Figure 9.13 (a) Basic insulated monopole. (b) Slot gap insulated monopole. (c) Choke sleeve insulated monopole.

Figure 9.13. Another type of insulated monopole is modified by applying a quarter-wavelength choke sleeve on the outer conductor, which is also used to prevent the propagated wave from traveling back along the coaxial line.

9.1.7 Dielectric Resonator Monopole Antennas

Recently, the monopole loaded with dielectric material was proposed to enhance the bandwidth and maintain the simplicity of design and cost as shown in Figures 9.14 to 9.17. The resonant frequency f_r of the dielectric resonator monopole in Figure 9.14 can be predicted as [24]:

$$f_r = \frac{c\sqrt{3.83^2 + (\pi/2)^2(a/h)}}{2\pi a\sqrt{\varepsilon_r + 2}} \tag{9.12}$$

where c and ε_r are the velocity of light and the dielectric constant of the dielectric material, respectively. Equation (9.12) is only valid for $0.4 \le a/h \le 6$.

The dielectric resonator monopole in Figure 9.15 was proposed by Ittipiboon et al. in 2002 [26]. A brief design guideline of the dielectric resonator monopole was proposed by Lapierre et al. in 2005 [27]. An improved guideline for the design

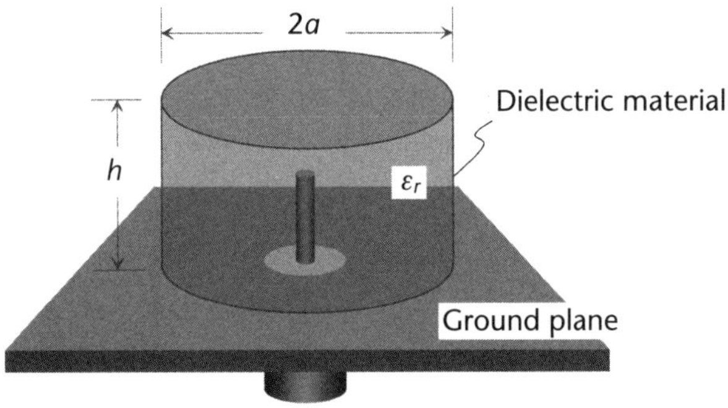

Figure 9.14 The basic dielectric resonator monopole antenna [25].

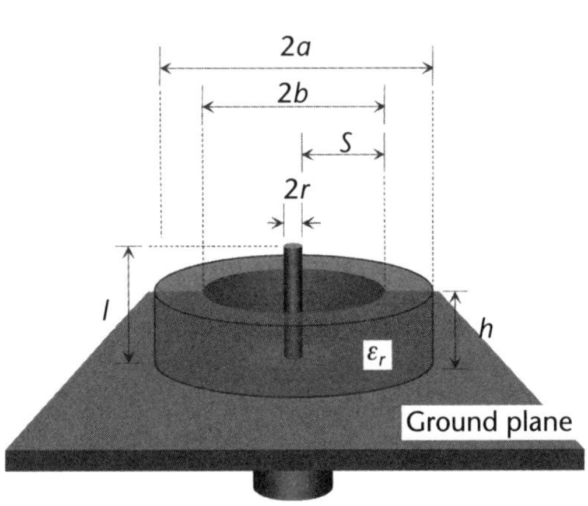

Figure 9.15 The monopole antenna surrounding by dielectric ring [26].

Table 9.7 The Improved Design Guideline for Dielectric Resonator Monopole in Figure 9.15

Optimum Dimensions	
Monopole part	*Dielectric ring part*
The optimum length of the monopole, $l = \lambda_{min}/4$, where $\lambda_{min} = c/f_{min}$ and c is the velocity of light in free space. The relationship between the radius a and the air gap space S must be obeyed: $S \geq r \geq S/2$	The optimum range of the dielectric ring dimensions are listed as: $0.016\ \lambda_{min} \geq S \geq 0.013\ \lambda_{min}$ $0.5l \geq h \geq 0.4l$ $b = r + S$ $a = b/0.3$ The dielectric constant ε_r of the dielectric ring can be estimated as [30]: $\varepsilon_r = [(\pi/2h)^2 + (4.71/a)^2]/(2\pi f_{max}/c)^2$

procedure was reported by Guha et al. [28]. A bandwidth ratio f_{max}/f_{min} of up to 2.5 can be achieved with this monopole. The design guideline is listed in Table 9.7 [28].

Normally, the dielectric material used in dielectric resonator monopole has lossless properties. However, lossy dielectric material was used in the monopole design studied by Zivkovic [29] and shown in Figure 9.16. The reflection coefficient, S_{11} of the antenna was measured up to 40 GHz and it was found that the lossy material can cause a larger bandwidth when compared to the lossless material in the same design configuration.

9.1.8 Negative Dielectric Resonator Monopole Antennas

The application of negative dielectric material (metamaterial) to a monopole antenna in order to reduce the size of the monopole was studied by Stuart and Pidwerbetsky [24], as shown in Figure 9.17.

In Figure 9.17(b), the small monopole is partially extended into the negative dielectric material. The fundamental resonant mode contributed by the negative permittivity sphere exhibits a uniform electric field within the sphere. However, the small monopole protruding into the center of the sphere provides a strong coupling between the coaxial transmission line mode and the resonant mode of the sphere [24]. The negative dielectric sphere acts like an inductor that tunes the capacitive reactance at a lower frequency, which is contributed by the monopole. Thus, the

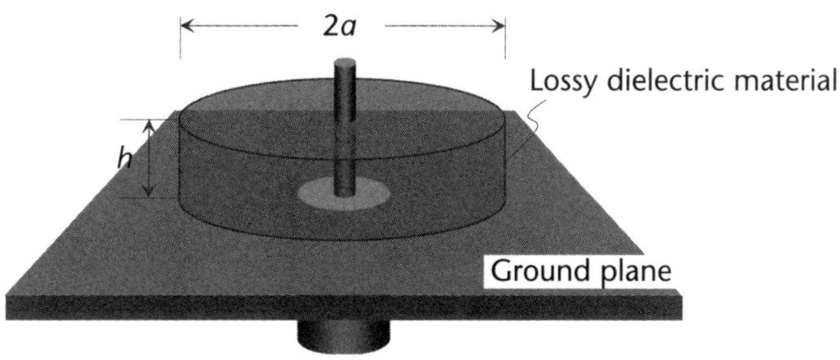

Figure 9.16 Cylinder-shaped dielectric loading around the monopole [29].

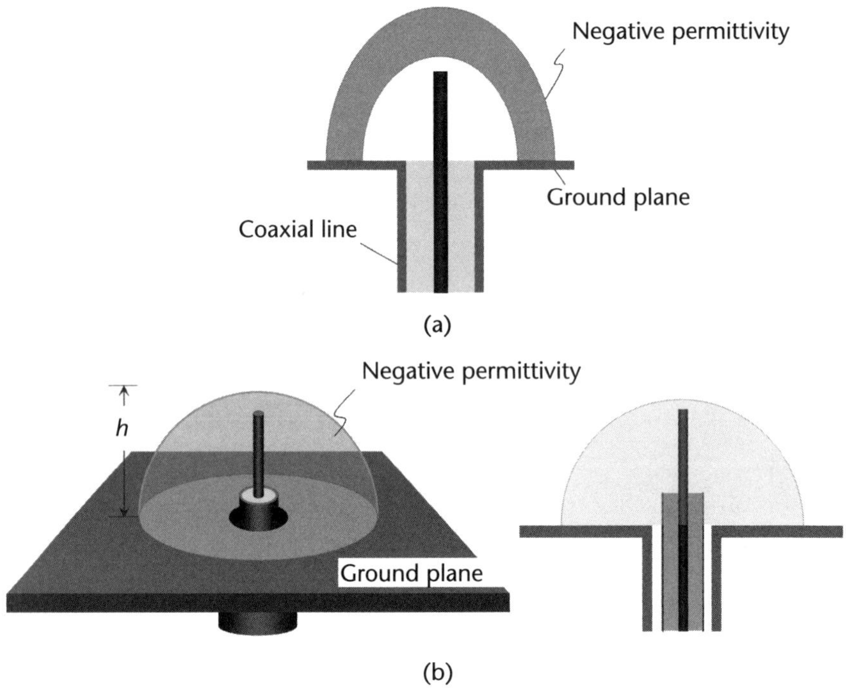

Figure 9.17 (a) General form of negative dielectric resonator monopole antennas. (b) The monopole driven from a coaxial line that is covered by a half-sphere of negative dielectric material on the ground plane [24].

negative dielectric resonator monopole is capable resonance at a very low frequency with the small size of the monopole.

9.1.9 Leaky Coaxial Cables

Wireless network access has become popular in public facilities, such as tunnels for subways, underground shopping arcades, train stations, long hotel corridors, and work offices. However, conventional antennas have several issues when they are used to provide wireless access in those areas. A number of antenna units have to be installed every several meters' distance in the areas. Besides, blind zones of the wireless signal will occur, which are out of the antenna coverage areas. Recently, leaky coaxial cables are being used to minimize the blind zones and provide a wider wireless access area with one single base station unit and one single frequency channel as shown in Figure 9.18. Besides wireless local area network (LAN) applications, the leaky coaxial cable can also be used for various sensing applications such as intruder sensor, linear cell MIMO, and position location sensor [31, 32]. In this section, the discussion of the theoretical background for the leaky coaxial cable in Chapter 8 will be extended in order to explain the application of the principles of the cable.

The propagation constant of the *m*th spatial harmonic in the leaky coaxial cable is given as [33]:

$$\beta_m d = \beta_o d + 2m\pi \quad (m = -1, -2, \ldots) \tag{9.13}$$

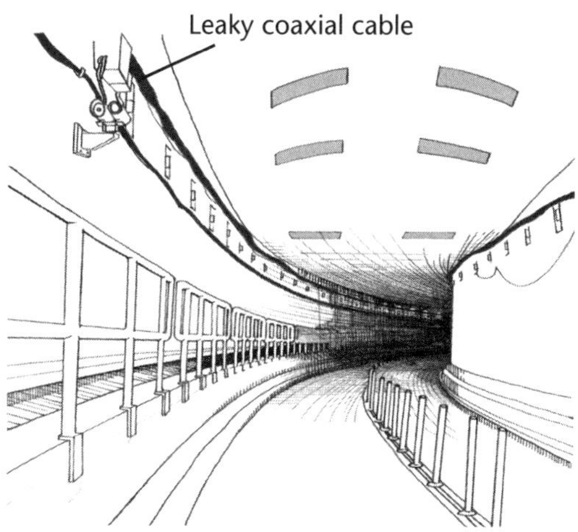

Figure 9.18 Leaky coaxial cable installation in New York City's Lincoln Tunnel.

where d is the period distance of the slot aperture on the leaky cable. However, $\beta_o(= 2\pi\sqrt{\varepsilon_r}/\lambda_o)$ is the wave propagation constant in the coaxial cable (neglecting the slot aperture effects on the outer conductor of the cable). The term k_o (= ω/c) is the propagation constant of free space. From (9.13), the angles ϕ_m of the radiation direction of the mth spatial harmonic, as shown in Figure 9.19, are determined by:

$$\sin\phi_m = \frac{\beta_m d}{k_o d} = \left(\frac{\beta_o d + 2m\pi}{k_o d}\right) = \left(\frac{\beta_o}{k_o} + \frac{m\lambda_o}{d}\right) = \sqrt{\varepsilon_r} + \frac{m\lambda_o}{d} \tag{9.14}$$

For a practical radiator, only a single radiated beam is preferred ($m = -1$), Thus, the angle, ϕ_{-1} of the radiation direction of the first spatial harmonic is expressed as:

$$\phi_{-1} = \sin^{-1}\left(\sqrt{\varepsilon_r} - \frac{\lambda_o}{d}\right) \tag{9.15}$$

Nishikawa et al. [32] proposed a new position detection method based on the time-domain measurement of both incident wave and reflected wave using a leaky coaxial cable. The generator produces incident wave into leaky coaxial cable. The incident wave is detected by the wireless terminal at position (x, y) through path

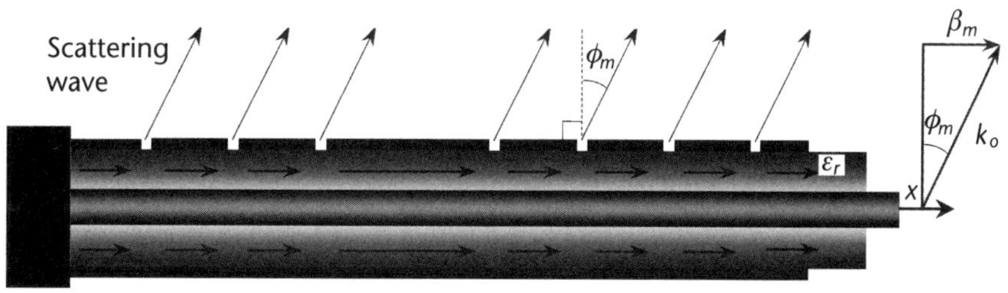

Figure 9.19 Radiated wave from the slot aperture of the leaky coaxial cable.

distance of $(l_1 + l_2)$ (in meters). The propagation time t_1 (in seconds) of the incident wave along the length of $(l_1 + l_2)$ can be estimated as:

$$t_1 \approx \frac{l_1}{v} + \frac{l_2}{c} \tag{9.16}$$

where $v = 3 \times 10^8/\sqrt{\varepsilon_r}$ m/s and $c = 3 \times 10^8$ m/s are the propagation velocity in leaky coaxial cable and free space, respectively. The partial incident wave is transmitted to the end of the cable and the remaining partial incident wave will reflect and travel back due to the discontinuity at the end of the cable. The reflected wave is detected by the wireless terminal after travel distance of $(l_3 + l_4 + l_5)$ as shown in Figure 9.20. Thus, propagation time t_2 of the reflected wave along the length of $(l_3 + l_4 + l_5)$ is estimated as:

$$t_2 \approx \frac{l_3 + l_4}{v} + \frac{l_5}{c} \tag{9.17}$$

where l_1, l_2, l_3, l_4, and l_5 in (9.16) and (9.17) are given as [32]:

$$l_1 = x + y\tan\theta_{-1} \tag{9.18a}$$

$$l_2 = l_5 = \frac{y}{\cos\theta_{-1}} \tag{9.18b}$$

$$l_4 = l_3 - x - y\tan\theta_{-1} \tag{9.18c}$$

From (9.16) to (9.18), after some mathematical arrangements, the coordinate position (x, y) of wireless terminal can be determined as:

$$x = \left[\frac{2l_3}{v} - (t_2 - t_1)\right]\frac{v}{2} \tag{9.19a}$$

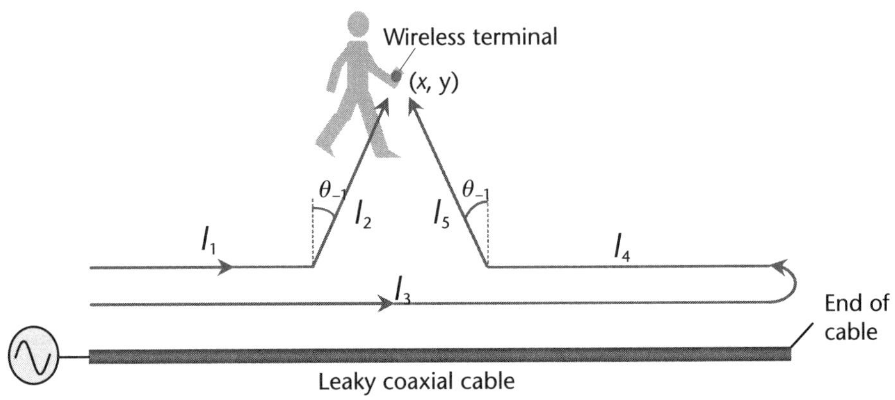

Figure 9.20 Principle of position detection.

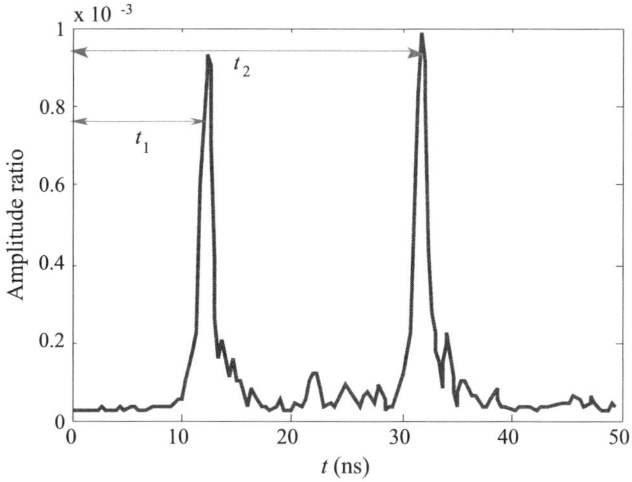

Figure 9.21 Time-domain impulse measurements with 0.083 ns of time resolution [32].

$$y = \left[\frac{2l_3}{v} - \left(t_2 + t_1\right)\right]\left[2\left(\frac{\tan\phi_{-1}}{v} - \frac{1}{c\cos\phi_{-1}}\right)\right]^{-1} \tag{9.19b}$$

The radiation angle ϕ_{-1} in (9.19b) is calculated from (9.15). Figure 9.21 shows the impulse response data between the generator and the wireless terminal (antenna), which is extracted from Nishikawa et al. [32]. Clearly, the propagation time (t_1 and t_2) of the incident wave and reflected wave are observed.

Higashino and Okada [31] used the similar time arrival location principle for 1-D position location detection as shown in Figure 9.22. The propagation time (t_1 and t_2) of each received signal at R_1 and R_2 is estimated using channel estimation functions. The estimated position at x-axis can be calculated as [31]:

$$x = \left(t_1 - t_2\right)\frac{v}{2} \tag{9.20}$$

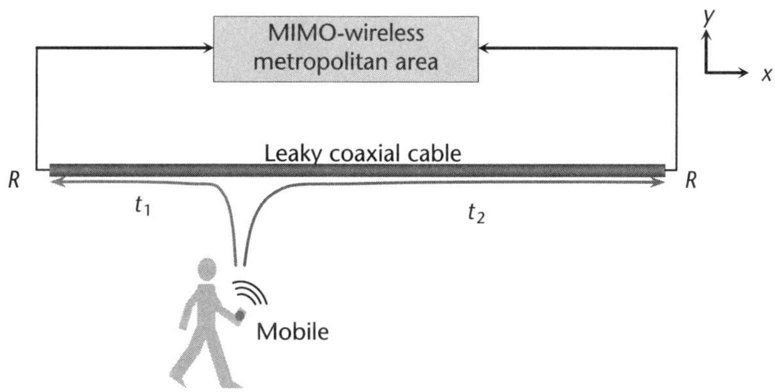

Figure 9.22 Operation of positioning system.

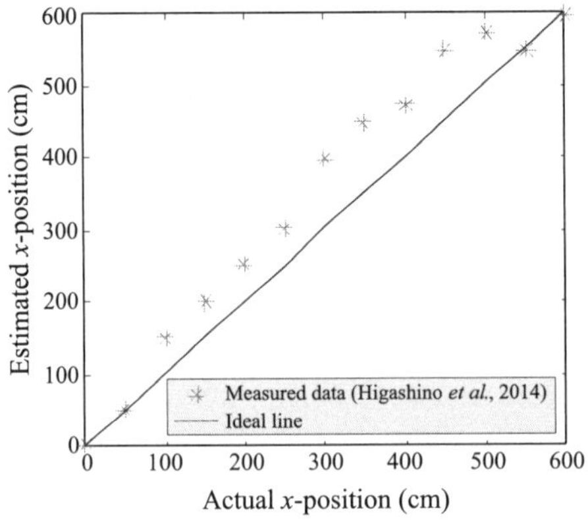

Figure 9.23 The relationship between the actual and estimated x-position distances [31].

The comparison between actual and estimated x-position distances using leaky coaxial cable at 500 MHz [31] is shown in Figure 9.23.

9.2 Health and Biomedical Applications

9.2.1 Dielectric Database for SAR Evaluation in Human Heads and Hyperthermia Treatment Simulation

Studies of interactions between electromagnetic waves and biological samples typically involve references to values of relative permittivity ε_r and ionic conductivity σ for the biological samples. Figure 9.24 shows a dielectric constant ε_r' and conductivity σ of an ideal biological tissue with four stages of dispersion (α, β, δ, and γ-dispersions) from dc to 70 GHz:

1. The α-dispersion is determined by the tangential flow of ions across the cell surface at a very low frequency (10^2–10^3 Hz).

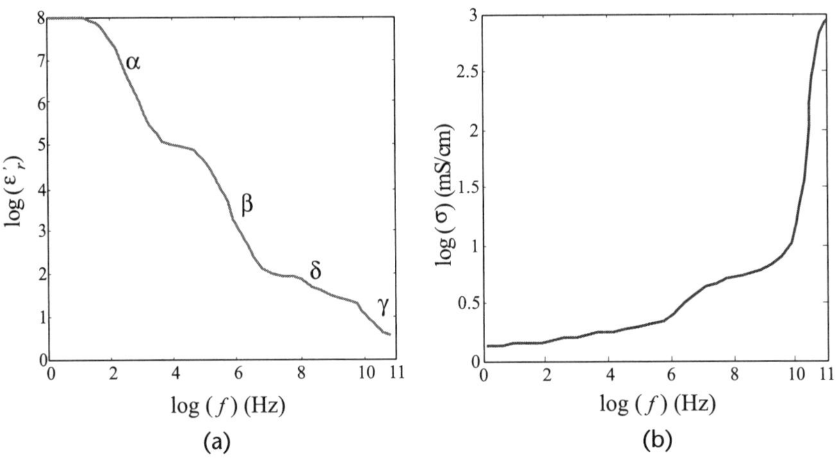

Figure 9.24 Dependence of (a) dielectric constant ε_r' and (b) conductivity σ of the typical biological tissues on the operating frequency.

2. The β-dispersion is determined by the charge at cell membranes caused by the Maxwell-Wagner effect (10^4–10^6 Hz).
3. The δ-dispersion is determined by the rotation of macromolecular side-chains and bound water (10^8–10^{10} Hz).
4. The γ-dispersion is determined by the dipolar rotation of small molecules, particularly water (>10^{10} Hz).

The intensive study of the dispersion mechanism in biological tissue can be found in [34]. The values of ε_r and σ are used as databases for SAR characterization, hyperthermia radiator simulation, or cancer cell detection.

Normally, an open-ended coaxial probe is used to measure the dielectric properties of the biological tissue that cover a wide frequency band in order to create a reliable database. For instance, the dielectric constant ε_r' and conductivity σ for several types of body tissues are listed in Table 9.8.

9.2.2 Hyperthermia Applicators

Hyperthermia is a cancer treatment in which body tissue is exposed to high temperatures up to 42°C as described in Chapters 5 and 7. Several types of hyperthermia applicators using coaxial line probes have been designed for this application.

Table 9.8 The Dielectric Constant ε_r' and Conductivity σ of Body Tissues at 37°C

	433 MHz		*915 MHz*		*2.45 GHz*		
Tissues	ε_r'	σ *(S/m)*	ε_r'	σ *(S/m)*	ε_r'	σ *(S/m)*	*Reference*
Artery	—	—	—	—	43	1.85	[37]
Blood	66	1.27	62	1.41	60	2.04	[38]
Bone with marrow	5.2	0.11	4.9	0.15	4.8	0.21	[39]
Brain							
White matter	48	0.63	41	0.77	35.5	1.04	[40]
Gray matter	57	0.83	50	1.0	43	1.43	
Fat	15	0.26	15	0.35	12	0.82	[41]
Kidney	60	1.22	55	1.41	50	2.63	[39, 41]
Liver	47	0.89	46	1.06	44	1.79	[39, 41]
Lung, deflated	35	0.71	33	0.78	—	—	[38]
Muscle	57	1.12	55.4	1.45	49.6	2.56	[42]
Ocular tissues							
Choroid	60	1.32	55	1.4	52	2.3	[43]
Cornea	55	1.73	51.5	1.9	49	2.5	
Iris	59	1.18	55	1.8	52	2.1	
Lens cortex	55	0.8	52	0.7	48	1.75	
Lens nucleus	31.5	0.29	30.8	0.5	26	1.4	
Retina	61	1.5	57	1.55	56	2.5	
Skin	47	0.84	45	0.97	44	—	[38]

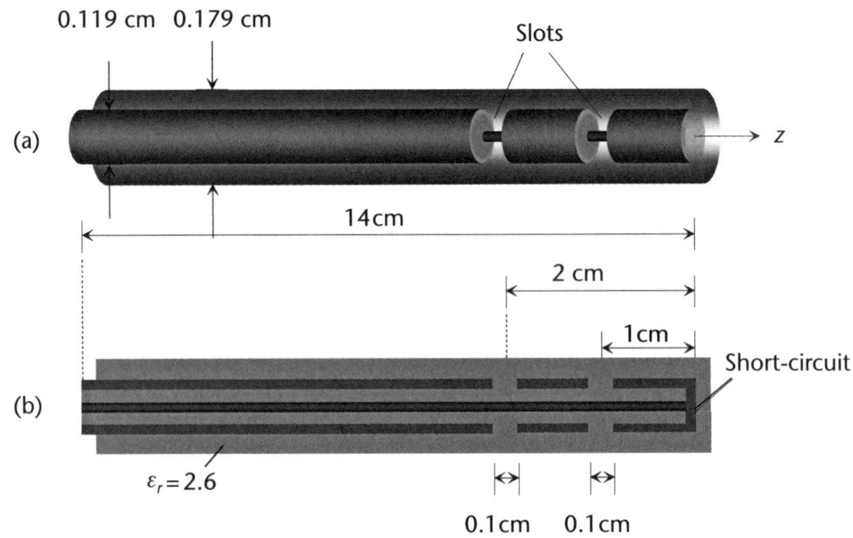

Figure 9.25 (a) The coaxial-slot applicator. (b) Cross-sectional view of the applicator with two slots [35].

In general, the applicator is designed to match the complex impedance of the particular tissue at the operating frequency (433 MHz, 9.15 GHz, or 2.45 GHz) to the impedance of the radiating applicator in order to ensure the RF energy is efficiently delivered into the target tissue. Normally, this kind of applicator is used for brain, liver, cardiac catheter ablation, and breast cancer due to its long and thin size. Different structures are modified at the end of the applicator in order to give different forms of heat distribution based on the size, shape, and location of the cancer tissue (or tumor).

For instance, the dimension of the applicator designed by Saito et al. [35] is shown in Figure 9.25. The applicator has two slot sections in order to concentrate the heat distribution around the slot areas. Normally, an array of four applicators is implemented in order to obtain an optimum heating region covering the cancer tissues or tumors as shown in Figure 9.26.

For brain tumors, the control of the heating environment is more important than other treatment cases [36]. Thus, a controllable heating applicator (sleeve choke applicator) is designed as shown in Figure 9.27, and it is used to replace the simple coaxial-slot applicator. The heat distribution along the z-axis of the applicator can be controlled by changing the length of the sleeves, $2L$. Figure 9.28 shows the SAR distribution using an applicator with $2L = 40$ mm versus the axial direction z. From Figure 9.28, the heating distribution area of applicator clearly depends on the length of the sleeves, $2L$.

9.2.3 Abnormal Tissues Detector

Tumor tissues typically have a high water content compared to normal tissues due to the hydration increases associated with rapid metabolism of cancer cells and surrounding angiogenic vasculature. Implicitly, it will cause differences between the dielectric properties of normal and cancerous tissue. Thus, it is possible to detect

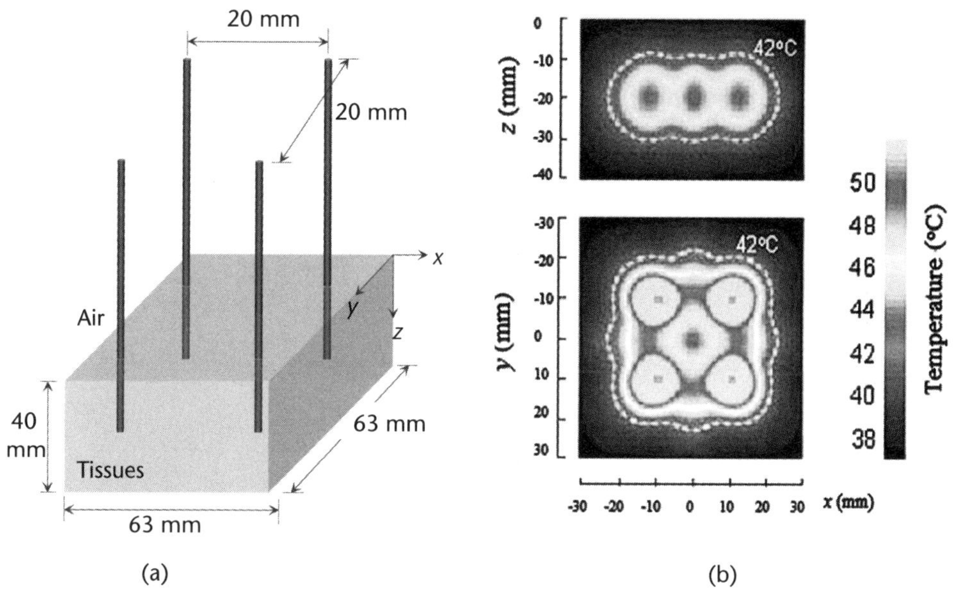

Figure 9.26 (a) Four parallel coaxial-slot applicators, forming a square surrounding a targeted tumor. (b) *z*-*x* and *y*-*x* cross-sectional views of the calculated heat distribution contour with input power, P_{in} = 20W, initial temperature, T_o = 37°C, and heating time, *t* = 300 seconds [36].

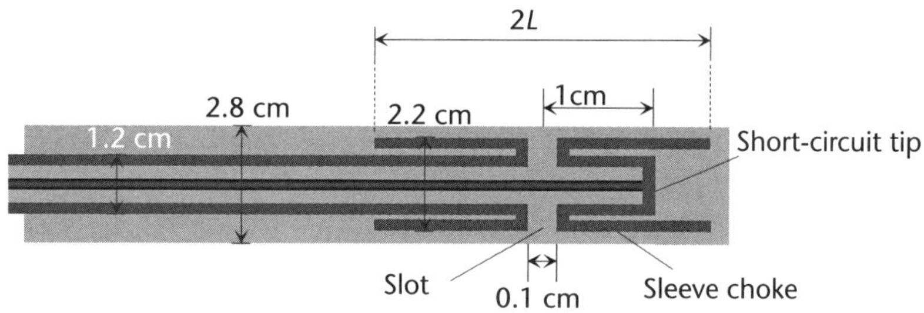

Figure 9.27 Cross-sectional view of the sleeve choke applicator [35].

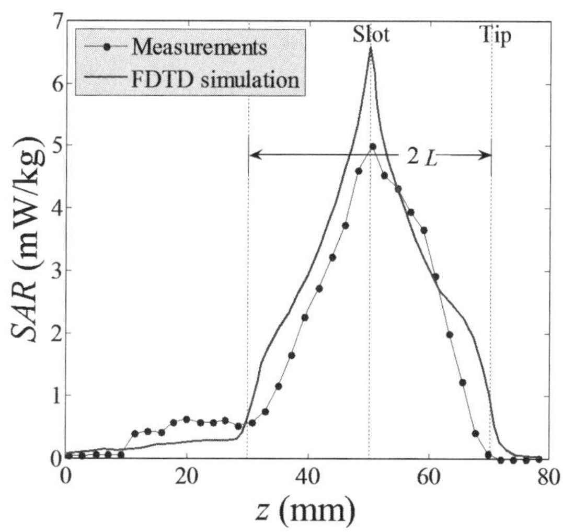

Figure 9.28 Simulated and measured SAR distributions for a sleeve choke applicator (2*L* = 40 mm) with input power, P_{in} = 30.4W [35].

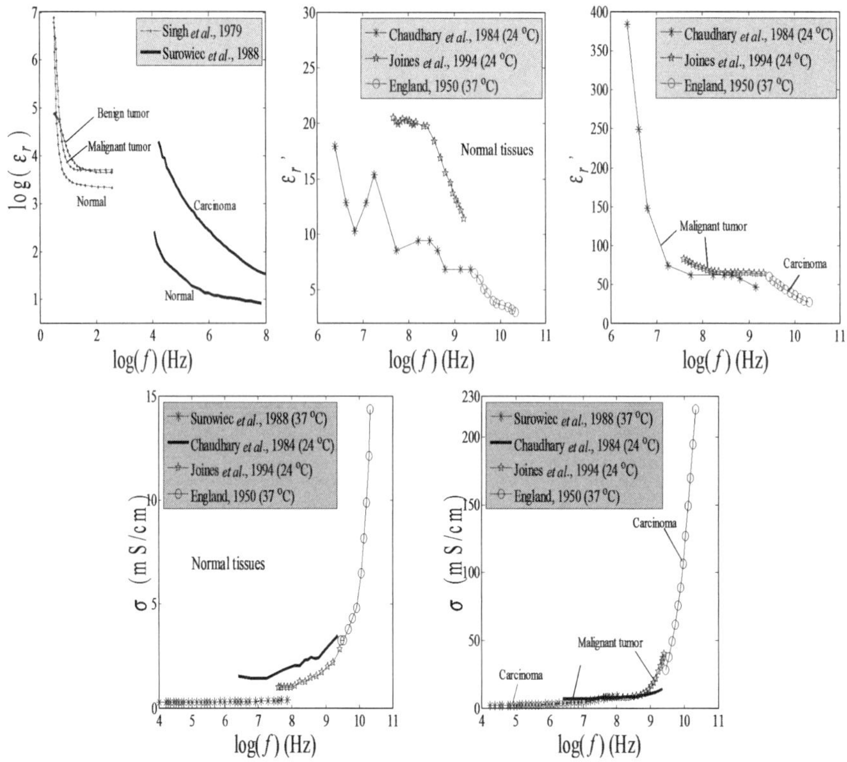

Figure 9.29 Comparison of the dielectric constant ε_r' and conductivity σ between the normal breast tissues and cancerous breast tissue from dc to 26 GHz.

the cancerous tissue based on dielectric properties or water content measurements using coaxial radiator. For instance, the normal breast tissue is dominated by adipose (fat) tissue, which has a low water content and low dielectric properties. The comparison of relative dielectric constant ε_r' and conductivity σ between the normal and cancerous breast tissue is shown in Figure 9.29. The dielectric database can be used as a guideline to identify whether a patient is diagnosed with cancer or not. The relationship among the dielectric constant, ε_r', conductivity, σ, and the percentage of water content in breast tissues is listed in Table 9.9. The deviation of the dielectric properties and the water content is higher in the measurement results. This is due to

Table 9.9 The Relationship Among Dielectric Constant ε_r', Conductivity σ, and Percentage of Water Content in Normal Breast Tissues and Cancerous Breast Tissues

Tissues Type	*Dielectric Constan, ε_r' at 3.2 GHz [45]*	*Conductivity σ (mS/cm) at 3.2 GHz [45]*	*Water Content (%) [45]*	*Water Content (%) (Age: 46–51) [44]*
Normal breast tissue	9.8–46	3.7–34	41–76	41–49
Benign breast tissue	15–67	7–49	62–84	53–56
Malignant breast tissue	9–59	2–43	66–79	9–75

Source: [44, 45].

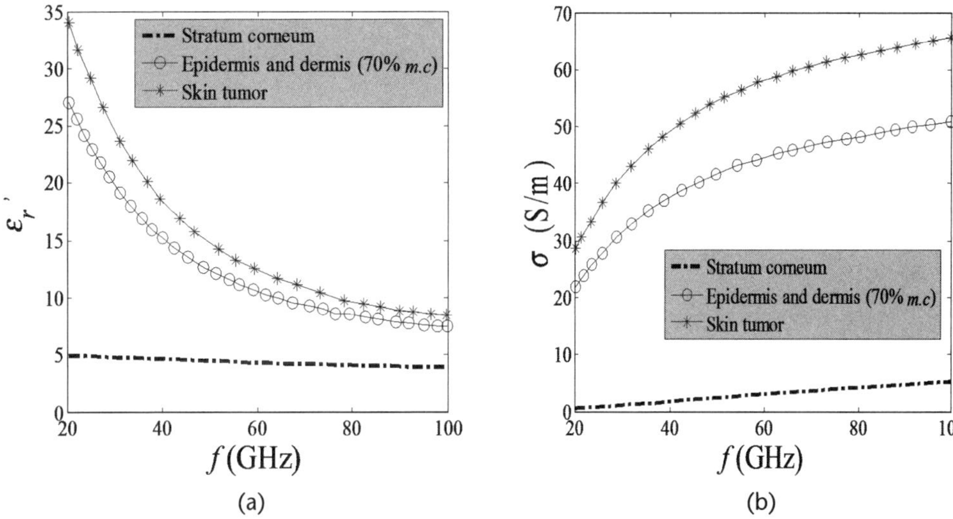

Figure 9.30 Comparison of the (a) dielectric constant ε_r' and (b) conductivity σ between the normal skin tissues and a skin tumor [46].

the adipose component, which is not uniformly distributed in the breast specimen. Furthermore, adipose content is different depending on a person's age and race.

Apart from breast cancer, other types of tissues, such as skin and liver, also showed a significant difference in the dielectric properties between the normal and cancerous tissue. In general, human skin consists of three layers, namely, stratum corneum, epidermis, and dermis. The stratum corneum is the outermost layer of the skin. Meanwhile, the second and third layers are the epidermis and dermis, respectively. The comparison of relative dielectric constant ε_r' and conductivity σ between normal skin tissues and a skin tumor from 20 to 100 GHz is shown in Figure 9.30. The percentages of water content and the dry component in the stratum corneum, epidermis, dermis, and skin tumor, respectively, are listed in Table 9.10.

Dielectric measurement of an excised liver tumor using an open-ended coaxial probe has been reported by Rourke et al. [47]. The comparison of relative dielectric constant ε_r' and conductivity σ between the normal and malignant liver tissues from dc to 20 GHz is shown in Figure 9.31. From Figure 9.31, it is found that the measured values of dielectric constant ε_r' and conductivity σ using in vivo and ex vivo techniques are different.

Table 9.10 Water Content and Dry Component Fractions of the Normal Skin Tissues and Skin Tumor

Skin Tissues	*Water Content (%)*	*Dry Component (%)*
Stratum corneum	20	80
Epidermis and dermis	65–70	30–35
Skin tumor	81.6	18.4

Source: [46].

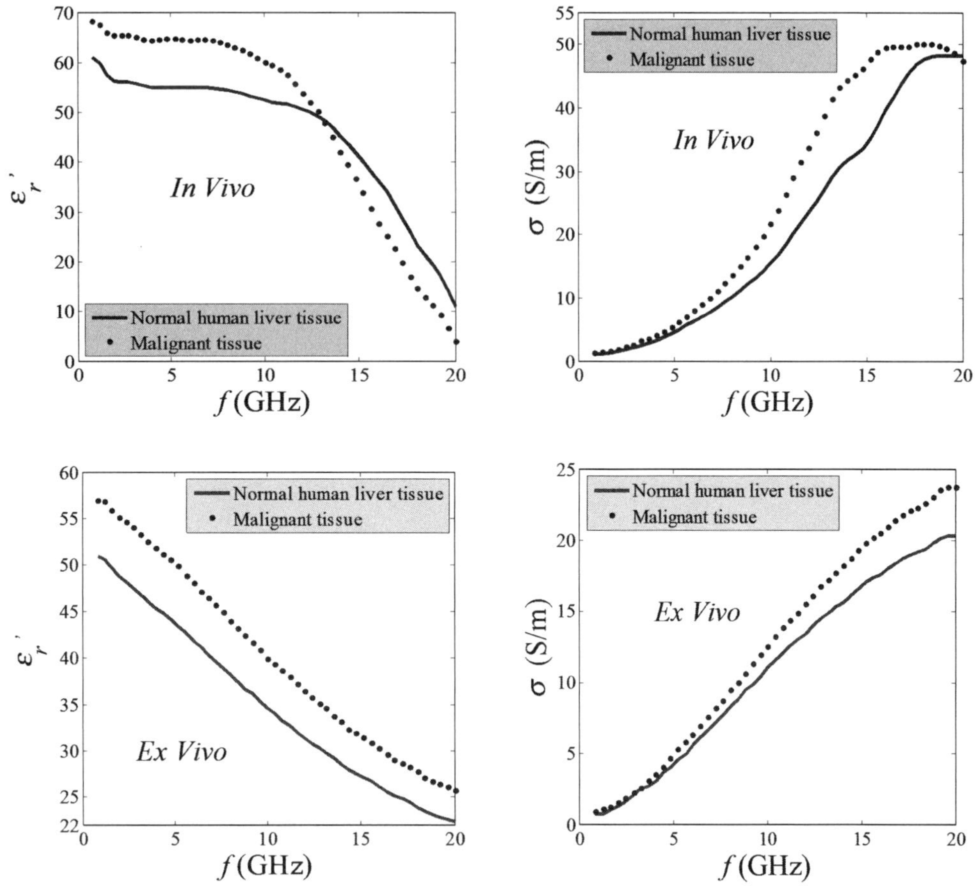

Figure 9.31 Dielectric measurement of in vivo and ex vivo normal and malignant human liver tissues from 0.5 to 20 GHz [47].

9.3 Agriculture and Food Processing

Most of the microwave measurements in agricultural and food processing are used to monitor the moisture content, *m.c.*, in agricultural products and foods under test. The reason is due to the polarization of water molecules contained in the agri-foods products, which are sensitive and showed a significant response when exposed to microwaves. The tendency of water is to absorb microwave energy, allowing the microwave device to be used as a measuring technique to sense the moisture content, *m.c.*, in agri-foods products containing water. In fact, the mechanisms of water molecules in agri-food specimens are different when exposed to different ranges of operating frequency f as shown in Figure 9.32. However, only three states of water contained in the specimen are typically analyzed and discussed in the measurements, namely, ionic conductivity state, bound water state, and free water state.

Figure 9.33 shows the relationship between the loss factor ε_r'' mechanism of water molecules and the operating frequency, f, which has provided some measurement guidelines. For instance, if the ionic conductivity σ of a specimen is desired to be measured, the low operating frequency ($f < 1$ GHz) must be used. Mechanisms

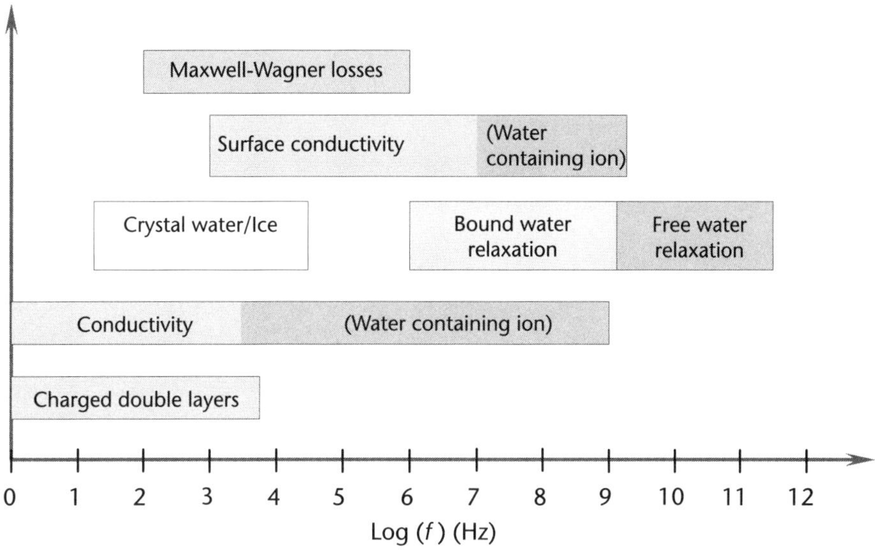

Figure 9.32 The mechanisms of water molecules when exposed to different range of operating frequency.

in water molecules are also dependent on the total volume of water contained in the agri-food specimen. Typically, bonding between water molecules and other substances in the specimen exists when the total volume of water content in the specimen is less than 30%.

9.3.1 Soil Moisture Monitoring

Soil moisture content, *m.c.*, is one of the most essential physical characteristics in agriculture, civil engineering, landscaping, irrigation engineering, and hydrology, since the consistency and workability of a clay-like soil strongly depend on its moisture content, *m.c*. The change in electrical properties can be directly correlated

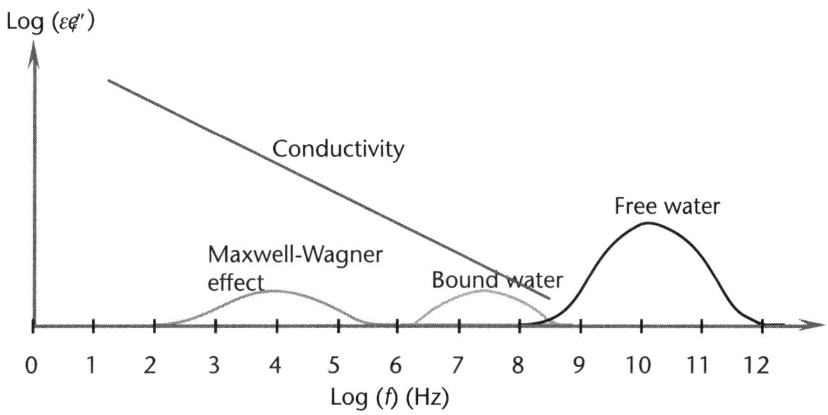

Figure 9.33 Relationship between the loss factor ε''_r mechanism of water molecules and the operating frequency, *f*.

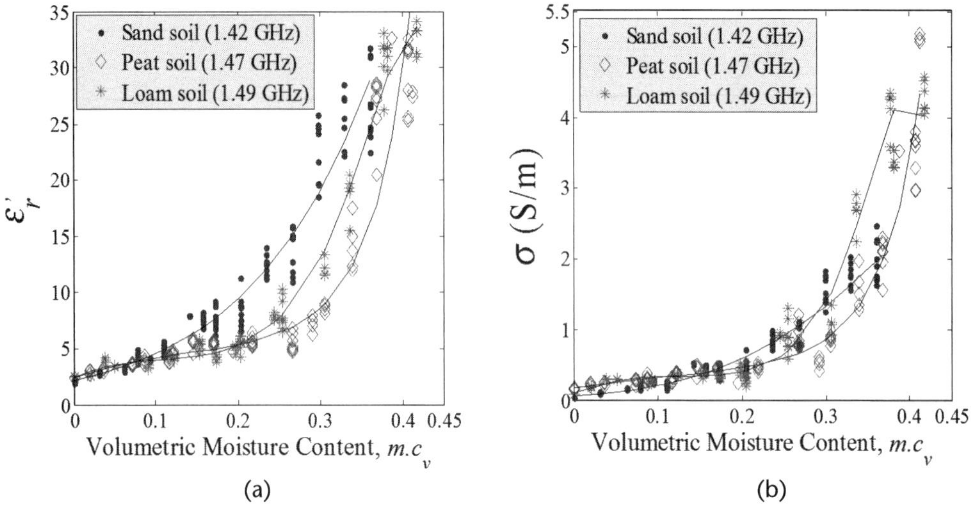

Figure 9.34 Variations in (a) relative dielectric constant ε_r' and (b) conductivity σ of soil with volumetric moisture content, $m.c_v$ at (25 ± 1)°C [48].

with a change in the actual moisture content, $m.c.$, of the soil obtained from oven drying method (direct method). The variations in the relative dielectric constant ε_r' and ionic conductivity σ of the soil with volumetric moisture content, $m.c_v$, are plotted in Figure 9.34 including a solid fitting line for measurement data. The values of ε_r and σ in Figure 9.34 are obtained by the commercial Keysight 85070D dielectric probe [48].

In general, the water contained in the soil can be in bound water or free water conditions. The bound water fraction is not the same for different kinds of soil, which depends on its texture. Loam is a soil that has a high water-holding capacity as compared to other soils, since it contains a lot of organic matter. The organic matter in loam soil can hold more water than mineral substances. Thus, the water-holding capacity or maximum fraction of bound water can be improved by adding organic matter. The measured reflected voltage V_r (in volts) using a monopole sensor for various soil types in the range from 0 $m.c_v$ to 0.45 $m.c_v$ is plotted in Figure 9.35(b). The small monopole driver from coaxial line is buried into the soil, and its reflected voltage V_r is measured and correlated with corresponding soil moisture content, $m.c_v$, which obtained from the oven dying method.

In Figure 9.35, peat soil is the soil has a high water-holding capacity up to 0.45 $m.c_v$; in fact, it contained a lot of organic matter (0.643%) as compared to loam soil (0.335%) and sand soil (0.058%). The organic matter in peat soil can hold more water than mineral substances. Thus, the sand soil is saturated with water above 0.3 $m.c_v$.

9.3.2 Rubber Latex Quality Measurements

Natural rubber latex is a complex biological product containing 50–80% water, 15–45% rubber hydrocarbon, and approximately 2–4% nonrubber constituents [49]. In standard manufacturing, the latex product is required to have dry rubber content (DRC) of 60% and approximate 39% of moisture content, $m.c$. Thus, the

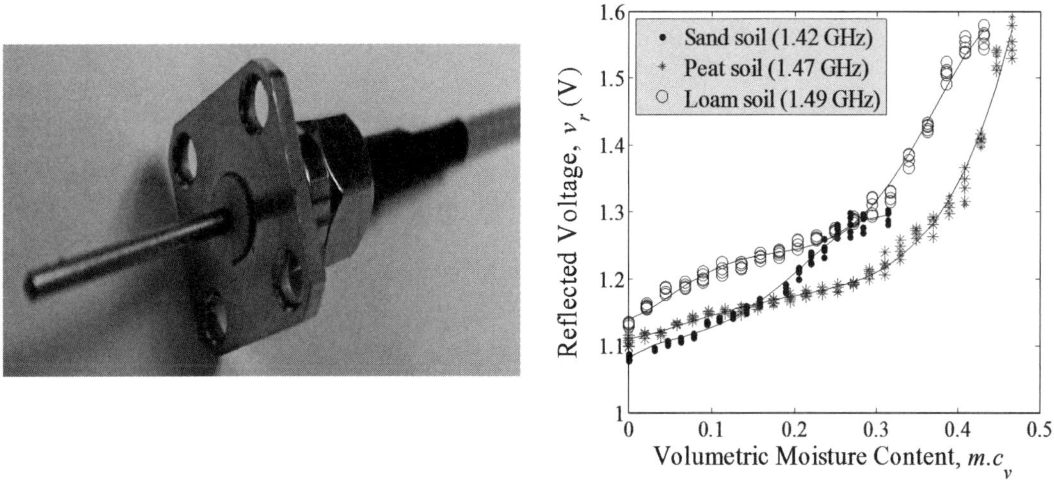

(a) (b)

Figure 9.35 (a) Monopole sensor. (b) Variations in reflected voltage V_r with volumetric moisture content $m.c_v$ of three kinds of soils at (25 ± 1)°C [48].

dry rubber content (DRC) in the latex can be indirectly estimated from moisture content measurement, since there are only two major components in the product. The variations in relative dielectric constant ε_r' and loss factor ε_r'' of the rubber latex with frequency f for various moisture content $m.c.$ values are plotted in Figure 9.36.

Ansarudin et al. [49] used an insulated monopole radiator [Figure 9.37(a)] to monitor the moisture content in the rubber latex. The insulated monopole with 15 mm of length is fully immersed in the rubber sample and the moisture content, $m.c.$ in the sample is predicted based on the change of the measured phase shift

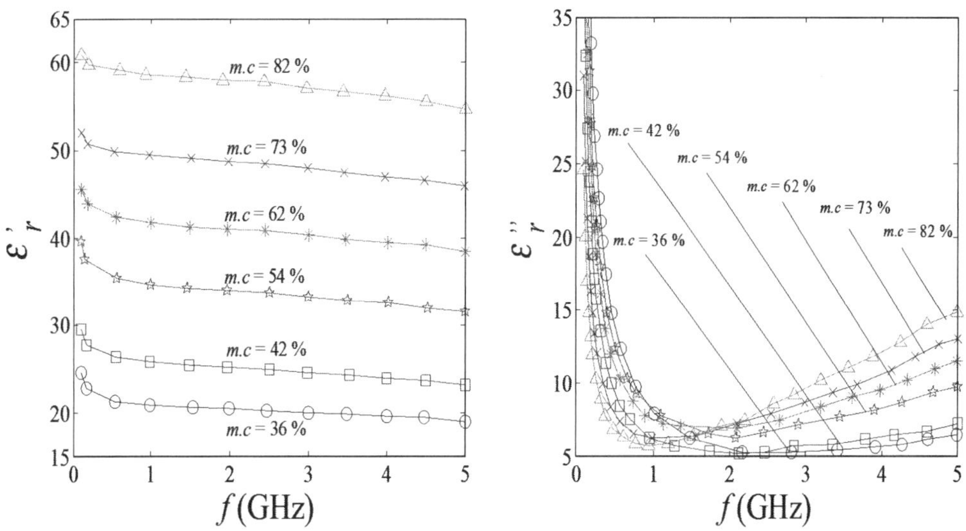

(a) (b)

Figure 9.36 Variation in (a) dielectric constant ε_r' and (b) loss factor ε_r'' with frequency at room temperature (25 ± 1)°C for various values of water content, $m.c.$ in rubber latex [49].

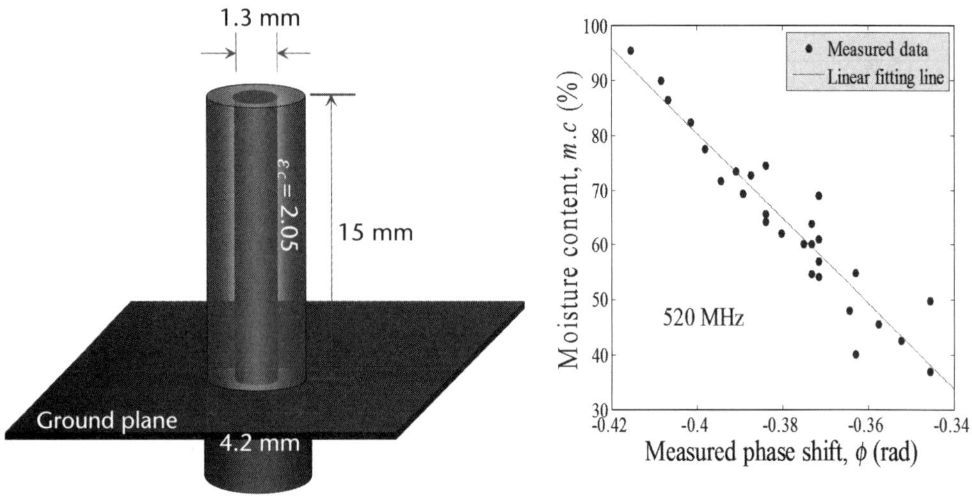

(a) (b)

Figure 9.37 (a) Dimensions of the insulated monopole radiator. (b) The moisture content *m.c.* in the rubber latex versus measured phase shift f using an insulated monopole [49].

ϕ. The variation in moisture content *m.c.* with the corresponding measured phase shift ϕ (in radians) is shown in Figure 9.37(b). Clearly, the phase shift ϕ is linearly correlated with the moisture content *m.c* in the rubber latex.

9.3.3 Oil Palm Fruit Ripeness Detector

Palm oil is obtained from the mesocarp of the oil palm fruits. Normally, a palm tree takes about three years to produce its first fruit bunch. In Ariffin et al. [50], it was found that the amount of moisture content *m.c.* is higher at the early stage of fruit development. The excess water in the mesocarp is reduced during fruit ripening, which coincides with oil accumulation occurring approximately 14 to 15 weeks after anthesis. The amount of water in fresh mesocarp is about 80%, which decreases rapidly to about 30% to 40% in the ripe fruit at 20 to 24 weeks after anthesis. The decrease in moisture content *m.c.* from 15 to 20 weeks after occurs in relation with the increases accumulation of oil in the mesocarp as shown in Figure 9.38(a) [51]. When the oil content reaches the maximum, the fruit becomes loose and falls to the ground. This is the optimum time for oil palm harvesting. In addition, Hartley [1] proposed that the relationship between oil and moisture content *m.c.* in a mesocarp may be expressed as:

$$\text{Oil} = 87.38 - 1.08\ \text{Water} \tag{9.21}$$

where *Oil* and *Water* are the percentage of oil per mesocarp and percentage of water per mesocarp, respectively. Therefore, the close relationship between the moisture and oil contents in mesocarp gives a possibility of using a percentage of moisture content *m.c.* as a parameter to gauge the mesocarp ripeness. The relationship

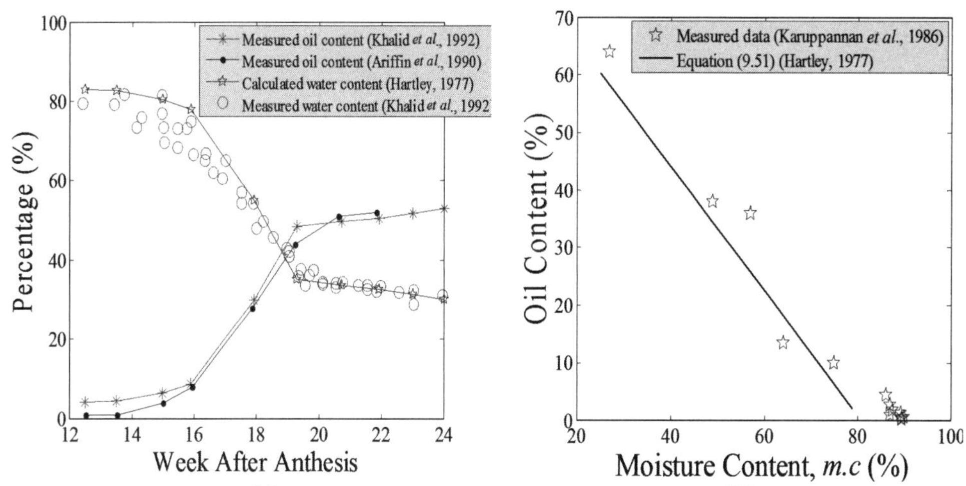

Figure 9.38 (a) Variation in moisture content *m.c.* and oil content with weeks after anthesis. (b) The relationship between the oil content and moisture content *m.c.* in oil palm mesocarp.

between the oil content and moisture content *m.c.* in oil palm mesocarp is shown in Figure 9.38(b).

Figure 9.39 shows the moisture measurement of oil palm fruits using an open-ended coaxial radiator. Part of the fresh mesocarp of each fruit was sliced in a longitudinal direction to ensure good contact between the surface of the mesocarp and the open-ended coaxial radiator.

The relationship between both dielectric constant ε_r' and loss factor ε_r'' of oil palm mesocarp and frequency for various percentage of water content *m.c* was shown in Figure 9.40. The dielectric constant ε_r' is measured using a commercial Agilent 85070E dielectric probe. From Figure 9.40, dielectric properties are quite linear at a moisture content below 40%, suggesting that at low moisture, there is

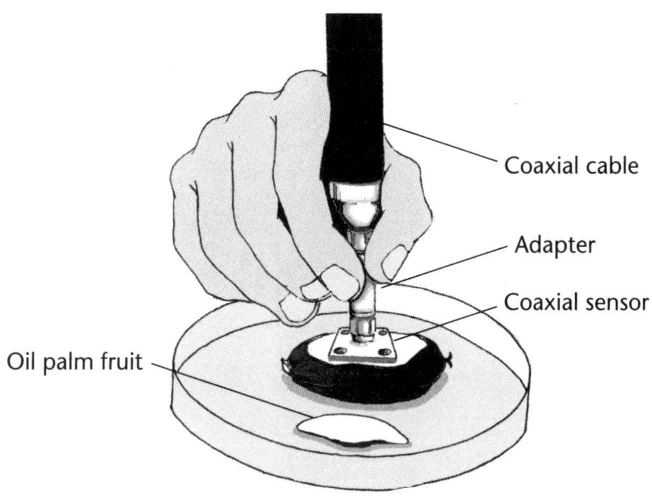

Figure 9.39 Moisture measurement of oil palm fruit using a coaxial radiator [52].

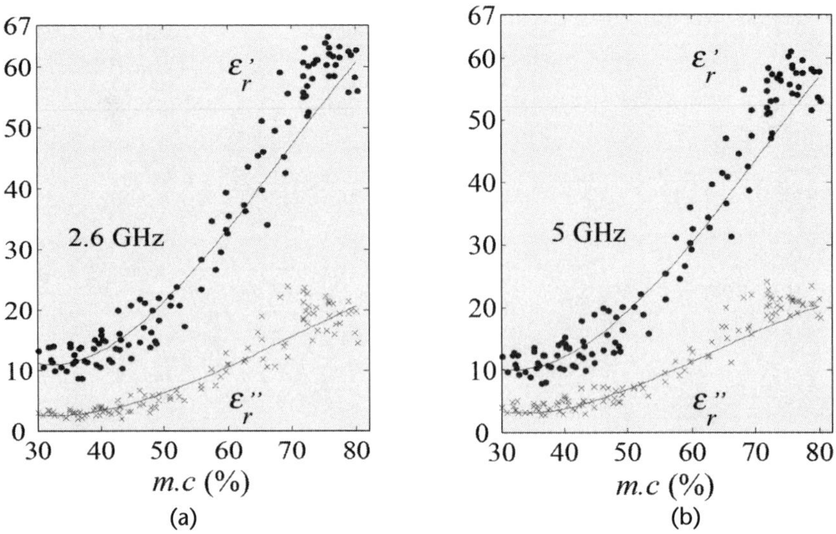

Figure 9.40 Variation in dielectric constant ε_r' and loss factor ε_r'' with moisture content *m.c.* of 25°C [51].

higher oil-water emulsion concentration. The higher uncertainty measurements appeared due to the water accumulated on the surface between sensor and fruit when the sensor is pressed firmly into the sample fruits that contained moisture above 70%. Based on Figure 9.41, the relationship between the dielectric constant ε_r' and the percentage of moisture content in mesocarp is listed in Table 9.11. In all measured fruits, the loss factor ε_r'' decreased at a low frequency, which reached a broad minimum between 1 and 3 GHz and then increased again, as shown in Figure 9.41. This behavior is influenced by ionic conductivity σ at lower frequencies, since dielectric properties in agriculture are believed to be primarily dependent on water activity and ionic conductivity of fluids contained in their cellular structures. The oil palm mesocarp has a higher relaxation time τ below a frequency of 10 GHz for different water contents *m.c*, while the relaxation of oil palm mesocarp

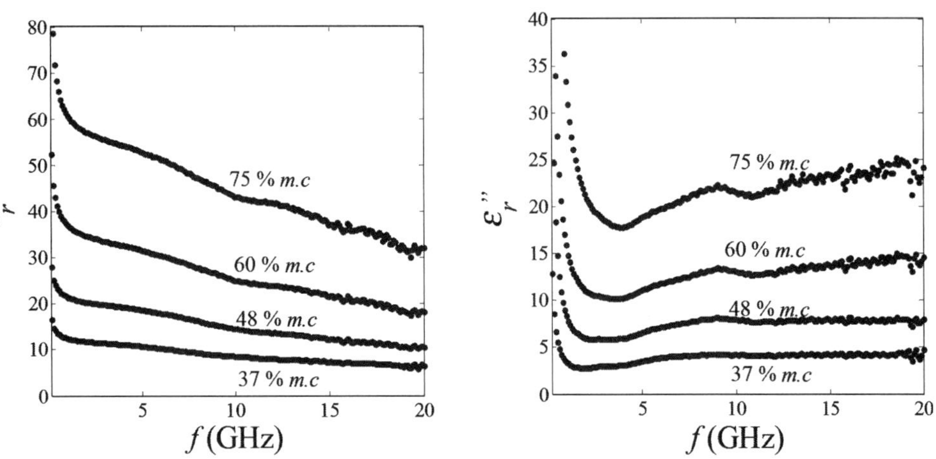

Figure 9.41 Variation in dielectric constant ε_r' and loss factor ε_r'' with frequency at room temperature (25 ± 1)°C for various values of water content *m.c.* in oil palm fruits [52].

Table 9.11 Relationship Between the Percentage of Water Content, *m.c.*, the Dielectric Constant, ε_r', and Conductivity, σ, of Oil Palm Mesocarp for 3 GHz at (25 ± 1)°C

Water Content, m.c. (%)	*Dielectric Constant,* ε_r'	*Conductivity,* σ $(\Omega^{-1}m^{-1})$
30–40%	10–20	0.11–0.17
40–50%	20–25	0.17–0.38
50–60%	25–35	0.38–0.7
60–70%	35–55	0.7–1.1
70–80%	55–65	1.1–1.7

Source: [52].

is approximately 8.377 ps relaxation of free water for frequencies above 10 GHz. The high value of relaxation time τ of oil palm mesocarp at low frequencies is due to the polarization orientation of polar molecules, which is slowed down by the binding of water molecules in the mixtures, such as the water absorption by a fiber and the capillary water between the fibers. However, the bound molecules are broken down above 10 GHz. In fact, the overall effect on water structure in oil palm mesocarp is extremely complex.

9.3.4 Palm Oil Quality Monitoring

The oil palm is a unique plant that yields two different vegetable oils, namely, the red crude palm oil (CPO) and palm kernel oil (CPKO). The CPO is extracted from the mesocarp of the palm fruit, while the yellow CPKO is obtained from the kernel as illustrated in Figure 9.42. Both palm oils are varied in their chemical compositions and physical properties as well as their industrial applications. The CPO with low free fatty acid (FFA) is less susceptible to oxidation, good bleachability, low trace metals, and low insoluble impurities and the moisture content *m.c.* is labeled as a good quality oil. In general, the standard quality CPO is defined at 5% FFA (maximum), 0.2% moisture (maximum), and 0.05% impurities (maximum). Hence, it is known that the water content in the oil influences the rate of FFA formation. This is due to the formation of FFA involving the hydrolysis reaction, which can only be carried out in the water or at the oil-water interface. When the water content in oil is greater than the solubility, the water will exist in droplets or an emulsion state that influences the rate of the hydrolysis reaction.

Figure 9.42 Red crude palm oil (CPO) is derived from the mesocarp of the palm fruit while the palm kernel oil (CPKO) is obtained from the kernel of the palm fruit [53].

However, the solubility of water in palm oil is very low, as in neutralized palm oil it is about 0.15% (maximum) at room temperature, 25°C. For palm oil containing 2.43% and 6.44% of FFA, the water solubility will increase to 0.2% and 0.25%, respectively. However, the total moisture content within the palm oil over 26 days is about 0.60%. In general, the rate-controlling factor means that the water content must be kept below the solubility of water in palm oil, in order to reduce the hydrolysis of the oil. Therefore, the water content in the oil is preferably to be kept as low as possible to maintain its stability during storage.

CPKO is obtained from the kernel inside the nut of the palm fruit and it is lauric-rich oil, which contains no carotene. It is saturated oil and more resistant to oxidation. The mechanisms of chemical hydrolysis and micro-organism activity of CPKO are mainly influenced by the water content as in CPO. The rate of FFA formation in CPKO is slightly higher due to the shorter hydrocarbon chains in CPKO. In addition, the water content below 7% inhibits the microbial activity in CPKO. Moreover, by keeping water content below 7%, the formation of FFA will be kept below 1.5% after 6 months of storage of CPKO. Generally, the standard quality of CPKO is defined at 5% FFA (maximum), 7% moisture (maximum), and 6% impurities (maximum). The CPO and CPKO quote requirements are listed in Tables 9.12 and Table 9.13, respectively.

In order to monitor the moisture content in the palm oil (*Elaeis guinensis*), a rapid and nondestructive dielectric measurement using commercial a coaxial probe and dielectric mixture modeling is analyzed macroscopically in this study. The

Table 9.12 Standard Quality CPO Based on Malaysian Oil Palm Growers' Council (MOPGC), Malaysian Standard (MS), and Palm Oil Research Institute of Malaysia (PORIM)

	Standard quality of crude palm oil (CPO)		
Factors	*Malaysian Oil Palm Growers' Council (MOPGC)*	*Malaysian Standard (MS)*	*Palm Oil Research Institute of Malaysia (PORIM, 97/98)*
FFA (maximum)	5%	2.5%	3.2%
Moisture (maximum)	0.25%	0.2%	0.18%
Impurities (maximum)	0.25%	0.05%	0.03%

Table 9.13 Standard Quality CPKO Based on Malaysian Oil Palm Growers' Council (MOPGC), Malaysian Standard (MS), and Standard Institute of Malaysia (SIM)

	Standard Quality of Crude Kernel Palm Oil (CPKO)			
			Standard Institute of Malaysia (SIM)	
Factors	*Malaysian Oil Palm Growers' Council (MOPGC)*	*Malaysian Standard (MS)*	*Grade 1*	*Grade 2*
FFA (maximum)	5%	5%	3%	6%
Moisture (maximum)	7%	7%	0.35%	0.5%
Impurities (maximum)	6%	6% (minimum)	0.35%	0.5%

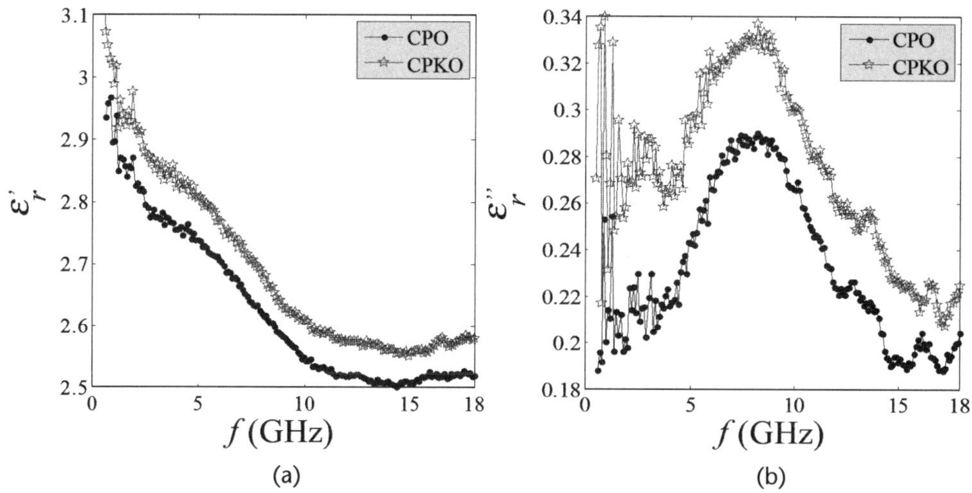

Figure 9.43 Variation in (a) dielectric constant ε'_{r_oil} and (b) loss factor ε''_{r_oil} with frequency at (25 ± 1)°C for pure crude palm oil (CPO) and crude palm kernel oil (CPKO) [53].

measured dielectric constant ε'_{r_oil} and loss factor ε''_{r_oil} of CPO and CPKO versus operation frequency f are plotted in Figure 9.43. Obviously, the CPKO naturally has higher values of relative permittivity compared to the CPO for entire operating frequency ranges. In Figure 9.44, we have plotted the variation in effective dielectric constant ε'_{eff} with frequency for various volumes of water v_{water} in CPO. As expected, the values of the effective dielectric constant ε'_{eff} increase with increasing volume of water in CPO over the frequency range.

The minor differences between the results obtained from the measurement and calculation may due to the following factors:

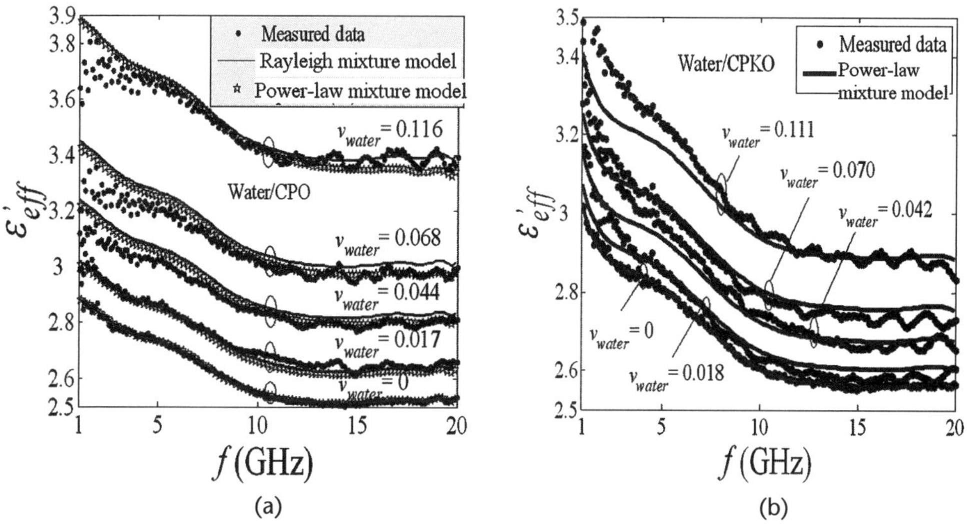

Figure 9.44 Variation in effective dielectric constant ε'_{eff} with frequency f at room temperature (25 ± 1)°C for various values of water volume in (a) CPO and (b) CPKO [53].

1. The direction of polarized electric field is not fully normal to the layer.
2. The distribution of water inclusion in the mixture is not uniform because it tends to sink in an oil environment.
3. The nonuniform distribution of water inclusion in oil environment around the sensing area.
4. The overlapping water inclusion can form the connected sets; thus, the size of water inclusion distributed in oil environment will not be uniform.

9.3.5 Rice Grain Moisture and Broken Rate Measurements

Microwave measurement techniques have gained an increased interest in the determination of moisture content *m.c.* (or dielectric properties) of bulk rice grain at microwave frequency range [54, 55]. This is because the percentage of *m.c.* (range from 9% *m.c.* to 26% *m.c.*) in the rice grain has played an important role in the aspects of marketing and storage. In marketing, the price of rice is dependent on the weight of the rice bulk, and more water content in the rice grain will increase the price of rice. For storage, it is the highest moisture content (*m.c.* < 14%) that is present in the grain mass that determines the storage duration of the rice grains, in which *m.c.* < 13% indicates a storage duration of more than 60 days [56, 57].

Recently, You et al. [57] used a millimeter, slim, open-ended coaxial probe to measure the moisture in the small grain as shown in Figure 9.45(a). The coaxial probe was fabricated from a 0.86-mm outer diameter (OD) semirigid coaxial cable with a male-type SMA plug connector. In addition, a specific probe holder was customized for the measurement of a single rice grain. The customized holder has

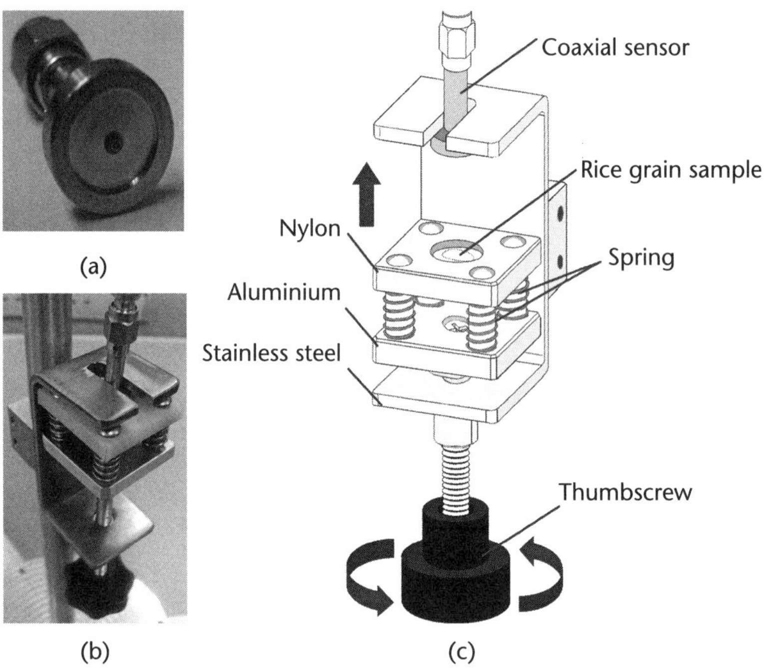

Figure 9.45 (a) The small open-ended coaxial radiator. (b) Actual configuration and (c) description of the experimental setup [57].

a movable nylon platform that was mounted on a retort stand as shown in Figure 9.45(b, c).

Figure 9.46 shows that the reflection coefficient Γ for 10 selected single rice grains from bulk rice grain, where the bulk has certain average water content, was measured using the millimeter coaxial probe. As is well known, the *m.c.* for a bulk rice grain (which consists of thousands of single rice grains) is a statistical mean value due to the slightly different $m.c_{\text{single}}$ in each single grain. Thus, the deviation of the actual $m.c_{\text{single}}$ distribution in a single rice grain is higher as compared to the

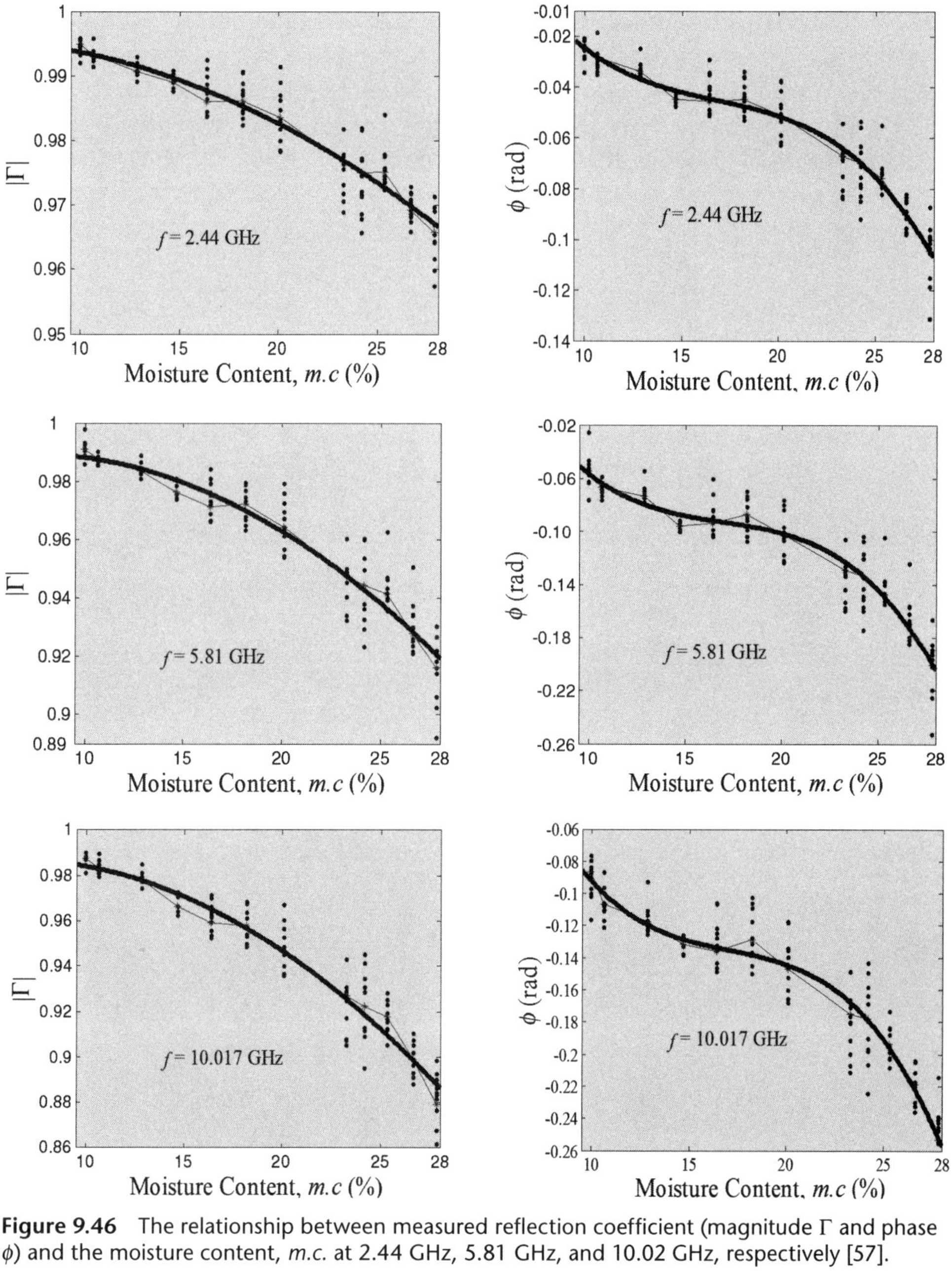

Figure 9.46 The relationship between measured reflection coefficient (magnitude Γ and phase ϕ) and the moisture content, *m.c.* at 2.44 GHz, 5.81 GHz, and 10.02 GHz, respectively [57].

average $m.c.$ of the entire bulk rice sample. This will lead to scattered measurement of the reflection coefficient Γ of the single grain in referring to the average $m.c.$ of the bulk sample as shown in Figure 9.46. However, Figure 9.46 does show a significant change in the measured reflection coefficient Γ of a single rice grain compared with the $m.c$ of bulk rice grain. The black solid line in Figure 9.46 is a regression fitting line from the average of measured reflection coefficient data.

Besides moisture content, the grade and price of the rice grain also depend on the percentage of broken grain. The rice grain that has a length less than three-fourths but more than one-fourth of the average head grain length is categorized as a broken rice. Normally, bulk rice containing broken grain for 0%–5%, 5%–20%, 20%–35%, and 35%–50% is classified as a premium grade, grade 1, grade 2, and grade 3, respectively. The single and coupling slot radiators with one input signal are introduced in You et al. [58] to measure the percentage of the broken rice. The slot radiators are constructed using commercial SMA stub panel with the radius of slot a = 0.65 mm and the length of slot h = 16 mm. The slots were driven from aluminum ground and covered by acrylic as shown in Figure 9.47. For the coupling

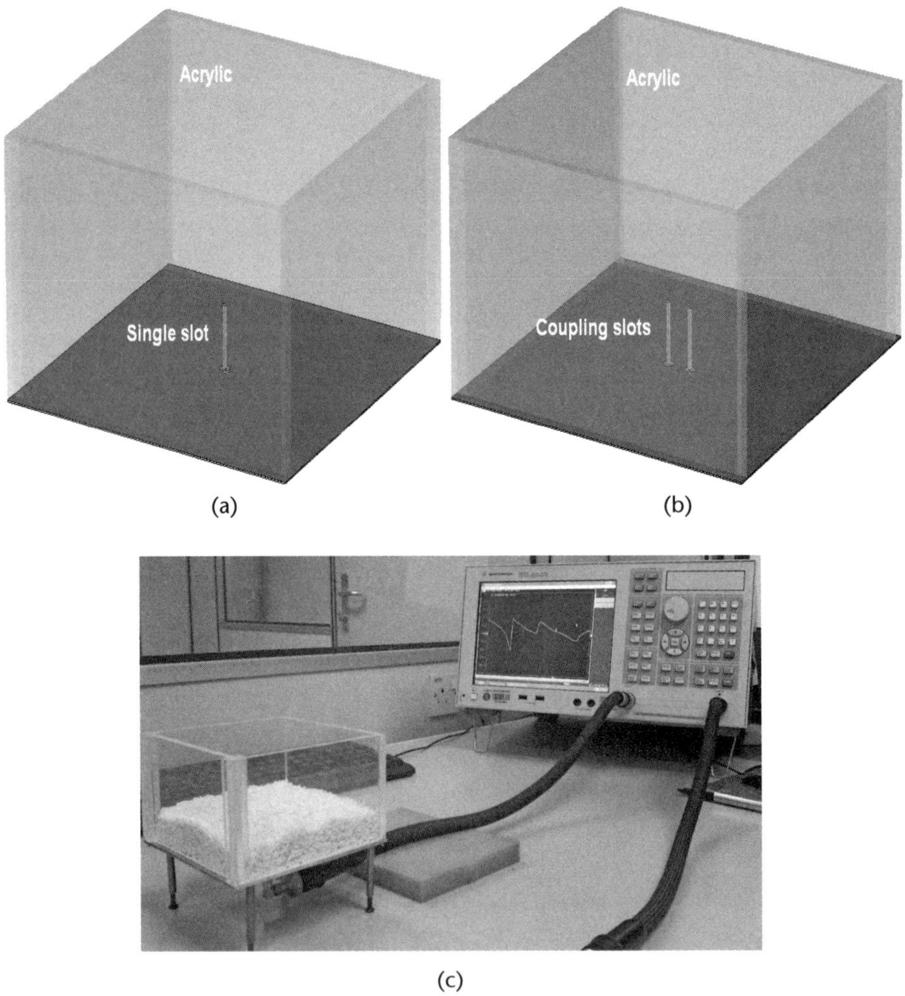

Figure 9.47 Configurations of (a) single and (b) coupling slots radiators. (c) Experimental setup [58].

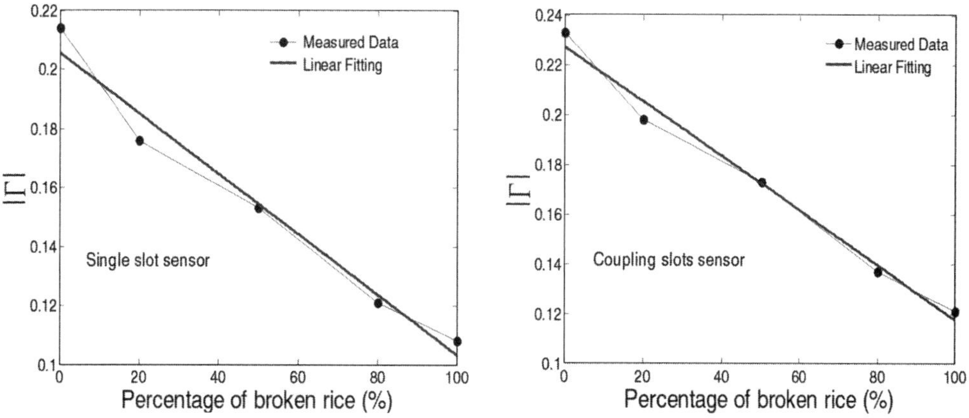

Figure 9.48 Variation in magnitude reflection coefficient $|\Gamma|$ with the percentage of broken grain in the rice sample at 13.5 GHz at room temperature 25°C [58].

slots radiator, the separation distance between the coupling slots is 9 mm and one of the slots was terminated. The density of sensing field was strengthened using the coupling slots radiator.

In measurements, the measured reflection coefficient will vary with changes in the density of rice grains in the sample. The electromagnetic wave is generated from coaxial radiator and radiates into the grain sample. The measurement technique is based on the change of air gap density in the bulk rice. Since the air gap between broken grains is smaller as compared to the normal rice grains, the measurement technique is based on the change of air gap density in the bulk rice. Thus, the density of rice grains in the mixed sample is high. A short wavelength signal (high operating frequency) is required to enhance the air gap sensitivity for broken rice measurement and reduces the penetration of energy into the rice grains. In fact, the air gap density in the grain sample is inversely proportional to the density of radiated field which is covering inside the grain sample. Figure 9.48 shows that the measured reflection coefficient $|\Gamma|$ of the rice grain varies with the percentage of broken rice in the samples at 13.5 GHz.

9.3.6 Fruit Sweetness Detector

The correlation between the dielectric properties and fruit sweetness (or fruit maturity) using an open-ended coaxial probe has been intensively studied by Nelson et al. [59–61] and Guo et al. [62, 63] as shown in Figure 9.49.

The major component of soluble solids content (SSC) in fruit is the sugar content. Moreover, the SSC is linearly correlated to the moisture content $m.c.$ in the fruit as shown in Figure 9.50; thus, the sugar content in the sweetness of the fruit can be indirectly determined by measuring the moisture content of the fruit. From Figure 9.50, it appears that most of the watermelons are sweeter than the honeydew melons, since the average of the SSC for watermelon is almost 12%, while the average of the SSC for honeydew melon is approximately 8%. The relationship between the loss tangent, tan δ, and the SSC for the watermelon at 1.8 GHz is given in Figure 9.51.

(a) (b)

Figure 9.49 Dielectric measurements of the (a) honeydew melon and (b) watermelon surfaces using an open-ended coaxial probe [61, 62].

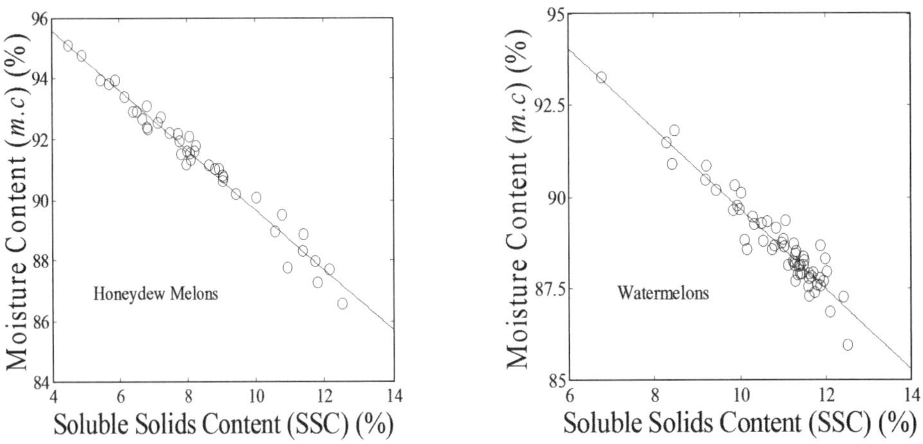

(a) (b)

Figure 9.50 Relationship between the moisture content *m.c.* and SSC for (a) honeydew melon and (b) watermelon [61, 62].

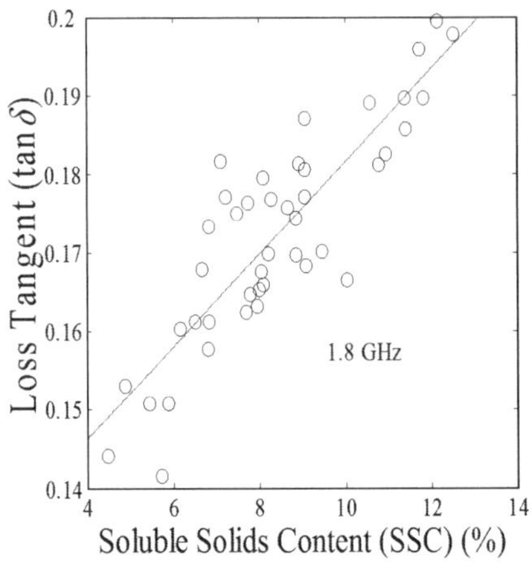

Figure 9.51 The linear relationship between the loss tangent, tan δ, and SCC of the watermelons [61].

9.4 Industrial and Scientific Materials Processing

9.4.1 Metal Crack Detector

Fatigue cracks occurring on metal are mainly due to repeatedly applied loads on the metal. If the loads are above a certain threshold, microscopic cracks will begin to form on the metal surface. When the crack on the surface metal reaches a critical size, the crack will propagate suddenly and the metal will fracture. Recently, an open-ended coaxial probe was used to detect fatigue cracks on metal surfaces [64–66], since its reflection measurement is very sensitive to any small air gap that exists between the probe aperture and the test metal surface as shown in Figure 9.52. The small coaxial probe aperture can be used to detect very tight cracks on the metal surface. In addition, the radiation field at the probe aperture has no specific polarization (omnidirectional radiation pattern); thus, the reflection measurement using a coaxial probe is independent of the relative orientation of the crack [64]. Detection of the cracked metal can be correlated to the magnitude or phase of the reflected signal (reflection coefficient, reflected power, or reflected voltage). The measurement normally operates at a high frequency (up 10 GHz) in order to increase the sensitivity of the measurement.

In the measurement, the coaxial probe aperture has scanned through the surface metal and the crack is detected when an evident change of the reflected signal occurs. In order to analyze conveniently, the crack below the center of the inner conductor is considered to be located at the relative position x = 0 mm as shown in Figure 9.53.

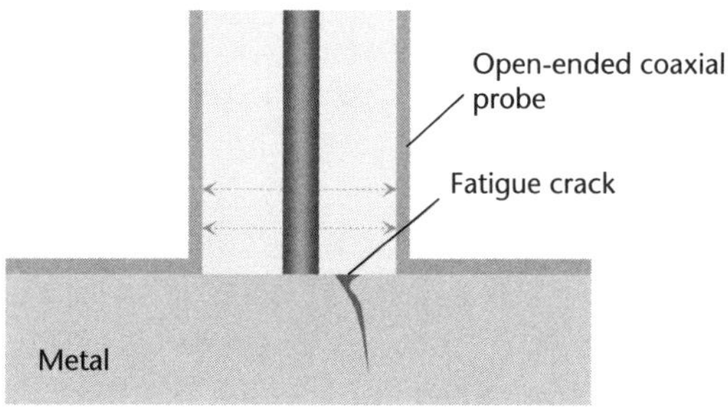

Figure 9.52 Cross-sectional view of the junction between an open-ended coaxial probe and a surface crack in a metal.

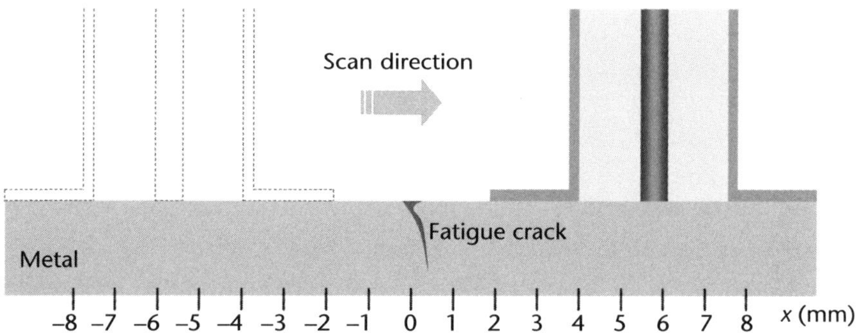

Figure 9.53 The scan direction of the coaxial probe and the relative position x of a crack from the inner conductor of the coaxial probe.

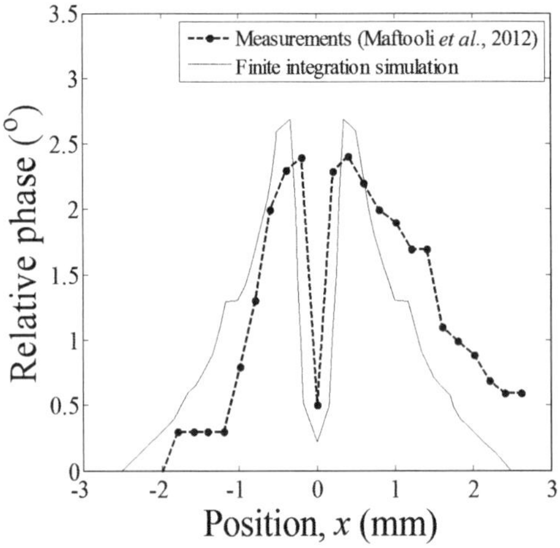

Figure 9.54 Variation in measured phase with relative position x of the crack in the metal at 10 GHz [65].

Figure 9.54 shows the phase of the reflection coefficient at the probe aperture versus scanning the relative distance x over a crack. The phase measurement is done using a coaxial probe with a radius of an inner conductor, a = 1.47 mm, and an outer conductor, b = 4.4 mm, at 10 GHz and size of the crack given as width = 20 mm, length = 0.8 mm, and depth = 10 mm. A high response of the reflected phase can be observed when the crack surface is exactly contacted with the inner conductor of the probe at x = 0 mm.

In fact, the response of the reflected signal is dependent on the size of the crack and also the size of the coaxial probe. For instance, the comparison of the use between a large and a small coaxial probe to detect the crack with width = 34 μm and depth = 12 mm at 12 GHz is shown in Figure 9.55 [66]. Clearly, it is shown that the probe with a large size (diameter of an inner conductor, $2a$ = 3.2 mm) gives two peaks on the measurement results when the probe is scanned through the crack as shown in Figure 9.55(a). When the crack is in the vicinity of the edge of the inner conductor, it causes a peak in the measured reflected voltage. The second response will occur when the probe aperture is moved over the crack and reaches another edge of the inner conductor. However, only one peak of the measured reflected voltage near the location of x = 0 mm, has occurred when the small coaxial probe (diameter of an inner conductor, $2a$ = 0.85 mm) is used as shown in Figure 9.55(b). In fact, the two distinct peaks have combined with each other due to the small size of inner conductor of the coaxial probe.

9.4.2 Electric Field and Magnetic Field Probes

The electric field and magnetic field probes (near-field probes) are commonly used in electromagnetic compatibility (EMC) testing, such as SAR measurement (Figure 9.56) and adjacent electromagnetic field distribution measurement of the IC package.

Those testing probes are mostly created by the coaxial radiator with an inherent electrostatic shield. Most of the EMC test equipment manufacturers have their own probe products. Normally, the electric field probes have a configuration which

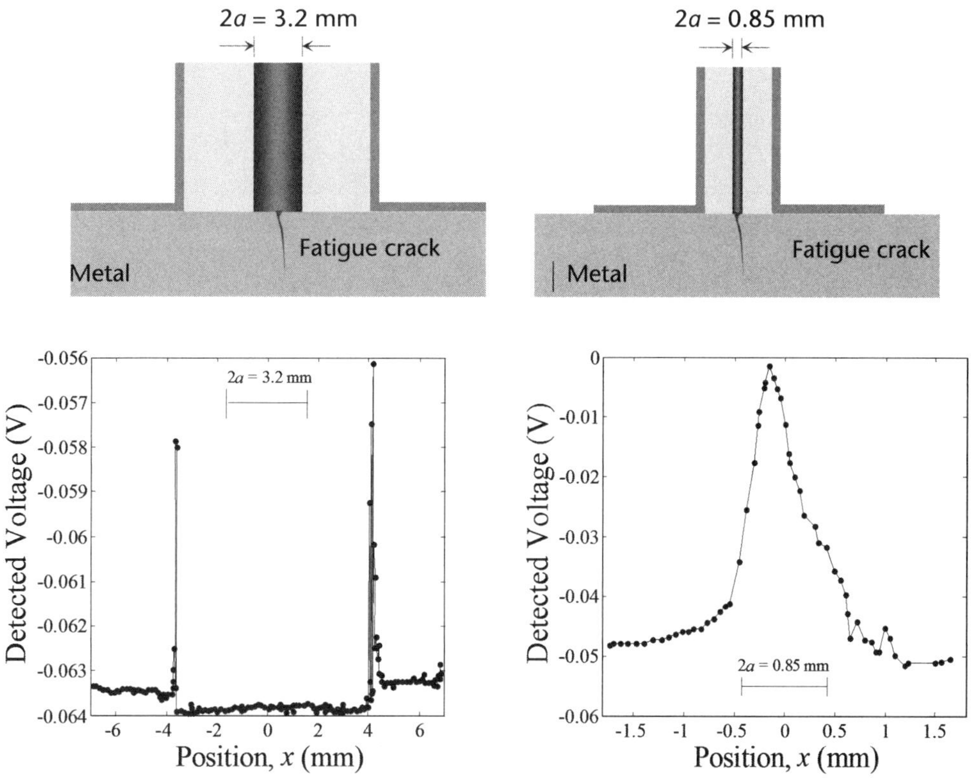

(a) (b)

Figure 9.55 Measured reflected voltage for a 34 μm of crack using (a) large coaxial probe and (b) small coaxial probe [66].

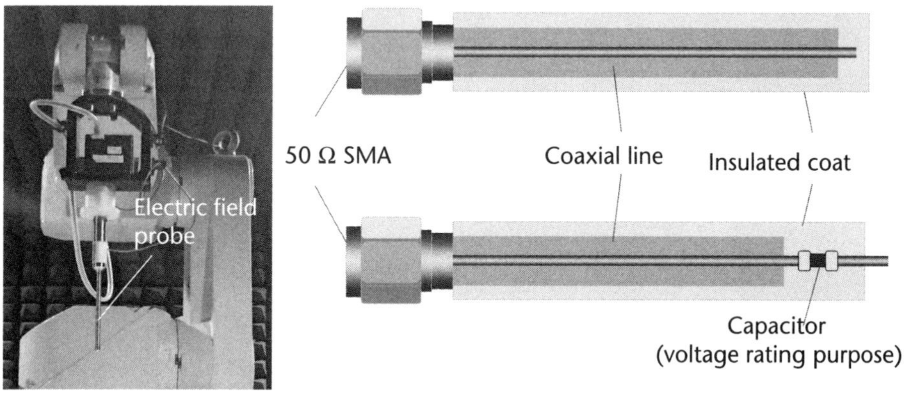

(a) (b)

Figure 9.56 (a) The electric field probe attached to the DASY4 robot system for SAR distribution measurements. (b) Two examples of a simple electric field probe.

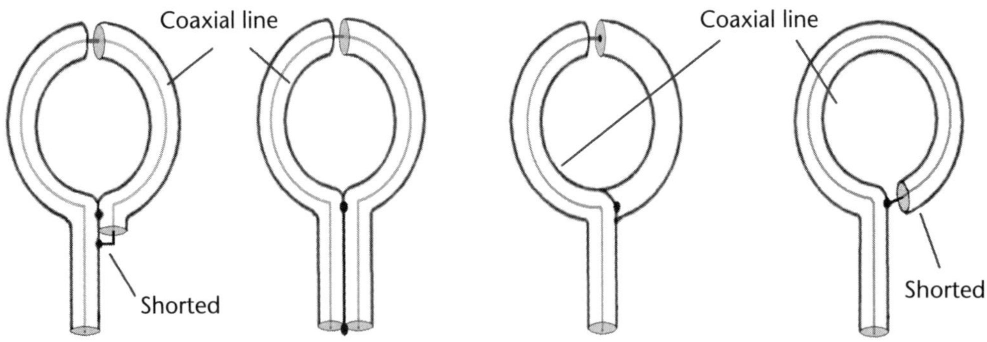

Figure 9.57 Four types of magnetic field probes.

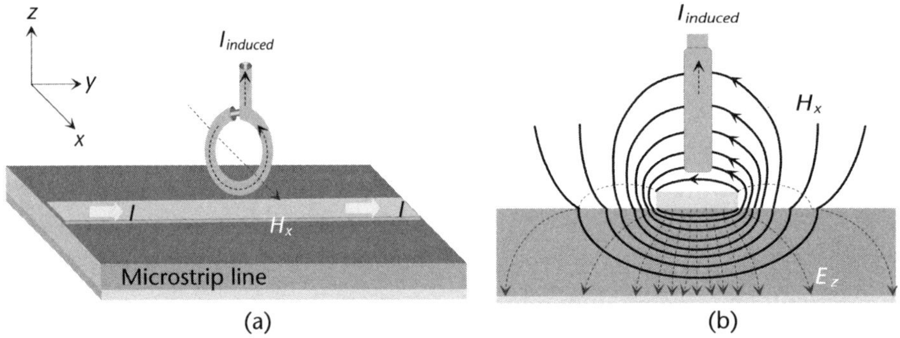

Figure 9.58 (a) The magnetic field distribution measurement of PCB circuit using magnetic field probe. (b) Cross-sectional view of the electromagnetic field distribution for the circuit.

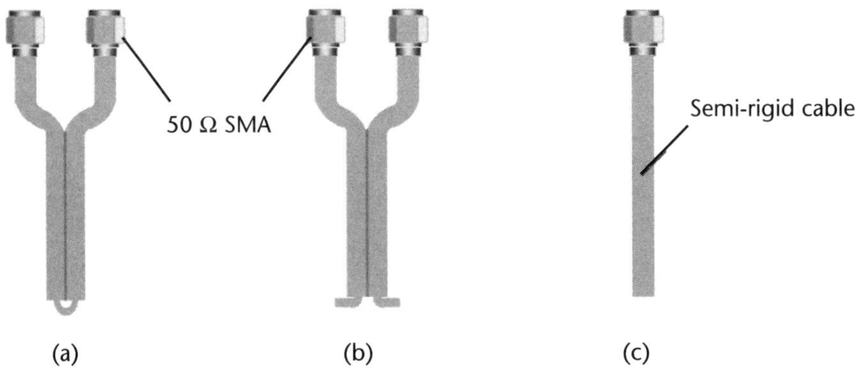

Figure 9.59 (a) Magnetic field probe. (b) Dipole-type electric field probe. (c) Open-ended coaxial-type electric probe [67].

is similar to insulated monopole, dipole or open-ended coaxial probe as shown in Figure 9.56(b). However, magnetic field probes are created in the form of loop as shown in Figure 9.57.

For instance, a current that flows in the microstrip line will produce a magnetic field H_x that surrounds it as shown in Figure 9.58. The magnetic field H_x passing through the probe loop and generating an induced voltage V_{induced} (or an induced current, $I_{induced}$) in the probe is proportional to the rate of change of magnetic flux through a circuit loop, according to Faraday's law. Finally, the induced current $I_{induced}$ can be measured by a spectrum analyzer or other EMF instrument.

In fact, there are many research efforts and technical reports describing the design of the simple and low-cost electric and magnetic field probes [67, 68]. For instance, Figure 9.59 shows the field testing probes designed using semirigid coaxial cable, which is described by Baudry et al. [67].

References

[1] Hartley, C. W. S., *The Oil Palm*, London: Longman Group Limited, 1977.

[2] Sokol, L., "JIB Antennas Will Form Array for GPS III Satellites," *Microwave & RF*, 2014, p. 24.

[3] Sokol, L., "Self-Deploying Monopole Antennas Seize Greater Coverage of the Seas," *Microwave & RF*, 2014, p. 24.

[4] Fujimoto, K., and H. Morishita, *Modern Small Antennas*, New York: Cambridge University Press, 2013.

[5] Yeh, C. Y., and R. Zoughi, "A Novel Microwave Method for Detection of Long Surface Cracks in Metals," *IEEE Trans. on Instrum. Meas.*, Vol. 43, 1994, pp. 719–725.

[6] Schantz, H., *The Art and Science of Ultrawideband Antennas*, Norwood, MA: Artech House, 2005.

[7] DuHamel, R. H., and D. E. Isbell, "Broadband Logarithmically Periodic Antenna Structures," *IRE National Convention Record*, Part I, 1957, pp. 119–128.

[8] Yagi, H., and S. Uda, "Projector of the Sharpest Beam of Electric Waves," *Proceedings of the Imperial Academy*, Vol. 2, 1926, pp. 49–52.

[9] Barbano, N., "Log Periodic Yagi-Uda Array," *IEEE Trans. on Antennas Propagat.*, Vol. 14, 1966, pp. 235–238.

[10] Balmain, K. G., and J. N. Nkeng, "Asymmetry Phenomenon of Log-Periodic Dipole Antennas," *IEEE Trans. on Antennas Propagat.*, Vol. 24, 1976, pp. 402–410.

[11] Collins, B. S., "VHF and UHF Antennas for Communications and Broadcasting," Ch. 29 in *Antenna Engineering Handbook*, 4th ed., J. L. Volakis, (ed.), New York: McGraw-Hill, 2007.

[12] Lindenblad, N. E., "Wide Band Antenna," U.S. Patent 2,239,724, April 1941.

[13] Poggio, A. J., and P. E. Mayes, "Pattern Bandwidth Optimization of the Sleeve Monopole Antennas," *IEEE Trans. on Antennas Propagat.*, Vol. 14, 1966, pp. 643–645.

[14] Carter, P. S., "Shortwave Antenna," U.S. Patent 2,175,252, October 1939.

[15] Papas, C. H., and R. King, "Input Impedance of Wide-Angle Conical Antennas Fed by a Coaxial Line," *Proceedings of the IRE*, Vol. 37, 1949, pp. 1269–1271.

[16] Shen, Z. X., and J. P. Wang., "Top-Hat Monopole Antenna for Conical-Beam Radiation," *IEEE Trans. on Antennas Propagat. Letters*, Vol. 10, 2011, pp. 396–398.

[17] Simpson, T. L., "The Disk Loaded Monopole Antenna," *IEEE Trans. on Antennas Propagat.*, Vol. 52, 2004, pp. 542–550.

[18] Stöhr, W., "Broadband Ellipsoidal Dipole Antenna," U.S. Patent 3,364,491, January 1968.

[19] Kraus, J. D., "Helical Beam Antenna," *Electronics*, Vol. 20, 1947, pp. 109–111.

[20] Kraus, J. D., "The Helical Antenna," *Proceedings of the IRE*, Vol. 37, 1949, pp. 263–272.

[21] Wheeler, H. A., "Helical Antenna for Circular Polarization," *Proceedings of the IRE*, Vol. 35, 1947, pp. 1484–1488.

[22] Wikipedia, "Helical Antenna," Wikimedia Foundation, January 2015.

[23] Drozd, B., and W. T. Joines, "Comparison of Coaxial Dipole Antennas for Applications in the Near-Field and Far-Field Regions," *Microwave Journal*, Vol. 47, 2004, pp. 160–176.

[24] Stuart, H. R., and A. Pidwerbetsky, "Electrically Small Antenna Elements Using Negative Permittivity Resonators," *IEEE Trans. on Antennas Propagat.*, Vol. 54, 2006, pp. 1644–1653.

[25] Mongia, R. K., and P. Bhartia, "Dielectric Resonator Antennas, A Review and General Design Relations for Resonant Frequency and Bandwidth," *International Journal of Microwave and Millimeter-Wave Computer-Aided Engineering*, Vol. 4, 1994, pp. 230–247.

[26] Ittipiboon, A., A. Petosa, and S. Thirakoune, "Bandwidth Enhancement of a Monopole Using Dielectric Antenna Resonator Loading," *Proc. ANTEM Conf.*, Montreal, QC, Canada, August 2002, pp. 387–390.

[27] Lapierre, M., et al., "Ultra Wideband Monopole/Dielectric Resonator Antenna," *IEEE Antennas and Wireless Components Letters*, Vol. 15, 2005, pp. 7–9.

[28] Guha, D., et al., "Improved Design Guidelines for the Ultra Wideband Monopole-Dielectric Resonator Antenna," *IEEE Antennas and Wireless Propagation Letters*, Vol. 5, 2005, pp. 373–376.

[29] Zivkovic, I., "Dielectric Loading for Bandwidth Enhancement of Ultra-Wide Band Wire Monopole Antenna," *Progress in Electromagnetics Research C*, Vol. 30, 2012, pp. 241–252.

[30] Mongia, R. K., et al., "Electric Monopole Antenna Using a Dielectric Ring Resonator," *Electron Lett.*, Vol. 29, 1993, pp. 1530–1531.

[31] Higashino, T., and M. Okada, "A Wireless Sensing Technique Based on Channel Estimation in Leaky Coaxial Cable Antenna System," *General Assembly and Scientific Symposium (URSI GASS)*, 2014, pp. 1–4.

[32] Nishikawa, K. I., et al., "A New Position Detection Method Using Leaky Coaxial Cable," *IEICE Electronics Express*, Vol. 5, 2008, pp. 285–290.

[33] Oliner, A. A., and D. R. Jackson, "Leaky-Wave Antennas," Ch. 11 in *Antenna Engineering Handbook*, 4th ed., J. L. Volakis, (ed.), New York: McGraw-Hill, 2007.

[34] Pethig, R., and D. B. Kell, "The Passive Electrical Properties of Biological Systems: Their Significance in Physiology, Biophysics and Biotechnology," *Phys. Med. Biol.*, Vol. 32, 1987, pp. 933–970.

[35] Saito, K., et al., "Clinical Trials of Interstitial Microwave Hyperthermia by Use of Coaxial-Slot Antenna with Two Slots," *IEEE Trans. on Microwave Theory Tech.*, Vol. 52, 2004, pp. 1987–1991.

[36] Ito, K., and K. Saito, "Antenna Issues in Microwave Thermal Therapies," Ch. 5 in *Antennas for Portable Devices*, Z. N. Chen, (ed.), New York: John Wiley & Sons, 2007.

[37] Brady, M. M., S. A. Symons, and S. S. Stuchly, "Dielectric Behaviour of Selected Animal Tissues in Vivo at Frequencies from 2 to 4 GHz," *IEEE Trans. on Biomed. Eng.*, Vol. 28, 1981, pp. 305–307.

[38] Schwan, H. P., "Biophysics of Diathermy," *Therapeutic Heat and Cold*, S. Licht, (ed.), New Haven, CT: E. Licht Publications, 1965, pp. 63–125.

[39] Stoy, R. D., K. R. Foster, and H. P. Schan, "Dielectric Properties of Mammalian Tissues from 0.1 to 100 MHz: A Summary of Recent Data," *Phys. Med. Biol.*, Vol. 27, 1982, pp. 501–513.

[40] Foster K. R., et al., "Dielectric Properties of Brain Tissue Between 0.01 and 10 GHz," *Phys. Med. Biol.*, Vol. 24, 1979, pp. 1177–1187.

[41] Burdette, E. C., F. I. Cain, and J. Seals, "In Vivo Probe Measurement Technique at VHF Through Microwave Frequencies," *IEEE Trans. on Microwave Theory Tech.*, Vol. 28, 1980, pp. 414–427.

[42] Schepps, J. L., and K. R. Foster, "The UHF and Microwave Dielectric Properties of Normal and Tumour Tissues: Variation in Dielectric Properties with Tissue Water Content," *Phys. Med. Biol.*, Vol. 25, 1980, pp. 1149–1159.

[43] Gabriel, C., R. J. Sheppard, and E. H. Grant, "Dielectric Properties of Ocular Tissues at 37°C," *Phys. Med. Biol.*, Vol. 28, 1983, pp. 43–49.

[44] Bindu, G., et al., "Microwave Characterization of Female Human Breast Tissues," *Proceedings of the 9th European Conference on Wireless Technology*, 2006, pp. 123–126.

[45] Champbell, A. M., and D. V. Land, "Dielectric Properties of Female Human Breast Tissue Measured In Vitro at 3.2 GHz," *Phys. Med. Biol.*, Vol. 37, 1992, pp. 193–210.

[46] Aminzadeh, R., M. Saviz, and A. A. Shishegar, "Dielectric Properties Estimation of Normal and Malignant Skin Tissues at Millimeter-Wave Frequencies Using Effective Medium Theory," *22nd Iranian Conference on Electrical Engineering (ICEE 2014)*, 2014, pp. 1657–1661.

[47] Rourke, A. P. O., et al., "Dielectric Properties of Human Normal, Malignant and Cirrhotic Liver Tissue: *In Vivo* and *Ex Vivo* Measurements from 0.5 to 20 GHz Using a Precision Open-Ended Coaxial Probe," *Phys. Med. Biol.*, Vol. 52, 2007, pp. 4707–4719.

[48] You, K. Y., et al., "Precise Moisture Monitoring for Various Soil Using Handheld Microwave-Sensor Meter," *IEEE Sensors Journal*, Vol. 7, 2013, pp. 2563–2570.

[49] Ansarudin, F., et al., "A Simple Insulated Monopole Sensor Technique for Determination of Moisture Content in Hevea Rubber Latex," *Measurement Science Review*, Vol. 12, 2012, pp. 249–254.

[50] Ariffin, A., et al., "Morphological Changes of the Cellular Component of the Developing Palm Fruit (Tenera: Elaeis guineensis)," *Palm Oil Research Institute of Malaysia Bulletin*, Vol. 21, 1990, pp. 30–34.

[51] Abbas, Z., et al., "Complex Permittivity and Moisture Measurements of Oil Palm Fruits Using an Open-Ended Coaxial Sensor," *IEEE Sensors Journal*, Vol. 5, 2005, pp. 1281–1287.

[52] You, K. Y., A. Zulkifly, and K. Kaida, "Application of Microwave Moisture Sensor for Determination of Oil Palm Fruit Ripeness," *Measurement Science Review*, Vol. 10, 2010, pp. 7–14.

[53] You, K. Y., et al., "Palm Oil Moisture Monitoring Based on Dielectric Properties at Microwave Frequencies," *International Journal of Microwave and Optical Technology*, Vol. 9, 2014, pp. 294–302.

[54] Kim, K. B., et al., "Measurement of Grain Moisture Content Using Microwave Attenuation at 10.5 GHz and Moisture Density," *IEEE Trans. on Instrum. Meas.*, Vol. 51, 2002, pp. 72–77.

[55] You, T. S., and S. O. Nelson, "Microwave Dielectric Properties of Rice Kernels," *Journal of Microwave Power and Electromagnetic Energy*, Vol. 23, 1988, pp. 150–158.

[56] Thakur, S. K. P., and W. S. Kolmes, "Permittivity of Rice Grain from Electromagnetic Scattering," *4th International Conference on Electromagnetic Wave Interaction with Water and Moist Substance*, 2001, pp. 203–210.

[57] You, K. Y., et al., "Small and Slim Coaxial Probe for Single Rice Grain Moisture Sensing," *Sensors*, Vol. 13, 2013, pp. 3652–3663.

[58] You K. Y., et al., "Cylindrical Slot Antennas for Monitoring the Quality of Milled Rice," *PIERS Proceedings*, Suzhou, China, September 12–16, 2011.

[59] Nelson, S. O., W. R. Forbus, Jr., and K. C. Lawrence, "Permittivities of Fresh Fruits and Vegetables at 0.2 to 20 GHz," *Journal of Microwave & Electromagnetic Energy*, Vol. 29, 1994, pp. 81–93.

[60] Nelson, S. O., W. R. Forbus, Jr., and K. C. Lawrence, "Assessment of Microwave Permittivity for Sensing Peach Maturity," *Transactions of the ASAE*, Vol. 38, 1995, pp. 579–585.

[61] Nelson, S. O., et al., "Dielectric Spectroscopy of Watermelons for Quality Sensing," *Meas. Sci. Technol.*, Vol. 18, 2007, pp. 1887–1892.

[62] Guo, W. C., et al., "Dielectric Properties of Honeydew Melons and Correlation with Quality," *Journal of Microwave & Electromagnetic Energy*, Vol. 41, 2007, pp. 41-2-44–41-2-54.

[63] Guo, W. C., et al., "Dielectric Properties of Fuji Apples from 10 to 4500 MHz During Storage," *Journal of Processing and Preservation*, Vol. 35, 2011, pp. 884–890.

[64] Donnell, K. M., A. McClanahan, and R. Zoughi, "On the Crack Characteristic Signal from an Open-Ended Coaxial Probe," *IEEE Trans. on Instrum. Meas.*, Vol. 63, 2014, pp. 1877–1879.

[65] Maftooli, H., et al., "Output Signal Prediction of an Open-Ended Coaxial Probe When Scanning Arbitrary-Shape Surface Cracks in Metals," *IEEE Trans. on Instrum. Meas.*, Vol. 61, 2012, pp. 2384–2391.

[66] Qaddoumi, N., S. Ganchev, and R. Zoughi, "A Novel Microwave Fatigue Crack Detection Technique Using an Open-Ended Coaxial Line," *Conference on Precision Electromagnetic Measurements*, 1994, pp. 59–60.

[67] Baudry, D., A. Louis, and B. Mazari, "Characterization of the Open-Ended Coaxial Probe Used for Near-Field Measurements in EMC Applications," *Progress in Electromagnetics Research*, Vol. 60, 2006, pp. 311–333.

[68] Whiteside, H., and R. W. P. King, "The Loop Antenna as a Probe," *IEEE Trans. on Antennas Propagat.*, Vol. 12, 1964, pp. 291–297.

Selected Bibliography

Adachi, S., "Conical Antennas," *Encyclopedia of RF and Microwave Engineering*, K. Chang, (ed.), New York: John Wiley & Sons, 2005, pp. 775–780.

Agilent Technologies, "Specifying Calibration Standards and Kits for Agilent Vector Network Analyzers," Application Note 1287-11, 2010.

Ahmed, J., H. S. Ramaswamy, and V. G. S. Raghavan, "Dielectric Properties of Indian Basmati Rice Flour Slurry," *Journal of Food Engineering*, Vol. 80, 2007, pp. 1125–1133.

Bone, S., P. R. C. Gascoyne, and R. Pething, "Dielectric Properties of Hydrated Proteins at 9.9 GHz," *J. Chem. Soc. Faraday Trans.*, Vol. 73, 1977, pp. 1605–1611.

Brown, G. H., and O. M. Woodward, Jr., "Experimentally Determined Radiation Characteristics of Conical and Triangular Antennas," *RCA Rev.*, Vol. 13, No. 4, 1952, p. 425.

Cebik, L. B., "Log Periodic Arrays," Ch. 10 in *The ARRL Antenna Book*, 21st ed., The American Radio Relay League, 2007.

Chaudharv, S. S., et al., "Dielectric Properties of Normal & Malignant Human Breast Tissues at Radiowave & Microwave Frequencies," *Indian Journal of Biochemistry & Biophysics*, Vol. 21, 1984, pp. 76–79.

Chow, H. C., *Methods for Measuring the Dry Rubber Content of Field Latex, RRIM Training Manual on Analytical Chemistry*, Rubber Res. Inst. Malaysia, Malaysia, 1979.

Colton, D., and Monk, P., "A New Approach to Detecting Leukemia Using Computational Electromagnetics," *IEEE Comput. Sci. Eng.*, Vol. 2, 1995, pp. 46–52.

De Loor, G. P., "Dielectric Properties of Heterogeneous Mixtures Containing Water," *J. Microw. Power*, Vol. 3, 1968, pp. 67–73.

Emerson, D. T., "The Gain of the Axial-Mode Helix Antenna," *Antenna Compendium*, Vol. 4, 1995, pp. 64–68.

England, T. S., "Dielectric Properties of the Human Body for Wave Lengths in the 1–10 cm Range," *Nature*, Vol. 166, 1950, pp. 480–481.

Federal Communications Commission (FCC), "First Report and Order in the Matter of Revision of Part 15 of the Commission's Rules Regarding Ultrawideband Transmission Systems," ET-Docket 98-153, FCC 02-48, 2002, pp. 3–4.

Greenblum, C., "Notes on the Development of Yagi Arrays," *QST*, Part 1 and Part 2, 1956.

Harrison, C. W., Jr., and C. S. Williams, Jr., "Transients in Wide-Angle Conical Antennas," *IEEE Trans. on Antennas Propagat.*, Vol. 13, 1965, pp. 236–246.

Hasted, J. B., *Aqueous Dielectric*, London, U.K.: Chapman and Hall, 1973.

Isbell, D. E., "Log Periodic Dipole Arrays," *IRE Trans. on Antennas Propagat.*, Vol. 8, 1960, pp. 260–267.

Joines, W. T., "Frequency-Dependent Absorption of Electromagnetic Energy in Biological Tissue," *IEEE Trans. on Biomedical Engineering*, Vol. 31, 1984, pp. 17–20.

Karuppannan, I., O. Omar, and A. Ariffin, "Profiles of Biochemical Changes During the Development of Oil Palm Mesocarp (Elaesis Guineensis)," *Proc. Mal. Biochem. Soc. Conference*, Vol. 11, 1986, pp. 129–134.

Khalid, K., and Z. Abbas, "A Microstrip Sensor for Determination of Harvesting Time for Oil Palm Fruits (Tenera: Elaeis Guineesis)," *J. Microwave Power and Electromag. Energy*, Vol. 27, 1992, pp. 3–10.

King, H. E., and J. L. Wong, "Characteristics of 1 to 8 Wavelength Uniform Helical Antennas," *IEEE Trans. on Antennas Propagat.*, Vol. 28, 1980, pp. 291–296.

King, R. W. P., "The Conical Antenna as a Sensor or Probe," *IEEE Trans. on Electromagnetic Compatibility*, Vol. 25, 1983, pp. 8–13.

Mohammadi, A., et al., "Development of Microcontroller-Based Microwave System to Measure Solid Content of Hevea Rubber Latex," *Measurement*, Vol. 45, No. 3, 2012, pp. 571–579.

Nyfors, E., and P. Vainikainen, *Industrial Microwave Sensors*, Norwood, MA: Artech House, 1989.

Papas, C. H., and R. King, "Radiation from Wide-Angle Conical Antennas Fed by a Coaxial Line," *Proceedings of the IRE*, Vol. 38, 1950, pp. 49–51.

Roy, M. N., "Constructing Normal-Mode Helical Antennas," *IERE Symposium on Electronics in Industry*, 1968, pp. 119–123.

Surowiec, A. J., et al., "Dielectric Properties of Breast Carcinoma and the Surrounding Tissues," *IEEE Trans. on Biomedical Engineering*, Vol. 35, 1988, pp. 257–263.

Tang, J., H. Feng, and M. Lau, "Microwave Heating in Food Processing," in *Advances in Bioprocessing Engineering*, X. Young and J. Tang, (eds.), New York: Scientific Press, 2002, pp. 1–43.

Tang, T. S., "Palm Kernel and Palm Kernel Oil Quality Today," *PORIM Bulletin*, Vol. 23, 1990, pp. 28–45.

Tan, Y. A., et al., "PORIM Crude Palm Oil Survey 97/98 Quality and Identity Characteristics," *PORIM Technology, Palm Oil Res. Inst. Malaysia*, Kuala Lumpur, Vol. 22, 2000, pp. 1–5.

United States Department of Agriculture (USDA), *United States Standards for Rice*, Washington, D.C.: United States Department of Agriculture, 2009.

Wikipedia, "Log-Periodic Antenna," Wikimedia Foundation, November 2014.

Wong, J. L., and King, H. E., "Empirical Helix Antenna Design," *Proc. IEEE Antennas Propagat. Int. Symp.*, 1982, pp. 366–368.

Xing, Z., and X. Yadong, "A Novel Electrically Small Monopole Load by Open-Circuited Transmission Line," *IEEE APS International Symposium*, Vol. 44, 2006, pp. 615–618.

Yagihara, S., et al., "Dielectric Relaxation Measurement and Analysis of Restricted Water Structure in Rice Kernels," *Meas. Sci. Technol.*, Vol. 18, 2007, pp. 983–990.

About the Author

K. Y. You obtained his B.Sc physics (Honours) degree from Universiti Kebangsaan Malaysia (UKM) in 2001. He pursued his M.Sc. in microwave at the Faculty of Science in 2003 and his Ph.D in wave propagation at the Institute for Mathematical Research in 2006 at Universiti Putra Malaysia (UPM). Recently, he has been a senior lecturer in the Communication Engineering Department, Faculty of Electrical Engineering, Universiti Teknologi Malaysia (UTM). His research interests include microwave moisture sensors and the theoretical and numerical modeling of microwave radiators. He has been a member of the IEEE and the IEEE Microwave Theory and Techniques Society since 2009. He also has been an overseas member of the IEICE since 2011.

Index

B

C

D

N

Recent Titles in the Artech House Antennas and Electromagnetics Analysis Library

Christos Christodoulou, Series Editor

Adaptive Array Measurements in Communications, M. A. Halim

Advanced Computational Electromagnetic Methods and Applications, Wenhua Yu, Wenxing Li, Atef Elsherbeni, Yahya Rahmat-Samii, Editors

Advances in Computational Electrodynamics: The Finite-Difference Time-Domain Method, Allen Taflove, editor

Advances in FDTD Computational Electrodynamics: Photonics and Nanotechnology, Allen Taflove, editor; Ardavan Oskooi and Steven G. Johnson, coeditors

Analysis Methods for Electromagnetic Wave Problems, Volume 2, Eikichi Yamashita, editor

Antenna Design with Fiber Optics, A. Kumar

Antenna Engineering Using Physical Optics: Practical CAD Techniques and Software, Leo Diaz and Thomas Milligan

Antennas and Propagation for Body-Centric Wireless Communications, Second Edition, Peter S. Hall and Yang Hao, editors

Antennas and Site Engineering for Mobile Radio Networks, Bruno Delorme

Analysis of Radome-Enclosed Antennas, Second Edition, Dennis J. Kozakoff

Applications of Neural Networks in Electromagnetics, Christos Christodoulou and Michael Georgiopoulos

The Art and Science of Ultrawideband Antennas, Second Edition, Hans G. Schantz

AWAS for Windows Version 2.0: Analysis of Wire Antennas and Scatterers, Antonije R. Djordjević, et al.

Broadband Microstrip Antennas, Girsh Kumar and K. P. Ray

Broadband Patch Antennas, Jean-François Zürcher and Fred E. Gardiol

CAD of Microstrip Antennas for Wireless Applications, Robert A. Sainati

The CG-FFT Method: Application of Signal Processing Techniques to Electromagnetics, Manuel F. Cátedra, et al.

Computational Electrodynamics: The Finite-Difference Time-Domain Method, Third Edition, Allen Taflove and Susan C. Hagness

Electromagnetic Modeling of Composite Metallic and Dielectric Structures, Branko M. Kolundzija and Antonije R. Djordjević

Electromagnetic Waves in Chiral and Bi-Isotropic Media, I. V. Lindell, et al.

Electromagnetics, Microwave Circuit and Antenna Design for Communications Engineering, Peter Russer

Engineering Applications of the Modulated Scatterer Technique, Jean-Charles Bolomey and Fred E. Gardiol

Fast and Efficient Algorithms in Computational Electromagnetics, Weng Cho Chew, et al., editors

Frequency-Agile Antennas for Wireless Communications, Aldo Petosa

Fresnel Zones in Wireless Links, Zone Plate Lenses and Antennas, Hristo D. Hristov

Handbook of Antennas for EMC, Thereza MacNamara

Handbook of Reflector Antennas and Feed Systems, Volume I: Theory and Design of Reflectors, Satish Sharma, Sudhakar Rao, and Lotfollah Shafai, editors

Handbook of Reflector Antennas and Feed Systems, Volume II: Feed Systems, Lotfollah Shafai, Satish Sharma, and Sudhakar Rao, editors

Handbook of Reflector Antennas and Feed Systems, Volume III: Applications of Reflectors, Sudhakar Rao, Lotfollah Shafai, and Satish Sharma, editors

Introduction to Antenna Analysis Using EM Simulators, Hiroaki Kogure, Yoshie Kogure, and James C. Rautio

Iterative and Self-Adaptive Finite-Elements in Electromagnetic Modeling, Magdalena Salazar-Palma, et al.

LONRS: Low-Noise Receiving Systems Performance and Analysis Toolkit, Charles T. Stelzried, Macgregor S. Reid, and Arthur J. Freiley

Measurement of Mobile Antenna Systems, Second Edition, Hiroyuki Arai

Microstrip Antenna Design Handbook, Ramesh Garg, et al.

Microwave and Millimeter-Wave Remote Sensing for Security Applications, Jeffrey A. Nanzer

Mobile Antenna Systems Handbook, Third Edition, Kyohei Fujimoto, editor

Multiband Integrated Antennas for 4G Terminals, David A. Sánchez-Hernández, editor

Noise Temperature Theory and Applications for Deep Space Communications Antenna Systems, Tom Y. Otoshi

Phased Array Antenna Handbook, Second Edition, Robert J. Mailloux

Phased Array Antennas with Optimized Element Patterns, Sergei P. Skobelev

Plasma Antennas, Theodore Anderson

Printed MIMO Antenna Engineering, Mohammad S. Sharawi

Quick Finite Elements for Electromagnetic Waves, Giuseppe Pelosi, Roberto Coccioli, and Stefano Selleri

Radiowave Propagation and Antennas for Personal Communications, Third Edition, Kazimierz Siwiak

Reflectarray Antennas: Analysis, Design, Fabrication and Measurement, Jafar Shaker, Mohammad Reza Chaharmir, and Jonathan Ethier

RF Coaxial Slot Radiators: Modeling, Measurements, and Applications, Kok Yeow You

Solid Dielectric Horn Antennas, Carlos Salema, Carlos Fernandes, and Rama Kant Jha

Switched Parasitic Antennas for Cellular Communications, David V. Thiel and Stephanie Smith

Ultrawideband Antennas for Microwave Imaging Systems, Tayeb A. Denidni and Gijo Augustin

Understanding Electromagnetic Scattering Using the Moment Method: A Practical Approach, Randy Bancroft

Wavelet Applications in Engineering Electromagnetics, Tapan Sarkar, Magdalena Salazar Palma, and Michael C. Wicks

For further information on these and other Artech House titles, including previously considered out-of-print books now available through our In-Print-Forever® (IPF®) program, contact:

Artech House
685 Canton Street
Norwood, MA 02062
Phone: 781-769-9750
Fax: 781-769-6334
e-mail: artech@artechhouse.com

Artech House
16 Sussex Street
London SW1V HRW UK
Phone: +44 (0)20 7596-8750
Fax: +44 (0)20 7630 0166
e-mail: artech-uk@artechhouse.com

Find us on the World Wide Web at: www.artechhouse.com